AF255920

Recent Progression in Metal-Organic Frameworks

Properties and Applications

Recent Progression in Metal-Organic Frameworks

Properties and Applications

Edited by
Dr. Tapan K. Pal
Dr. Prem Lama

CWP
Central West Publishing

For more information about the books published by Central West Publishing, please visit https://centralwestpublishing.com

Disclaimer
Every effort has been made by the publisher, editors and authors while preparing this book, however, no warranties are made regarding the accuracy and completeness of the content. The publisher, editors and authors disclaim without any limitation all warranties as well as any implied warranties about sales, along with fitness of the content for a particular purpose. Citation of any website and other information sources does not mean any endorsement from the publisher, editors and authors. For ascertaining the suitability of the contents contained herein for a particular lab or commercial use, consultation with the subject expert is needed. In addition, while using the information and methods contained herein, the practitioners and researchers need to be mindful for their own safety, along with the safety of others, including the professional parties and premises for whom they have professional responsibility. To the fullest extent of law, the publisher, editors and authors are not liable in all circumstances (special, incidental, and consequential) for any injury and/or damage to persons and property, along with any potential loss of profit and other commercial damages due to the use of any methods, products, guidelines, procedures contained in the material herein.

A catalogue record for this book is available from the National Library of Australia

ISBN (print): 978-1-922617-46-0

Preface

Over the past 2 decades, significant development has been done in the field of metal-organic frameworks. MOFs have shown important properties such as magnetic, luminescence, gas adsorption, proton conduction etc. and applications in the field of sensing, separation, catalysis. This book consists of eight chapters where various important properties, utilization and applications are discussed. Chapter 1 presents the recent advancement in energetic material using metal organic frameworks and describes about some of the interesting and important properties of an energetic material. In chapter 2, solid stated [2+2] photo cycloaddition reaction is discussed in detail along with examples. Chapter 3 deals with the magneto-structural analysis of MOFs. Here, the change of magnetic properties due to the coordination of oxygen and nitrogen atoms in various axial and equatorial position to metal ions is elaborated. Chapter 4 presents one of an important and fascinating thermal expansion behavior in MOF. The chapter briefly discussed about the MOF design, characterization and calculation of various thermal expansion behavior that generally occurred in MOFs. Its addition, different types of thermal expansion behavior are explained with mechanisms. Chapter 5 describes the fluoro-switchable MOFs and its sensing application for variety of analytes and molecules. In chapter 6, different types of secondary building unit in MOFs are explained. It also shows the use of MOF as catalyst in reduction reactions, C-C and C-X (X= N, B) bond formation, oxidation reactions, tandem reactions and asymmetric catalysis. Chapter 7 deals with proton conducting MOFs and their application in various filed such as fuel cells, bio-inspired light harvesting, proton batteries, sensing etc. The final chapter 8 discussed about the conversion of CO_2 to valuable chemicals using MOFs. In this chapter, the cycloaddition reaction between CO_2 and epoxide is presented and the photoreduction of CO_2 by various components of MOFs and its hybrid composite are explained.

About the Editors

Tapan K. Pal received his PhD in Inorganic Chemistry from the Indian Institute of Technology Kanpur, India in 2016. He joined National chemical Laboratory, India as a National Postdoctoral Fellow (SERB-NPDF) in 2016. Currently, he is an Assistant Professor at Pandit Deendayal Energy University, India. His research focuses on the synthesis of metal–organic frameworks for sensing and other potential applications.

Dr. Prem Lama obtained his Bachelor in Science (2004) and Master in Science (2006) from St. Xaviers's College, Kolkata and Calcutta University respectively. He received his PhD degree in 2012 under the supervision of Prof. Parimal K. Bhradwaj at Indian Institute of Technology Kanpur, India in. In 2013, he joined as a postdoctoral research fellow in Prof. Leonard J. Barbour's group at Stellenbosch University,South Africa. In 2018, he moved to Goa University (India) as a DST-Inspire faculty. After working in Goa University for 20 months, he then joined CSIR-Indian institute of Petroleum (India) in September 2019 as a Scientist. At present, his current research interests focus on the synthesis of metal oxides, nanomaterials, porous coordination polymers for the catalytic conversion to value added products. In addition, he is involved in the field of crystallography for obtaining a basic understanding of the structure–property relationship of complexes at the molecular level in order to explore such materials for new applications in the field of smart materials.

Table of Contents

High-Energy Metal-Organic Frameworks: A Recent Approach to Energetic Materials

Saona Seth

Department of Applied Sciences, Tezpur University, Assam, India

1. Introduction

Metal-organic frameworks (MOFs) have emerged as highly potential functional materials since the beginning of this century, and find applications in different aspects of synthetic chemistry, medicinal chemistry as well as in materials sciences. It is important to note that majority applications of MOFs are based on their ordered porous structures and host-guest chemistry therein. However, there are few applications for which the porous structures are not essential.[1,2] Rather densely packed frameworks are preferred. One such application is developing energetic materials, where high-energy materials with high density are desirable.

The area of energetic materials is as old as the subject 'Chemistry'.[3] Energetic materials are compounds or formulations that typically contain intimately mixed oxidizing and fuel components, such that they can undergo self-sustained combustion with release of energy in the form of heat or light when initiated by suitable external stimuli. Energetic materials can be categorized as explosives, propellants and pyrotechnics, based on their nature of decomposition. Explosives have very high energy density so that the decomposition leads to huge energy release. Decomposition of propellants is associated with formation of gaseous products leading to rapid and large increase in volume or pressure in the system. Pyrotechnic materials undergo deflagration with flames that can have variable colors depending on the ingredients. Energetic materials are enormously important because of their usefulness in several military and civilian applications, such as munitions, mining, etc. Developing environmentally green energetic materials, those produce nonhazardous decomposition products, or decomposition products having biocidal activities, have been a major focus of recent time. In addition to designing new organic energetic materials, their crystal engineering through discovery of novel polymorphs and cocrystals, synthesizing salts or metal complexes of azole-based energetics, and developing high-energy composites are some of the strategies of developing novel energetic materials. High-energy MOFs (HE MOFs) are recent addition in this area.[1,4-6] MOFs can have tunable properties based on the constituent metal ions and organic linkers, which provides an opportunity to investigate the structure-property relationship. Recent developments on He MOFs focus on the use of environmentally benign transition metal ions, such as, Zn^{2+}, Cu^{2+}, Cu^+, Co^{2+}, Ag^+, and main group metal ions, such as Na^+, K^+ and Ca^{2+}.[1,7-11] Some of these HE MOFs can potentially replace the traditionally used lead and mercury-based energetics, such as, lead azide, mercury fulminate, etc. Furthermore, energetic MOFs containing Ag^+ ions or iodine-based guests, are also being investigated where the decomposition products show biocidal effects on the environment.[12]

In this chapter, we shall discuss the key properties of energetic materials and attributes of high-energy MOFs with the aim of understanding structure-property relationship.

2. Key Properties of Energetic Materials

The desirable properties of energetic materials depend on the particular applications they are designed for. For example, high energy density and low sensitivity is desirable for explosives, whereas high sensitivity is essential for energetic materials to be used as initiators. Therefore, it is important to identify their key properties.[1,5] The key properties that determine the power of an energetic material are heat of detonation, detonation velocity and detonation pressure. The energy released upon detonation of an energetic material is called heat of detonation, which can be measured by calorimetric methods. However, experimental values of heat of detonation for energetic MOFs are rarely reported, presumably due to large amount of sample requirement to carry out the experiments and potential safety-hazard therein. Usually, the heat of detonation for these materials are estimated by using density functional theory (DFT), which computes the energy of detonation (ΔE_{det}), and from which heat of detonation (ΔH_{det}) is calculated by using a linear correlation equation. It should be noted that detonation energy depends on the composition of detonation products, and therefore, it is important to characterize the detonation products to estimate ΔE_{det} and ΔH_{det} correctly. Furthermore, high heat of formation of energetic compounds ensures high heat of detonation. Detonation velocity and detonation pressure are two other vital parameters. Detonation velocity refers to the velocity at which reaction process propagates in the mass of the explosive, and therefore, indicates the rate of energy release. Detonation pressure is an estimate of the energy of shock wave of the detonation products. Both of these parameters can be estimated from molar enthalpies of formation (determined experimentally) of the energetic compounds by using empirical Kamlet formula, or the Kamlet–Jacobs equation, or CHEETAH thermochemical code. It should be noted that all these energetic parameters are functions of density. For a given explosive, detonation velocity is proportional to density of the material and detonation pressure is proportional to the square of its density. Therefore, high density is desirable for formidable energetic properties.

Another crucial parameter is oxygen balance (OB) that estimates the internally available oxidant content in an energetic material. The ideal composition of an energetic material should be such that OB is 0%, meaning that the material has balanced amount of fuel and oxidant components, and therefore, during detonation/deflagration, it is able to undergo complete selfsustained combustion leading to maximum energy release. OB of an energetic material is estimated by the following equation:

$$OB\% = \frac{-1600}{molecular\ wt\ of\ the\ compound} \left(2x + \frac{y}{2} + M - z\right)$$

where, x = number of carbon atoms
y = number of hydrogen atoms
z = number of oxygen atoms
M = number of atoms of metal (assuming that metallic oxide is produced with formula MO).

OB is calculated based on the assumption that each element in the compound tend to form the most stable molecular state as terminal decomposition product upon self-combustion process. Therefore, it is assumed that carbon component will be converted to CO_2, hydrogen component to H_2O and nitrogen component to elemental N_2 molecules due to very low heats of formation of these molecules. It should be noted that there are a few reports where OB is calculated by assuming CO (not CO_2) as the terminal decomposition product of the carbon content as it is experimentally observed to be formed upon decomposition of many energetics.[10] Except for a few energetic oxidants, most energetic materials have negative OB, meaning they are deficient in oxidant content. For example, the traditionally known energetic material trinitrotoluene (TNT) has a highly negative OB of -74%. However, it should be noted that the energy released upon detonation depends on both OB and heat of formation, and therefore, TNT exhibits formidable detonation properties (heat of detonation: 4564 J g^{-1} and detonation velocity: 6.9 km s^{-1}) despite having a highly negative OB.[13]

Sensitivity to different stimuli is another key factor for energetic materials as for safety is concerned. Based on sensitivity, energetic materials are categorized as primary and secondary explosives.[3] Primary explosives are the class of energetic materials that can be initiated with a small stimulus and undergo characteristically fast transitions from deflagration to detonation upon initiation. Usually, primary explosives have impact sensitivities less than 4 J and for secondary explosives are greater than 4 J.[14,15] For example, lead azide is a traditionally used primary explosive (impact sensitivity: 0.089 J), which is a coordination polymer.[16] Primary explosives are used as initiators to trigger explosions. In contrast, secondary explosives have high power (high energy-density) but low sensitivity. Some of the state-of-the-art organic secondary explosives are HMX (7.4 J), RDX (7.5 J), TNT (15 J) and that are used in munitions.[1] In this regard, it should be noted that impact sensitivity is often represented by the parameter 'drop height' ($D_{50}h$), which represents the height from where when a certain drop weight falls on the sample, the probability of detonation is 50%. When impact sensitivity is reported in terms of $D_{50}h$, pentaerythritol tetranitrate, (PETN) is used as the reference ($D_{50}h$ ~12±5 cm).[1] If the impact sensitivity of an energetic material is greater than or equal to that of PETN, the material is categorized as primary explosive, and materials that exhibit higher $D_{50}h$ than PETN, are categorized as secondary explosives. In general, energetic materials (secondary explosives in particular) should be stable under ambient conditions, should exhibit sufficient thermal stability, and be sufficiently insensitive to external factors, such as, impact, friction, heat, shock, electrostatic discharge, etc. so that they can be transported safely. However, they should not be too insensitive to be initiated by any stimuli. In addition to all these properties, the terminal decomposition products of energetic materials should be non-hazardous or environmentally benign.

3. High-Energy MOFs

3.1 Linker Design for High-Energy MOFs

Based on their tunable properties, high-energy MOFs have gained significant contemporary research interest. Several strategies have been investigated to develop MOF-based energetic materials. Use of energetic linkers as the organic building blocks is the straightforward way to construct high-energy MOFs. As an example, the organic energetic compound 4,4′-azo-1,2,4-triazole (ATRZ) is a secondary explosive with remarkably high detonation velocity of 7.52 km/s and detonation pressure of 23.5 GPa.[17] Furthermore, ATRZ is a multitopic linker with six potential sites for metal coordination. Therefore, several energetic MOFs have been developed with this linker that demonstrate interesting properties. It should be noted that despite metal-carboxylate linkage constitutes majority of MOFs compared to other metal-coordination bonds, there are only a few high-energy MOFs known with metal-carboxylate functionality.[5,15,18] The reason behind this is that the carboxylate group does not contribute much to the heats of formation of these MOFs, which in turn reduce their heats of detonation. Majority of HE MOFs are synthesized from nitrogen rich linkers containing azole scaffolds and N-donor sites because of high heats of formation of these molecules. For example, heats of formation for tetrazole, 1,2,4-triazole and pyrazole are 333 kJ mol^{-1}, 182 kJ mol^{-1}, 160 kJ mol^{-1}, respectively.[19] Another advantage of nitrogen rich systems is that the decomposition products are less hazardous because the nitrogen component tend to form highly stable elemental nitrogen molecules (N_2) upon decomposition, which is a natural component of air. Therefore, the energetic MOFs reported so far are mostly composed of triazole and tetrazole based heterocyclic linkers. In addition to heat of detonation, OB may also be significantly impacted by the constituent organic linkers. Therefore, employing linkers of high OBs can be useful. One such interesting linker is 1,5-di(nitramino)tetrazole (DNAT), which has a positive OB of 8.4%. The HE MOFs synthesized from this linker also have positive OBs.[8] There are a few other heterocyclic systems that have also been employed as linkers to develop HE MOFs. For example, Shreeve and coworkers employed 4,4'-bis(dinitromethyl)-3,3'-azofurazanate as linker to synthesize an energetic potassium MOF.[10] It should be noted that the heat of formation for furazan moiety is also very promising (217 kJ mol[-1]) to be used in energetic materials.[19] Often, energetic MOFs are synthesized by mixed-linker based approach, where azide, hydrazine, etc. are introduced as colinkers to further increase energy-density (by increasing the heats of formation) of the materials.[1] Both of these ligands can also be present as terminal ligands. Azide is an interesting ligand for high-energy MOFs based on its nitrogen and energy content, and modes of coordination. Azide is 100% nitrogen-containing ligand with heat of formation of 355 kJ mol[-1].[1] It can bridge metal ions in $\mu_{1,1}$ and/or $\mu_{1,3}$ coordination modes.[1] It is observed that the coordination modes of azide play significant role on the energetic properties of the MOFs. For example, HE MOFs containing terminal azides are usually more sensitive to external stimuli and exhibit less thermal decomposition temperatures relative to HE MOFs formed through bridging azide groups. Organic linkers can become thermally stabilized or sensitized in the structures of MOFs depending on coordination environment and rigidity of the systems. Therefore, Linker design plays a crucial role on the properties of HE MOFs. Structures of some representative organic linkers have been shown in figure 1.

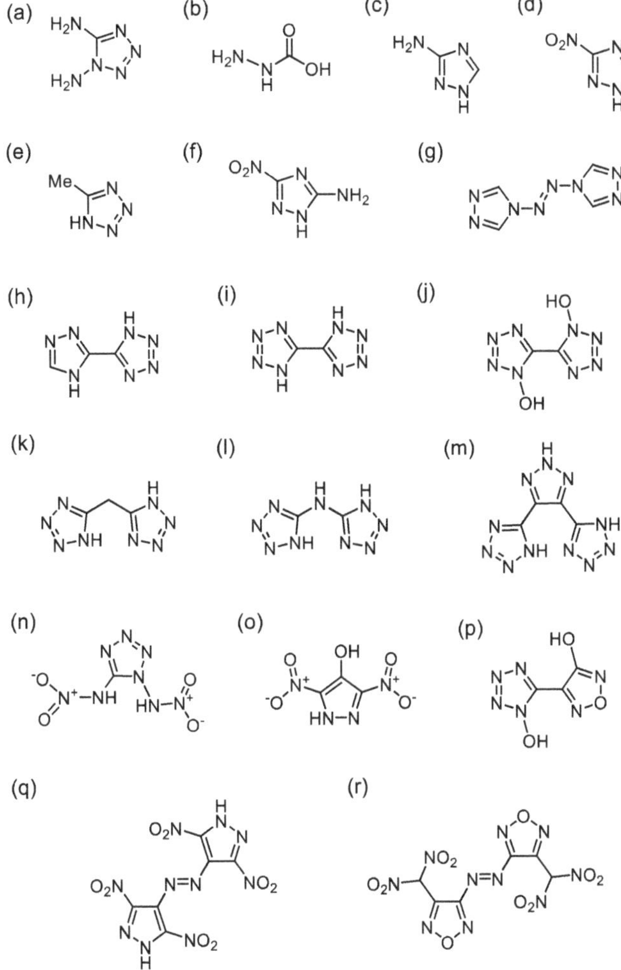

Figure 1. Chemical structures of the organic energetic molecules that have been employed as building blocks to synthesize energetic MOFs. (a) 1,5-diaminotetrazole, (b) hydrazinecarboxylic acid, (c) 3-amino-1*H*-1,2,4-triazole (atrz), (d) 3-nitro-1*H*-1,2,4-triazole (Hntz), (e) 5-methyltetrazole, (f) 5-amino-3-nitro-1*H*-1,2,4-triazole (ANTA), (g) 4,4′-azo-1,2,4-triazole (ATRZ), (h) 3-(tetrazole-5-yl)-1,2,4-triazole (H2tztr), (i) 5,5′-bistetrazole (j) 5,5′-bistetrazole-1,1′-diol (BTO), (k) bis(1*H*-tetrazol-5-yl)methane (H2btm), (l) *N,N*-bis(1*H*-tetrazole-5-yl)-amine (H2bta), (m) 4,5-di(1*H*-tetrazol-5-yl)-2*H*-1,2,3-triazole (H3dttz), (n) 1,5-di(nitramino)tetrazole (DNAT), (o) 3,5-dinitro-1*H*-pyrazol-4-ol (p) 3-(1-hydroxytetrazol-5-yl)furazan-4-ol (BTF), (q) 1,2-bis(3,5-dinitro-1*H*-pyrazol-4-yl)diazene, (r) 4,4′-bis(dinitromethyl)-3,3′-azofurazan.

3.2 High-Energy MOFs and their Properties

In this section, we shall discuss some selective examples of energetic MOFs and their key properties. Although a number of MOFs based on triazole and tetrazole containing organic linkers have been investigated for properties, such as, luminescence, magnetism, carbon capture, etc., energetic properties of such MOFs are being investigated only recently.[1,5] Notably, unlike most other applications of MOFs, where 3D porous structures are desirable, 2D or 1D metal-organic polymers are more important for energetic materials because those are usually less porous than the 3D analogues, and therefore, have relatively higher densities. Nonetheless, several energetic MOFs with 3D network structures have also been reported, and porous 3D frameworks have judiciously been utilized to further improve the material density and energetic properties by incorporating suitable guest species. The materials with 1D structures are out of the scope of this chapter as those 1D coordination polymers cannot be categorized as MOFs.[20] Therefore, only 2D and 3D frameworks will be discussed.

3.2.1 Energetic 2D MOFs

One early report on energetic 2D MOFs was by Tang et al. by employing the mixed liker-based approach.[21] A 2D energetic MOF $[Cd(DAT)_2(N_3)_2]_n$ was synthesized by employing highly nitrogen rich 1,5-diaminotetrazole (DAT) (N% = 63.42) and azide as colinkers. The 2D framework is constructed through azide bridging of the metal ions and the DAT coordinates the metal ion as a terminal ligand (Figure 2a). The MOF has a very high nitrogen content of 53.98%. It exhibits moderate density of 2.144 g cm^{-3}, and thermal stability up to 208 °C. Subsequently, several other 2D energetic MOFs have been reported containing azide as bridging linker and terminal ligand.[5,15,22]

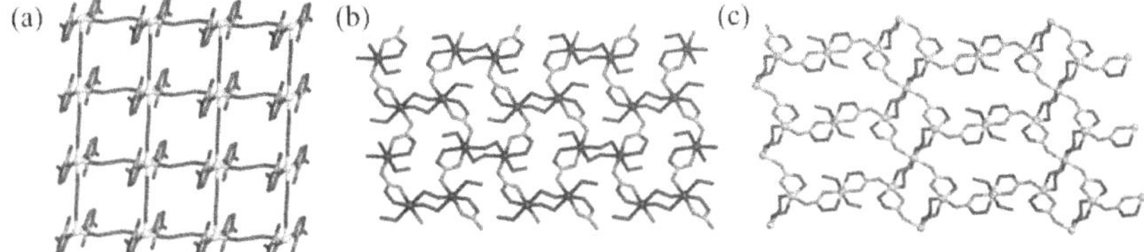

Figure 2. (a) 2D network of $[Cd(DAT)_2(N_3)_2]_n$ MOF, where azide group serves as the bridging linker and DAT coordinates the octahedral Cd(II) ion as a pendant ligand. 2D networks of (b) cobalt hydrazine hydrazinecarboxylate perchlorate (CHHP) and (c) zinc hydrazine hydrazinecarboxylate perchlorate (ZnHHP) MOFs. Hydrogens and counterion guests are omitted for clarity (Cd: light yellow, Co: maroon; Zn: orange, C: gray, N: blue, O: red). Reproduced from Reference 1 with permission from American Chemical Society.

Hopeweek and coworkers synthesized a couple of energetic 2D MOFs, viz., cobalt hydrazine hydrazinecarboxylate perchlorate (CHHP) and zinc hydrazine hydrazinecarboxylate perchlorate (ZnHHP) from the reaction of corresponding metal perchlorates and hydrazinecarboxylate that is formed in situ from hydrazine and atmospheric CO_2.[23] Hydrazine serves as bridging as well as pendant linker in both frameworks. In CHHP, each Co(II) ion is observed to be octahedral in geometry and is chelated by one hydrazinecarboxylate through the terminal nitrogen and carboxylate oxygen atom, and is further coordinated by two bridging hydrazines, one oxygen atom from another hydrazinecarboxylate anion and one pendant hydrazine (Figure 2b). In ZnHHP, three crystallographically independent octahedral Zn(II) ions are present with different coordination environments (Figure 2c). Both MOFs are cationic in nature, and perchlorate ions are present as the counterion. The MOFs have crystallographic densities of 2.000 g cm^{-3} and 2.117 g cm^{-3}, for CHHP and ZnHHP, respectively. Thermal analysis indicates that the MOFs are stable up to 231 °C and 293 °C, respectively. Furthermore, CHHP is slightly more sensitive than ZnHHP to other stimuli. For example, ZnHHP is insensitive to electrostatic discharge unlike CHHP. In terms of impact sensitivity, ZnHHP has a drop height 50 cm, while CHHP has a drop height of 30 cm (for 2.5 kg wt). Furthermore, ZnHHP deflagrates upon exposure to flame, while CHHP detonates.

A series of 2D energetic MOFs were derived from the energetic linker 3-(tetrazole-5-yl)-1,2,4-triazole (H_2tztr), which exhibits variable co-ordination modes in MOFs.[14,24] In the [Cu(Htztr)]$_n$ MOF, both Cu(I) ion and the monodeprotonated linker act as 4-connecting units, resulting a 2D net of density 2.44 g cm^{-3} (Figure 3a). Two more HE 2D MOFs were reported with Pb(II) ions. In [Pb(Htztr)$_2$(H_2O)]$_n$, Htztr serves a 3- as well as 4-connecting units to bridge the 8-coordinated Pb(II) ions leading to a 2D MOF with density 2.519 g cm^{-3} (Figure 3b). In the [Pb(H_2tztr)(O)]$_n$ MOF, metal ion shows distorted pyramidal geometry and H_2tztr as a 3-connecting linker (Figure 3c). The metal-ligand co-ordination results in a high density (3.511 g cm^{-3}) 2D network. All of three MOFs exhibit excellent thermal stabilities in TGA; 355 °C for [Cu(Htztr)]$_n$, 340 °C for [Pb(Htztr)$_2$(H_2O)]$_n$ and 318 °C for [Pb(H_2tztr)(O)]$_n$. All the three MOFs demonstrate to serve as highly insensitive secondary explosives with IS =32 J for [Cu(Htztr)]$_n$, IS >40 J for the Pb-MOFs, and FS >360N for all three MOFs. The estimated detonation velocity and detonation pressure of the MOFs are 10.40 km s^{-1} and 56.5 GPa for [Cu(Htztr)]$_n$, 7.72 km s^{-1} and 31.6 GPa for [Pb(Htztr)$_2$(H_2O)]$_n$ and 8.12 km s^{-1} and 40.1 GPa for [Pb(H_2tztr)(O)]$_n$. The higher detonation velocity and detonation pressure of [Pb(H_2tztr)(O)]$_n$ than [Pb(Htztr)$_2$(H_2O)]$_n$ can be attributed to the higher density of the former.

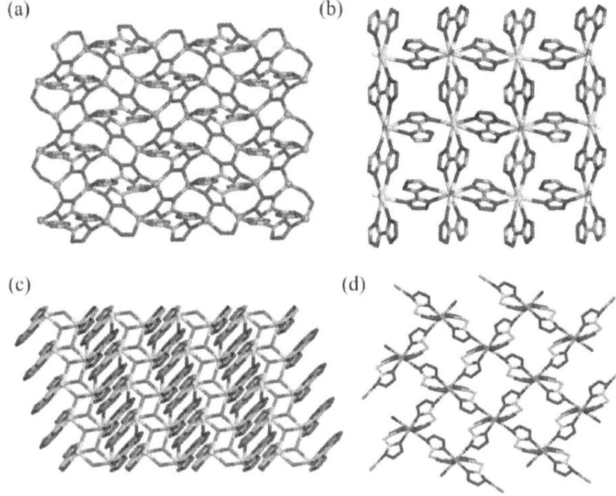

Figure 3. Crystal packing diagrams of (a) [Cu(Htztr)]$_n$, (b) [Pb(Htztr)$_2$(H_2O)]$_n$, (c) [Pb(H_2tztr)(O)]$_n$, and (d) [(AG)$_2$(Cu(btm)$_2$)]$_n$ MOFs. Hydrogens and guest counterions are omitted for clarity (Cu(I): dark turquoise, Cu(II): orange, Pb: light green, C: gray, N: blue, O: red). Reproduced from Reference 1 with permission from American Chemical Society.

Zhang and coworkers reported an energetic MOF $[(AG)_2(Cu(btm)_2)]_n$ (AG: aminoguanidinium, H_2btm: bis(1*H*-tetrazol-5-yl)methane) containing a 2D anionic framework (Figure 3d).[25] The deprotonated bis(tetrazole)methane) acts as both a chelating and a bridging linker that connects the octahedral Cu(II) ions to form the overall 2D network. The negative charge on the framework is balanced by the AG guests present in the pores of the MOF. Due to the presence of nitrogen rich linker and guest species, the MOF has a remarkably high nitrogen content of 65.4%. Heat of detonation of the MOF is estimated to be 5.41 kcal g^{-1}, detonation pressure as 53.9 GPa and detonation velocity as 10.97 km s^{-1}. In addition to azole-based linkers, energetic 2D MOFs have been reported with other heterocyclic linkers as well. For example, 3-(1-hydroxytetrazol-5-yl)furazan-4-ol (Figure 1p) form two 2D MOFs, viz., Pb-BTF and Ba-BTF.[26] The linker acts as a 6-connecting chelating and bridging unit in the Pb-MOF leading to a crystal density of 3.382 g cm^{-3}. In the Ba-MOF, the linker serves as a 4-connecting unit to form a framework of density of 2.336 g cm^{-3}. Both energetic MOFs have good calculated OBs of -15.56% (for Pb-BTF) and -16.97% (for Ba-BTF). The estimated detonation velocities and pressures are 9.21 km s^{-1} and 50.9 GPa for the Pb-MOF and 8.45 km s^{-1} and 36.5 GPa for the Ba-MOF. Both MOFs are stable beyond 250 °C and are fairly insensitive to impact and friction.

3.2.2 Energetic 3D MOFs

A number of energetic MOFs with 3D framework structures have been reported with moderate to good crystallographic densities, thermal stabilities and energetic properties. Among the polyazole-based linkers, the most widely used example is ATRZ that has been used as the organic building block for several energetic MOFs. One well-investigated 3D energetic MOF is known as ATRZ-1 $([Cu(ATRZ)_3(NO_3)_2]_n)$, synthesized by Pang and co-workers from the reaction of ATRZ linker with $Cu(NO_3)_2$ (Figure 4a).[9] The octahedral Cu^{2+} ions are bridged by the ATRZ linkers to form a doubly interpenetrated cationic 3D network structure. The MOF is porous in nature, and the pores are occupied with oxidizer guest nitrate anions to counterbalance the cationic framework. The porous nature of the MOF results in a moderate density of 1.68 g cm^{-3}. The same group reported another energetic 3D MOF ATRZ-2 $([Ag(ATRZ)_{1.5}(NO_3)]_n)$ from the same linker (Figure 4b). In this case the metal ion is tetrahedral in nature and ATRZ serves as 2- and 3-connecting linker leading to a denser (2.16 g cm^{-3}) 3D cationic network. For this MOF also the cationic charge on the framework is balanced by guest nitrate ion. Both MOFs have very high nitrogen contents (53.35% for ATRZ-1 and 43.76% for ATRZ-2). Despite low crystallographic density, which is detrimental for explosive materials, ATRZ-1 exhibits formidable energetic properties. It has high estimated heat detonation (3.62 kcal g^{-1}). Furthermore, the calculated detonation velocity (9.16 km s^{-1}) and detonation pressure (35.7 GPa) of ATRZ-1 are significantly higher than the precursor ATRZ linker (7.52 km s^{-1} and 23.5 GPa, respectively). These parameters for ATRZ-2 are 1.381 kcal g^{-1}, 7.77 km s^{-1}, and 29.7 GPa, respectively, which are less than ATRZ-1, but higher than ATRZ linker. Differential scanning calorimetric (DSC) study indicates that both MOFs are thermally more sensitive (decomposition temperatures are 243 °C and 257 °C for ATRZ-1 and ATRZ-2, respectively) than the pure ATRZ linker (313 °C). However, both MOFs exhibit relatively low impact and friction sensitivities than ATRZ and are resistant to electrostatic discharge and shock. The same group later reported another ATRZ-based 3D energetic MOF [Zn(ATRZ)$_3$(ClO$_4$)$_2$·2H$_2$O]$_n$, which is isostructural to ATRZ-1 and have perchlorate ions as oxidizing guests.[27] The energetic parameters of this MOF are similar to ATRZ-1.

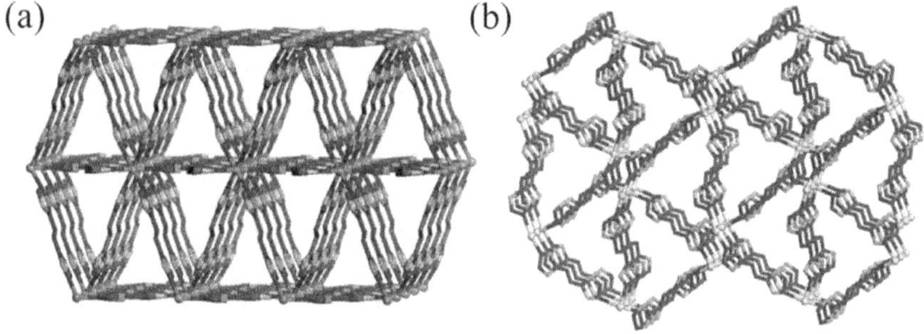

Figure 4. Porous 3D frameworks of (a) ATRZ-1, and (b) ATRZ-2. Hydrogen atoms of the frameworks and guest nitrate ions are omitted for clarity (Cu: dark turquoise, Ag: baby blue, C: grey, N: blue). Reproduced from Reference 1 with permission from American Chemical Society.

Pang and co-workers combined ATRZ with a Cu-graphene oxide complex to synthesize a 3D energetic MOF, which was referred to as GO/Cu(II)/ATRZ hybrid (Scheme 1).[28] In this material, ATRZ link up the 2D layers of Cu-graphene oxide complex to form the 3D network with a very high density of 2.85 g cm^{-3}. Furthermore, the material exhibits low sensitivity (impact sensitivity > 98 J; ESD of 1000 mJ) and extremely high thermal decomposition temperature of 456 °C unlike ATRZ-1 and ATRZ-2 where the linker ATRZ is thermally sensitized. Furthermore, the estimated detonation velocity of this material is 7.08 km s^{-1}.

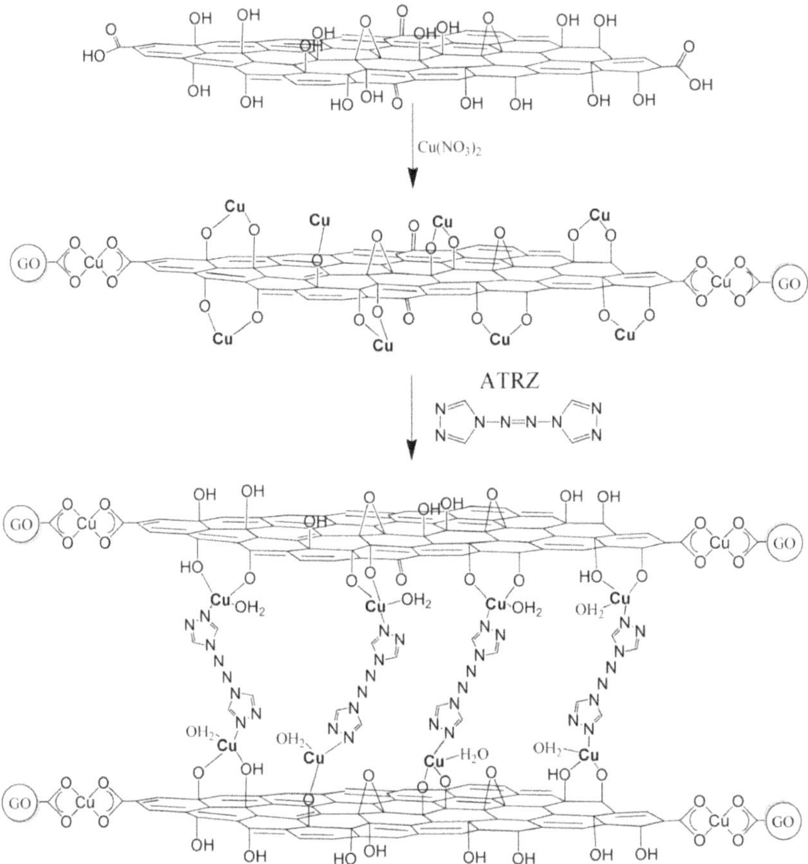

Scheme 1. Construction of GO/Cu(II)/ATRZ hybrid by grafting Cu(II) ions on the graphene oxide surfaces and subsequent treatment with ATRZ linker. Reproduced from Reference 28 with permission from American Chemical Society.

An energetic 3D MOF {[Cu(tztr)]·H$_2$O}$_n$ was reported by Gao and co-workers containing high-energy linker tztr.[24] In this MOF, the linker exhibits coordination mode of 5 and serves as both a chelating and a bridging unit to link up the distorted square pyramidal metal ions. Despite high connectivity between the metal ions and linkers, the MOF has a density of 2.36 g cm^{-3}, which is marginally less than the copper-based 2D MOF of the same linker, due to higher porosity of the 3D framework than the 2D structure. The MOF is stable up to 325 °C and is fairly insensitive to impact, electric discharge and friction. The MOF has an OB of -49.08%, and estimated heat of detonation of 1.32 kcal g^{-1}, detonation velocity of 7.92 km s^{-1} and detonation pressure of 32 GPa. Notably, these estimated energetic parameters are marginally lower than the analogous Cu-tztr MOF with 2D network, which can be attributed to slightly less density of the 3D MOF. Zhang and coworkers employed 5,5'-bistetrazole (Figure 1i) as linker (OB: -58%), which is structurally very similar to tztr, to obtain a 3D Cu-MOF.[29] Coordination of the octahedral Cu(II) ions with the 4-connecting bistetrazolate anions results in a 3D framework of density 2.505 g cm^{-3}. The MOF exhibits slightly higher OB (−36.76%) than the 3D Cu-tztr MOF, has high thermal stability (349.1 °C), and high estimated heat of detonation of 26.73 kJ g^{-1}. The parameters indicate that this MOF is superior energetic material to Cu-tztr. The 5,5'-bistetrazole linker was functionalized with N-hydroxyl groups to develop the linker 5,5'-bistetrazole-1,1'-diol (BTO) with significantly improved OB (-28.2%), and an energetic Pb-MOF was synthesized employing this linker.[30] In this MOF, each Pb(II) ion adopts slightly distorted mono-capped antiprismatic geometry, and each BTO acts as a hexadentate bridging linker to coordinate six metal ions to form a 3D framework. The MOF exhibits high thermal stability of 309.0 °C, and excellent estimated detonation parameters, such as, detonation pressure of 53.1 GPa, detonation velocity of 9.20 km s^{-1} and heat of detonation of 1.0 kcal g^{-1}. Qin et al. employed a tetrazole-functionalized triazole derivative, viz. 4,5-di(1*H*-tetrazol-5-yl)-2*H*-1,2,3-triazole (H₃dttz) as a linker to produce a 3D MOF {[Zn(Hdttz)]·DMA} (DMA: *N,N'*-dimethyl acetamide) upon reaction with zinc nitrate.[31] The linker is dideprotonated in the MOF structure. Each metal ion is tetrahedrally coordinated to four nitrogen atoms from three Hdttz^{2-} fragments to form zeolite-like network. The nitrogen content of the MOF is 47.26%. However, the MOF has less density (1.47 g cm^{-3}) than the pure linker (1.75 g cm^{-3}) due to porous 3D structure. Despite lower density than the organic precursor, the estimated heat of detonation of the MOF (5.62 kcal g^{-1}) is more than five times higher than the pure linker H₃dttz (1.06 kcal g^{-1}). Interestingly, the MOF shows very high thermal stability (up to 392 °C) and thermally more stable than H_3dttz (300 °C) linker. Furthermore, the MOF is insensitive to impact (>40 J), friction (>360 N) and electrostatic discharge.

Recently, several 3D energetic MOFs comprising environmentally benign sodium and potassium ions have been reported that are found to exhibit good thermal stability. As a common feature, these alkali metal-containing MOFs comprise linkers that are functionalized with nitro groups, which coordinate to the alkali metal ions. Therefore, the metal-linker bonds have significant ionic character, which presumably results in high thermal stability of these MOFs. As examples, Shreeve and co-workers reported 3D Na- and K-MOFs employing 3,5-dinitro-1*H*-pyrazol-4-ol as linker (Figures 5a, b).[11] The Na- and K-salts of the same organic compound were also prepared to compare their structures and energetic properties with the MOFs. In the crystal structures, both MOFs were observed to have higher coordination numbers (eight/nine) for the metal ions and shorter M-O bond distances, and therefore, higher densities (1.957 g cm^{-3} for the Na-MOF and 2.223 g cm^{-3} for the K-MOF at room temperature) than the respective metal salts. Furthermore, the MOFs demonstrate super-heat-resistant property with remarkably high thermal stabilities up to 395 °C and 376 °C, respectively. Both MOFs are fairly insensitive secondary explosives with formidable detonation velocities (6.84 km s^{-1} and 7.08 km s^{-1}, respectively) and pressures (21 GPa and 21.8 GPa, respectively). There are also energetic K-MOFs, which can be classified as primary explosives based on their sensitivity to impact, but exhibit high thermal stabilities. One example of such material is a 3D MOF potassium 4,4'-bis(dinitromethyl)-3,3'-azofurazanate.[10] Crystal structural analysis indicates that each K$^+$ ion is coordinated to five oxygen atoms of nitro groups on the linker. The density of the MOF is 2.039 g cm^{-3}. The authors calculated the OB for this MOF considering CO as the decomposition product for the carbon content, and the calculated OB is +10.66%. Interestingly, the K-MOF exhibits good thermal stability up to ~229 °C, which is not common for primary explosives. The impact sensitivity of the K-MOF is 2 J and the friction sensitivity is 20 N. The MOF has detonation velocity of 8.14 km s^{-1}, and detonation pressure of 30.1 GPa. Another K-MOF comprising deprotonated 1,2-bis(3,5-dinitro-1*H*-pyrazol-4-yl)diazene linker was reported by Zhou and coworkers as primary explosive.[32] The MOF contains a 3D framework comprising of high-connected K$^+$ ions coordinated by nitrate and pyrazolate groups (Figure 5c). The MOF has high OB (-15.3%). The estimated velocity and pressure of detonation are

8.28 km s^{-1}, 31.1 GPa, respectively. The MOF exhibits high thermal stability up to 315 °C and impact sensitivity of 1.5 J and friction sensitivity of 60 N. Recently, a series of bimetallic energetic MOFs with formula MK$_2$(DNAT)$_2$·4H$_2$O (M = Fe, Cu, Ni, Co, and Zn) were synthesized based on the linker 1,5-di(nitramino)tetrazole (DNAT) (Figure 5d), and can be categorized as primary explosives.[8] In these MOFs, K$^+$ ion exhibits eight-coordination mode, whereas six-coordinated disposition is adopted by the divalent metal ions. The linker coordinate

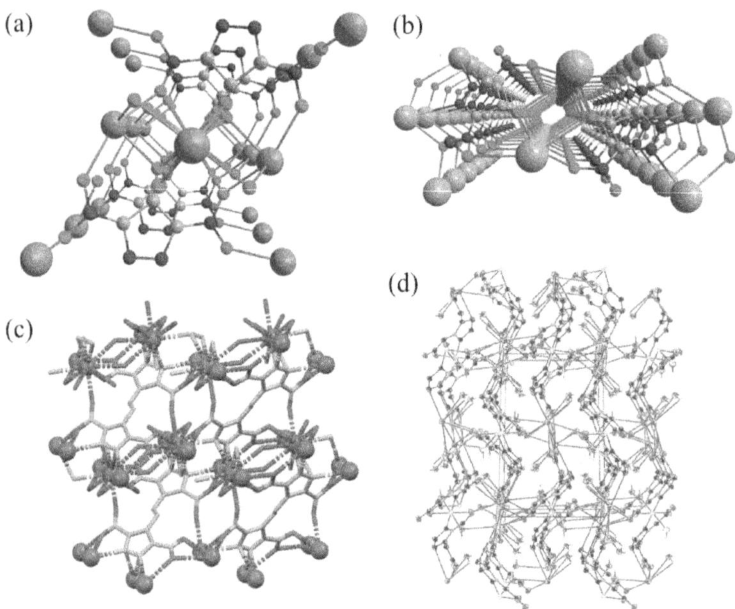

Figure 5. Crystal structures of (a) sodium 3,5-dinitro-4-hydropyrazolate (Na: turquoise), (b) potassium 3,5-dinitro-4-hydropyrazolate (K: green), (c) potassium 1,2-bis(3,5-dinitro-1*H*-pyrazol-4-yl)diazene (Cu: dark turquoise) and (d) CuK$_2$(DNAT)$_2$·4H$_2$O (K: light blue, Cu: turquoise) MOFs. Note that the coordination networks are formed through metal-nitrate coordination linkages. Hydrogen atoms and guest molecules are omitted for clarity (C: gray, N: blue, O: red). Reproduced from Reference 8 (Copyright © 2019 Li, Yu, Zhang, Hu, Chen, Wang and Xu) and from Reference 11 with permission from American Chemical Society.

to the metal ions through nitrogen and oxygen atoms from nitramino groups as well as through the tetrazole rings to form 3D coordination networks in all the cases. The MOFs exhibit acceptable thermal stabilities ranging from 153.8 °C to 208.9 °C. Interestingly, all these MOFs have positive OBs (5.4–5.8%). The detonation velocities of these MOFs range from 8.15 km s⁻¹ to 8.48 km s⁻¹ and detonation pressures vary from 29.7 GPa to 32.8 GPa. Furthermore, impact and friction sensitivities of these MOFs range within 3–6 J and 16–26N, respectively. Based on their formidable energetic parameters and environmentally benign decomposition products, they can be considered as a greener replacement of lead azide (D: 5.88 km s⁻¹, P: 33.4 GPa and OB: −11.0%).

One major drawback of 3D MOFs is that they are often porous, which results in low densities. Furthermore, the nonenergetic guests (usually solvent molecules) present in the porous frameworks reduce the energy density further. For example, [Ag$_7$(tza)$_3$(Htza)$_2$(H$_2$tza)(H$_2$O)] (H$_2$tza: 1*H*-tetrazole-5-acetic acid) is a porous energetic 3D MOF with density of 2.572 g cm⁻³ (Figure 6).[7] Three different modes of coordination are observed for the metal ions, (a) trigonal pyramidal, (b) trigonal planar geometries, and (c) distorted trigonal pyramidal (Figure 6b). The linker coordinates metal ions through both N- and O-donor sites, and serves as 3,4 and 5-coonecting units (Figure 6a). The pores of the MOF are occupied with water molecules in the as-synthesized material and can be removed to obtain the desolvated framework. The desolvated MOF demonstrate higher heat of detonation (2.02 kcal g⁻¹) than the as-synthesized solvated MOF (1.92 g⁻¹). Furthermore, decomposition temperature of the desolvated MOF is slightly less than the as-synthesized MOF, and the desolvated MOF is more sensitive to friction.

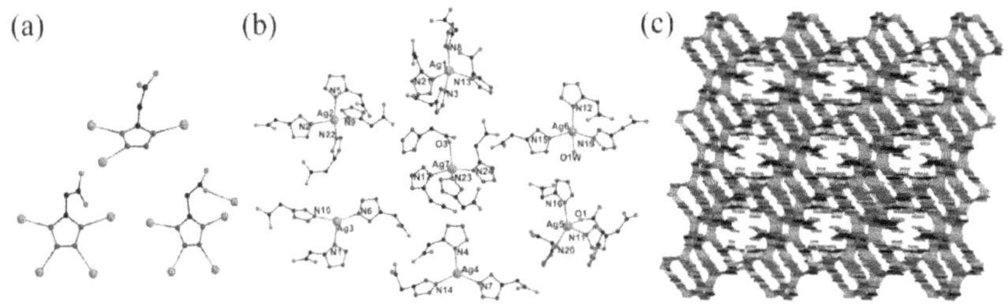

Figure 6. Coordination modes of (a) 1*H*-tetrazole-5-acetate linker and (b) Ag(I) ion, and 3D framework of [Ag$_7$(tza)$_3$(Htza)$_2$(H$_2$tza)(H$_2$O)] MOF (Hydrogen atoms are omitted for clarity). Reproduced from Reference 7 with permission from American Chemical Society.

4. Property Modulation of Energetic MOFs

One key advantage of MOFs is their modularity, which can be judiciously utilized to tune the properties of energetic materials. One way to modulate the properties of HE-MOFs is to substitute the existing guest species with more energetic ones or load the pores with energetic guests. Porous frameworks can be beneficial for this method, although porosity is usually detrimental for energetic properties. For example, the guest nitrate ion in ATRZ-1 MOF can be substituted with more energetic anion dinitramide $[N(NO_2)_2{}^{2-}]$ by treating the MOF crystals with aqueous solution of ammonium dinitramide at room temperature.[33] The anionic guest exchange results in drastic increase of heat of detonation from 4.56 kJ g^{-1} to 7.18 kJ g^{-1}, and decrease in impact sensitivity from 16 J to 9 J. Thus, secondary explosive ATRZ-1 is transformed into a primary explosive through this guest exchange. Furthermore, the relative ratio of nitrate and dinitramide in the MOF pores can be controlled to tune the heat of detonation and sensitivity. Similarly, the nitrate ion can also be replaced with oxidizing anion $IO_3{}^-$, which results in almost doubling (3.168 g cm^{-3}) of the MOF density.[12] Exchange of nitrate guest with iodate results in huge enhancement of OB of the MOF from −57.7% for pristine ATRZ-1 to −12.47% for the iodate incorporated MOF. Consequently, impact and friction sensitivities of the iodate included MOF are much higher (18 J) than ATRZ-1 (60 N). Furthermore, the decomposition products of the iodate included MOF are observed to exhibit excellent biocidal effects. Interestingly, when ATRZ-1 is soaked in a solution of potassium nitroformate $[KC(NO_2)_3]$, the pillaring ATRZ linkers are removed leading to a slightly denser (density: 1.76 g cm^{-3}) 2D MOF $\{Cu(ATRZ)_2[C(NO_2)_3]_2(H_2O)_2 \cdot ATRZ \cdot 2H_2O\}_n$ with a cationic framework structure (Scheme 2). The nitrate anions of the 3D precursor MOF are replaced with nitroformate and ATRZ guests in the resulting 2D MOF.[27] The same transformation occurs for the isostructural $[Zn(ATRZ)_3(ClO_4)_2 \cdot 2H_2O]_n$ MOF as well. In both cases, the estimated heats of detonation for the 2D MOFs (6.05 and 6.48 kJ g^{-1}, respectively for Zn-ATRZ and Cu-ATRZ) are found to be higher than the precursor 3D MOFs (5.68 kJ kg^{-1}; 2a, 4.56 kJ $kg{-1}$, respectively for Zn-ATRZ and Cu-ATRZ).

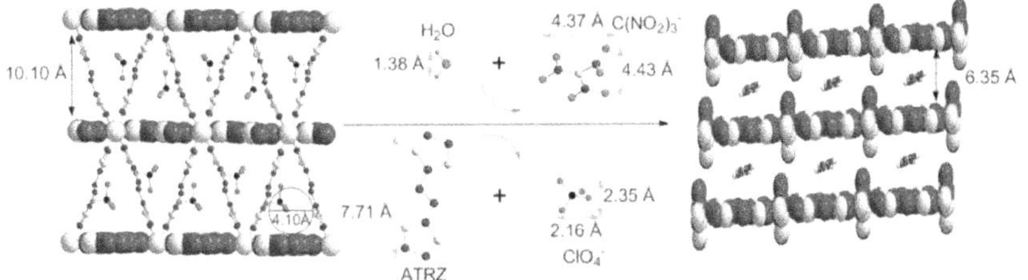

Scheme 2. (a) Tandem exchange of perchlorate guest anions with nitroformate and partial removal of ATRZ linkers leading to transformation of a 3D framework of $[Zn(ATRZ)_3(ClO_4)_2 \cdot 2H_2O]_n$ MOF to a denser MOF with 2D-layered structure (Zn: green, N: blue, O: red, C: light grey, Cl: black). Note that the distance between the neighbouring 2D planes decreases upon removal of the pillaring ATRZ linker. Reproduced from Reference 27 with permission from American Chemical Society.

Isostructural MOFs offer the advantage of property modulation through variation of constituent metal ions or organic linkers. However, the number of known isostructural energetic MOFs with different metal ions or linkers are only a few although several energetic MOFs have been reported to date. Qu et al. synthesized isostructural MOFs Ag-atrz and Ag-ntz by employing 3-amino-1H-1,2,4-triazole (Hatrz) and 3-nitro-1H-1,2,4-triazole (Hntz) as organic linkers, respectively to investigate the effect of amino and nitro pendant groups on the isoreticular frameworks.[34] It should be noted that Hntz (OB: -14%) has higher OB than Hatrz (OB: -38%). Both metal and linkers act as triconnecting units leading to dense 2D network structures. While the 2D nets in Ag-atrz are completely planner, Ag-ntz has corrugated 2D layers (Figure 7). The estimated energetic parameters for Ag-ntz MOF are marginally better than Ag-atrz. For examples, enthalpies of formation are (578.04 ± 2.31) kJ mol^{-1} and (637.04 ± 3.14) kJ mol^{-1}, detonation velocities are 6.52 km s^{-1} and 7.94 km s^{-1}, and detonation pressures are 24.1 GPa and 36.5 GPa, for Ag-atrg and Ag-ntz, respectively. The marginally better properties of Ag-ntz may presumably be attributed to the replacement of the amino groups with oxidizing nitro groups in the isostructural frameworks.

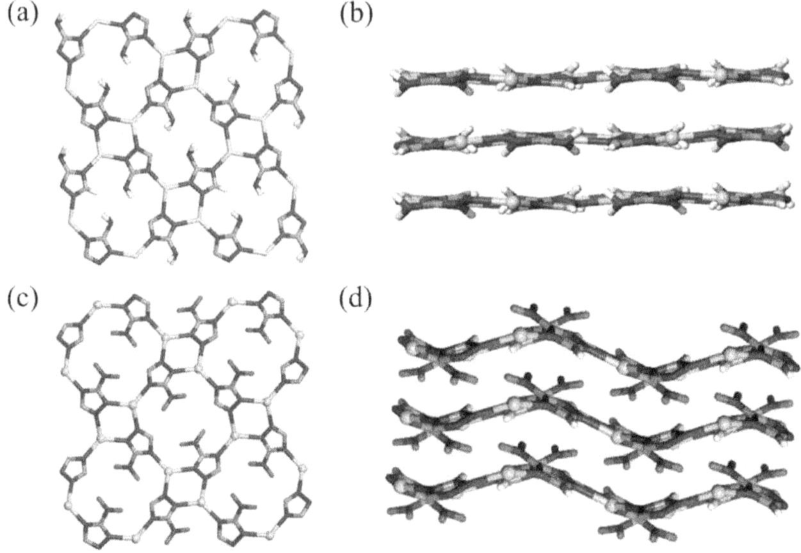

Figure 7. Isostructural 2D networks of (a) Ag-atrz and (c) Ag-ntz MOFs and packing of the 2D layers of (b) Ag-atrz and (d) Ag-ntz in their crystal lattices. Hydrogen atoms are omitted for clarity (Ag: baby blue, C: gray, N: blue, O: red). Note that the 2D polymeric layers of Ag-atrz are almost planner, while in case of Ag-ntz, the 2D layers are corrugated due to the presence of sterically bulky -NO$_2$ groups. Reproduced from Reference 1 with permission from American Chemical Society.

Matzger and coworkers reported two isostructural energetic MOFs with Zn^{2+} and Co^{2+} metal ions using 5-amino-3-nitro-1H-1,2,4-triazole (ANTA) as the organic building block.[35] The MOFs have 2D square-lattice networks where the metal ions exhibit tetrahedral coordination mode and are bridged with deprotonated ANTA and hydroxide anions (Figure 8a). The anions of ANTA are arranged almost perpendicular to the corrugated 2D planes of the coordination polymers, which results in high packing efficiency as well as high crystallographic densities for both MOFs (2.297 g cm^{-3} for Co-ANTA and 2.336 g cm^{-3} for Zn-ANTA at room temperature). The OBs of these MOFs were calculated to be -27.6% for Co-ANTA and -26.6% for Zn-ANTA. Notably, the OBs are significantly higher than the precursor ANTA molecule (-43.8%) and also comparable with the state-of-the-art energetic materials RDX and HMX. While the precursor organic compound ANTA decomposes at 225 °C, both MOFs are thermally stable beyond 300 °C. Zn-ANTA (decomposition temperature: 310 °C) has marginally higher decomposition temperature than Co-ANTA (decomposition temperature: 325 °C) (Figure 8b). Both MOFs behave as secondary explosives based on their sensitivities to impact. However, these MOFs exhibit strikingly different magnitude of impact sensitivities despite being isostructural in nature with comparable crystallographic densities, OBs and thermal stabilities. Similar to organic energetic ANTA, Zn-ANTA exhibits low impact sensitivity ($D_{50}h$ higher than 217 cm for 5 lb weight), whereas Co-ANTA has much higher sensitivity to impact ($D_{50}h$: 94 cm for 5 lb weight) (Figure 8c). The authors demonstrated high impact sensitivity of Co-ANTA as the result of unpaired electrons present on the metal ion, which presumably influence the decomposition pathway of the MOF and render it much more sensitive to impact than Zn-ANTA, containing metal ion of d^{10} electronic configuration.

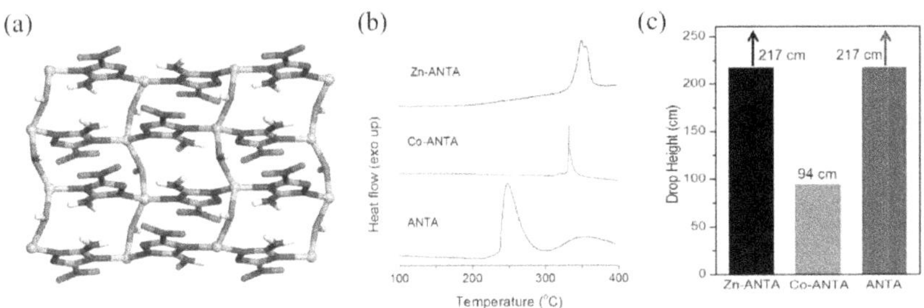

Figure 8. (a) 2D network (metal ion: baby blue, N: navy blue, O: red, H: white), (b) DSC profiles and (c) impact sensitivity of the isostructural MOFs containing ANTA linker. Reproduced from Reference 35 with permission from American Chemical Society.

Another pair of isostructural energetic MOFs are $[Co(N_3)_2(ATRZ)]_n$ and $[Cd(N_3)_2(ATRZ)]_n$ (ATRZ = 4,4′-azo-1,2,4-triazole) that were synthesized by employing azide as an energetic colinker with ATRZ.[36] In both MOFs, the metal ions exhibit octahedral disposition and are coordinated by two ATRZ and four N_3^- ions to form 2D networks. Both MOFs are stable up to 200 °C (Cd-MOF exhibits ~10 °C higher stability than Co-MOF). The estimated heats of formation are 4.21 kJ·g⁻¹ and 4.08 kJ g⁻¹, heats of detonation are 4.41 and 4.25 kJ g⁻¹, detonation velocities are 7.67 km s⁻¹ and 7.54 km s⁻¹, and detonation pressure are 28.5 GPa and 28.7 GPa, respectively, for the Co- and Cd-MOFs. Furthermore, their friction sensitivities are 5 and 12 N, and impact sensitivities are 1.2 J and 1.6 J, respectively. Based on the mechanical sensitivities, the Co-MOF is categorized as 'extremely sensitive' primary explosive, while the Cd-MOF is categorized as 'very sensitive' primary explosives. To understand the factors that cause the variation in the energetic properties of the isostructural MOFs, noncovalent interactions in the crystal structures were calculated, which indicates that the Co-MOF has stronger repulsive (steric) interactions among the adjacent azide ions than the Cd-MOF, and therefore, more sensitive than the later (Figure 9). These examples indicate that it is indeed possible to modulate the properties of energetic MOFs by the varying the metal ions or organic building blocks for a given framework structure, or by exchanging the guest species with more energetic ones.

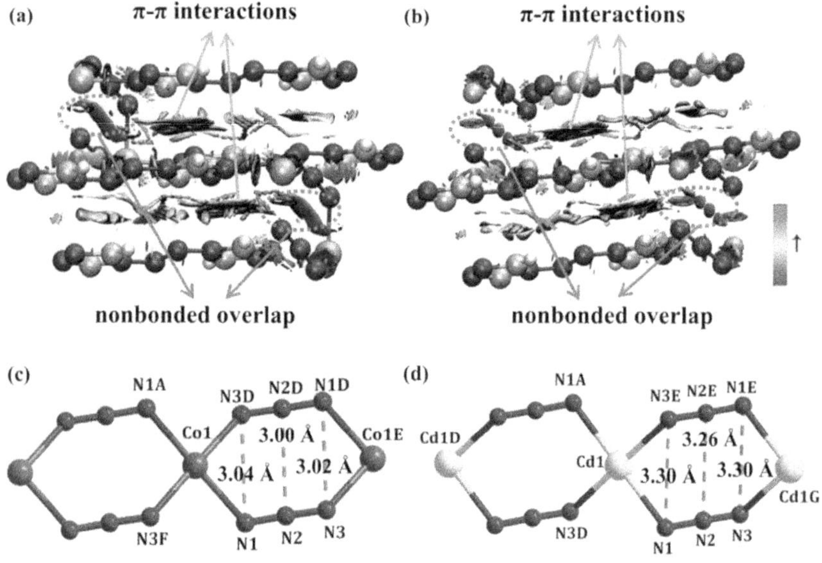

Figure 9. Gradient isosurfaces for noncovalent interactions in (a) $[Co(N_3)_2(ATRZ)]_n$ and (b) $[Cd(N_3)_2(ATRZ)]_n$. The extent of interactions is represented in a blue-green-red (BGR) scale (−0.04 to 0.02 au), where blue represents strong attractive interactions, green indicates weak attractive interactions, and red represents strong nonbonded overlap. (c,d) N⋯N distances between the adjacent azide ions in the Co-MOF and Cd-MOF, respectively (Co: brick red, Cd: baby blue, N: royal blue). It should be noted that nonbonded overlap between the adjacent azide ions in the Co-MOF is larger than the Cd-MOF, which is evident from the smaller average N⋯N distances between the adjacent azide ions in the Co-MOF than the Cd-MOF. Reproduced from Reference 36 with permission from American Chemical Society.

5. MOF-Oxidizer Composites

It has been discussed in the previous section 4 that exchange of counterionic guests with suitable species can systematically modulate the
energetic properties of MOFs. This approach can be further extended
to load an oxidizer guest (can be ionic or neutral) in the pores of 'activated' (guest evacuated) MOFs to the saturation level to obtain energetic composites. While the guests with high positive OBs act as the
oxidizer components, the carbon rich frameworks of MOFs serve as
fuels in such composites. Inclusion of the guest species in the micropores of MOFs results in intimate mixing of the oxidizing and fuel
components. Nelson et al. introduced this concept and developed several energetic composites by loading oxidizers in porous MOFs.[37] A
3D MOF with formula of [$Zn_2(NDC)_2DPNI$] was used as the porous
host, and a number or liquid organic energetic guests, viz., butanetriol
trinitrate (BTTN), tetramethylolethane trinitrate (TMETN), triethy-
leneglycol dinitrate (TEGDN), and n-butyl-2-nitratoethylnitramine
(BuNENA) were loaded. Typically, the DMF guest molecules present
in the as-synthesized MOF were exchanged with these energetic
guests by soaking them in the neat liquids. Another porous MOF, com-
mercially available as Basolite™Al00 was utilized to load solid energetic guest, viz., trinitrotoluene. The MOF was activated by heating at
250 °C under reduced pressure. Subsequently, the activated MOF was
mixed with TNT crystals, and the mixture was heated to 90 °C to allow
TNT melt and diffuse into the MOF pores. This method, however, resulted in a physical mixture rather a host-guest composite due to deposition of solid TNT on the MOF crystals in addition to diffusion into
the pores as revealed from DSC analysis. DSC profile of the resulting
material indicates endothermic peaks at 81 °C and 211 °C, respectively in addition to the exothermic decomposition signal of the composite above 300 °C. The endothermic peaks correspond to melting
and decomposition of the surface bound TNT. Matzger and coworkers introduced a modified method of sublimation under reduced
pressure to obtain consistent loading of solid organic oxidizer without any deposition on MOF crystals (Scheme 3).[38] They utilized activated MOF-5 as the host and hexanitroethane as the oxidizer guest.
Rather than starting with a physical mixture of the two components,
they used separate chambers for the compounds, and consistent
loading was obtained by controlling the sublimation temperature and
diffusion time. The same group synthesized another energetic composite MOF-5-TNM by loading liquid oxidizer tetranitromethane in

MOF-5 by vapor diffusion. It is observed that at saturation of loading, the MOF-5-TNM composite contains as much as 55 wt% of TNM, which translates to an OB of -14.6%. Similarly, up to 62 wt% oxidizer loading is possible in the case of MOF-5-HNE composite leading to an OB of -8.7%. It should be noted that the host MOF-5 has a highly negative OB of -93.6% and OB for the oxidizer guests are +57% for TNM and +42.7% for HNE. As expected, the heat released upon thermal decomposition (by DSC) of the composites is higher than the corresponding oxidant guests. The released heat for MOF-5-TNM composite is slightly inconsistent as the volatile guest tend to escape from the composite. The result is more consistent for MOF-5-HNE, which contains a less volatile guest. The released heat is 2.48 kJ g^{-1} compared to 2.14 kJ g^{-1} for pure solid HNE. Detonation velocities of the composites (5.24 km s^{-1} for MOF-5-TNM and 6.16 km s^{-1} for MOF-5-HNE) were calculated using thermochemical code Cheetah 7.0 and are found to be slightly smaller than pure oxidants. Based on the DSC thermograms, the composites are thermally more sensitive than the pure oxidants. MOF-5-TNM is stable up to 197 °C and MOF-5-HNE decomposes at 135 °C. Furthermore, both TNM and HNE become more sensitive to impact upon inclusion in MOF-5. Both of pure TNM and HNE behave as secondary explosives with $D_{50}h$ of 34 and 21 cm, respectively. The $D_{50}h$ values for MOF-5-TNM and MOF-5-HNE are 17 and 7 cm, respectively, which indicates that secondary explosive HNE behaves like a primary explosive when loaded in MOF-5. This concept of loading organic oxidizers in MOFs to develop energetic composites was further tested for other MOFs, such as, ZIF-8 (pore volume of 50%) and ZIF-70 (pore volume of 54%) for TNM oxidant.[39] 15.8 wt% (OB: -122%) and 46.7 wt% (OB: -26.5%) loading of TNM were achieved for ZIF-8-TNM and ZIF-70-TNM composites, respectively. It should be noted that TNM loading in ZIF-8 is significantly lower than ZIF-70 although pore volume of ZIF-8 is only marginally less than ZIF-70. The authors reported that the loading in ZIF-8 is underestimated due to easier escape of guests from the composite under ambient conditions. The ZIF-8-TNM ($D_{50}h$: 36 cm for 5 lb weight) acts as a secondary explosive and impact sensitivity of ZIF-70-TNM composite ($D_{50}h$: 17 cm for 5 lb weight) lies in the borderline of primary and secondary explosives.

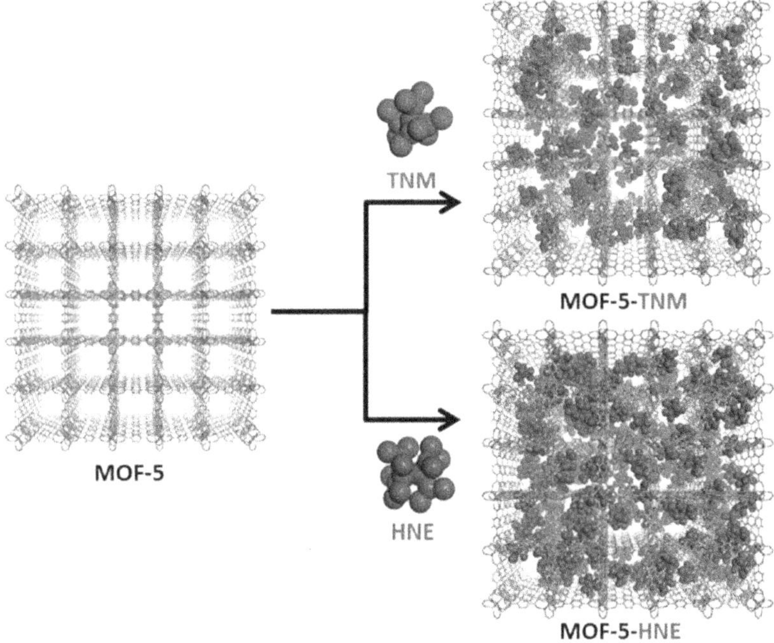

Scheme 3. Schematic representation for the loading of TNM (blue) and HNE (purple) oxidizer guests into MOF-5 to develop MOF-5-TNM and MOF-5-HNE energetic composites.

The same group developed impact insensitive energetic composites as secondary explosives by loading inorganic oxidizers in porous MOFs. In general, inorganic oxidizers (usually nitrate and perchlorate salts) are nonvolatile, and less sensitive to stimuli, such as impact and heat, compared to organic oxidizers. One challenge with them, however, is that they exhibit significantly high melting points and their vapor pressures are extremely low even at temperatures as high as 250 °C. Therefore, loading inorganic oxidizers in MOFs by diffusion in molten or vapor phases is extremely challenging, except for ammonium nitrate. Ammonium nitrate exhibits sublimation property, and a relatively low melting point of 169 °C. In fact, AN was successfully loaded in the MOF UiO-66 by heating a ground mixture of the MOF and salt at only 75 °C. For other oxidizer salts, attempt of loading from their aqueous solutions often result in physical mixtures with salt included in the MOF pores as well as deposited on outer surfaces of MOF crystals.[37] However, solvent-assisted grinding of MOF and salt in the presence of optimal amount of solvent has been successful to load $LiNO_3$ in UiO-66.[40] Matzger and coworkes were successful to load 32 wt% NH_4NO_3 and 38 wt% $LiNO_3$ in UiO-66 at saturation. The resulting NH_4NO_3-UiO-66 and $LiNO_3$-UiO-66 composites have densities 1.787 g cm^{-3} and 1.967 g cm^{-3}, respectively and OBs of -53.9% and -32.7%, respectively as compared to a density of 1.24 g cm^{-3} and OB of -88.0% for the pristine host MOF.[41] SEM-EDX analysis supports that complete inclusion of the salts in the pores of the MOF in these composites leads to molecular level mixing of oxidizer and fuel components (Figure 10). Both composites are insensitive secondary explosives and are stable up to 225 °C (for NH_4NO_3-UiO-66) and 340 °C (for $LiNO_3$-UiO-66) (Figure 11). The DSC profiles indicate that phase transitions and melting of the pristine salts are completely suppressed when they are loaded in the MOFs. It should be noted that phase transition is a major limitation for NH_4NO_3 (Figure 11a) to be used as energetic material, because the phase transitions are associated with significant change in crystal volume and density. Overall, loading of suitable oxidizers in MOFs can be utilized to obtain primary and secondary explosives with tunable properties.

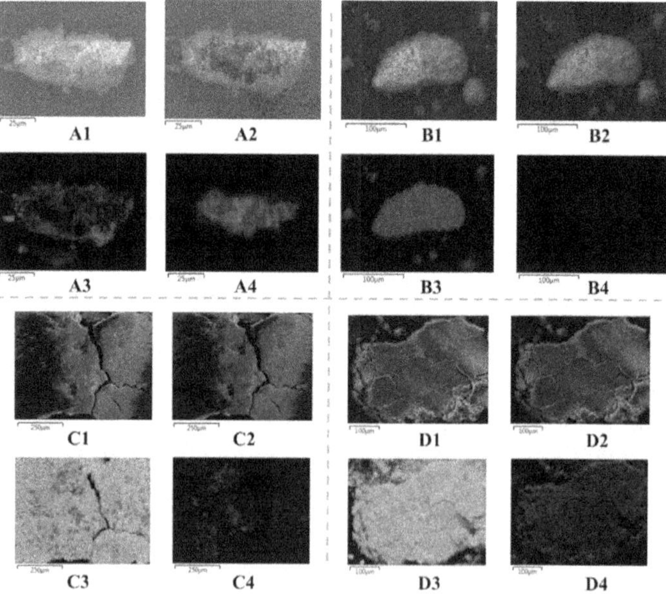

Figure 10. SEM image and elemental mapping of zirconium (green) and nitrogen (red) in the (A) physical mixture of NH4NO3 and UiO-66, (B) NH4NO3-UiO-66 composite, (C) physical mixture of LiNO3 and UiO-66 physical mixture and (D) LiNO3-UiO-66 composite based on EDS analysis. Reproduced from Reference 40, copyright Royal Society of Chemistry.

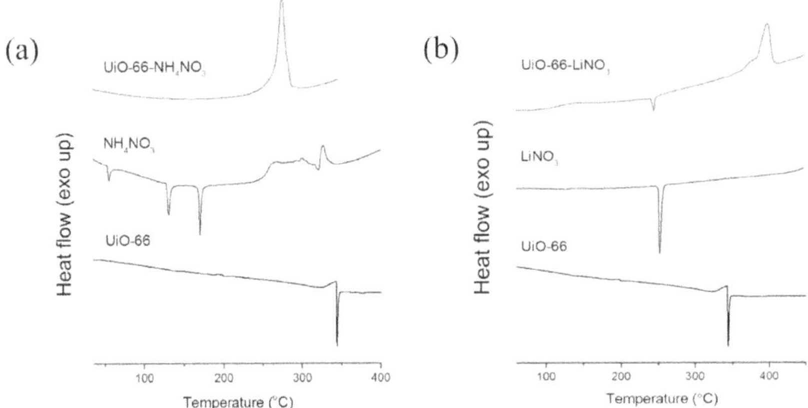

Figure 11. DSC profiles of (a) NH_4NO_3-UiO-66 and (b) $LiNO_3$-UiO-66 composites and their comparison with that of the pristine host UiO-66 and the oxidizer salts. Reproduced from Reference 40, copyright Royal Society of Chemistry.

6. Conclusions and Outlook

The area of high-energy MOFs has made significant advancement over the last fifteen years with the development of novel energetic MOFs and development of methods to estimate their explosive power. Based on the diversity of metal dispositions and linker coordination modes, it is possible to create diverse energetic MOFs with modular properties. Furthermore, isostructural frameworks provide the opportunity to tune the energetic properties through metal or linker substitution. Energetic MOFs can have variable sensitivities to be categorized as primary and secondary explosives. Particularly, several potassium MOFs with furazanate or pyrazolate linkers have recently been developed that exhibit high impact sensitivities and can be considered as green replacement of heavy metal containing primary explosives, such as, lead azide, mercury fulminate etc. In addition to primary or secondary explosive behaviors of energetic MOFs, pyrotechnic and hypergolic properties have also been observed in some cases. Energetic MOFs exhibit promising heats of detonation, detonation velocities and detonation pressures. For many MOFs, these parameters are significantly high and comparable with the state-of-the-art energetics (Figure 12). Furthermore, high-energy MOFs may have tunable OBs, and good OBs have been achieved in many of them (Figure 13).

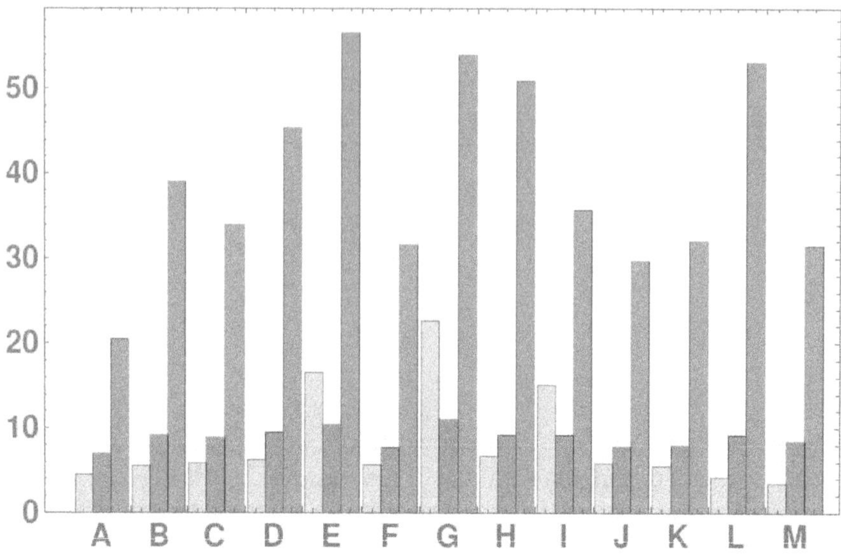

Figure 12. Comparison of estimated heat of detonation (in kJ g^{-1}, color code: orange), detonation velocity (in km s^{-1}, color code: brown) and detonation pressure (in GPa, color code: blue) of some promising high-energy MOFs with state-or-the-art energetic materials. A: TNT, B: HMX, C: RDX, D: CL-20, E: [Cu(Htztr)]$_n$ (2D MOF), F: [Pb(Htztr)$_2$(H$_2$O)]$_n$, G: [(AG)$_2$(Cu(btm)$_2$)]$_n$, H: Pb-BTF, I: ATRZ-1, J: ATRZ-2, K: {[Cu(tztr)]·H$_2$O}$_n$ (3D MOF) and L: Pb-BTO, M: CuK$_2$(DNAT)$_2$·4H$_2$O. Note that the estimated detonation properties of some high-energy MOFs are even better than the state-or-the art explosives.

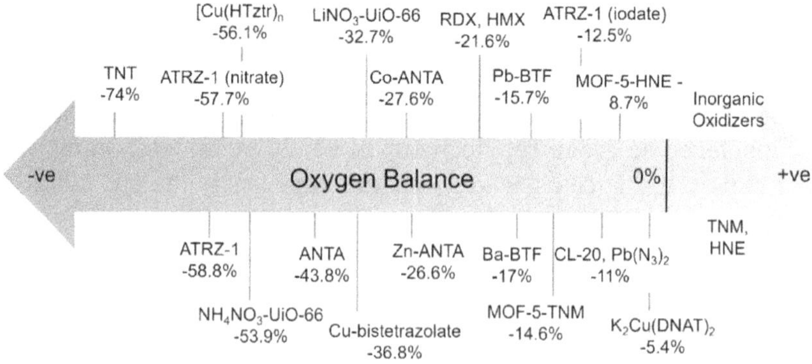

Figure 13. Calculated OBs of energetic MOFs and their comparison with OBs of some state-of-the-art energetic materials.

In addition to nonporous energetic MOFs, porous MOFs can also be utilized to obtain dense energetic composite materials by loading energetic guests in them. For ionic frameworks, counterionic guest species have been successfully substituted with desired oxidizer ions. The carbon rich frameworks of MOFs can be used as fuel hosts for oxidizer guest species. This method results in molecular level mixing of oxidizer and fuel, which is evident from different spectroscopic analyses. Properties of intimately mixed oxidizer-fuel composites, e.g., density, OB and sensitivity to external stimuli can be reliably varied based on the nature of the guest as well as extent of loading. By this approach, it is possible to increase the impact sensitivity so much so that the precursor secondary material can switch to a primary explosive.

Despite all these attributes of high-energy MOFs and MOF-based energetic materials, this area lacks a thorough and systematic investigation to understand structure-function relationship. For most cases, the energetic properties (heat of detonation, detonation velocity and pressure) are estimated by the help of theoretical calculations, where crystallographic density is used in the calculations. It should be noted that most of the MOFs have higher densities than the constituent organic linkers due to the presence metal ions, which may or may not take part in the energetic decomposition pathways of the MOFs. Therefore, the energetic parameters might be overestimated until the mechanism of decomposition is well-established. Furthermore, the estimated values have hardly been verified experimentally, except for a very few cases, the heats of detonations have been experimentally determined. Additionally, both heats of detonation and sensitivities are tested only in laboratory scale by using very small quantity of the materials. For real applications, safe methods for manufacturing these materials and property testing in larger scales are crucial.

This field of research is still in the developing stage with a lot of scope yet to be explored. In this regard, it should be noted that particle size and morphology play important roles on properties of materials. This is evident in the cases of small molecular organic energetic materials that polymorphic phases can have variable properties (density, sensitivity, etc.). However, morphological variation of energetic MOFs has not yet been reported. Overall, it is envisioned that high-energy MOFs have promising future prospects in the field of energetic materials.

Acknowledgements

SS gratefully acknowledges financial support from SERB-DST (grant no. SRG_2020_001014) and UGC-BSR Research Start-up-Grant from UGC India.

References

1. McDonald, K. A., Seth, S., and Matzger, A. J. (2015) Coordination Polymers with High Energy Density: An Emerging Class of Explosives. Crystal Growth & Design, 15(12), 5963-5972.
2. Tominaka, S., Hamoudi, H., Suga, T., Bennett, T. D., Cairns, A. B., and Cheetham, A. K. (2015) Topochemical conversion of a dense metal–organic framework from a crystalline insulator to an amorphous semiconductor. Chemical Science, 6(2), 1465-1473.
3. Klapötke, T. M. (2012) Chemistry of High-Energy Materials, 2nd edition, De Gruyter, Germany.
4. Zhang, J., and Shreeve, J. n. M. (2016) 3D Nitrogen-rich metal–organic frameworks: opportunities for safer energetics. Dalton Transactions, 45(6), 2363-2368.
5. Zhang, S., Yang, Q., Liu, X., Qu, X., Wei, Q., Xie, G., Chen, S., and Gao, (2016) S. High-energy metal–organic frameworks (HE-MOFs): Synthesis, structure and energetic performance. Coordination Chemistry Reviews, 307, 292-312.
6. Zhang, Q., and Shreeve, J. n. M. (2014) Metal–Organic Frameworks as High Explosives: A New Concept for Energetic Materials. Angewandte Chemie International Edition, 53(10), 2540-2542.
7. Ma, X., Cai, C., Sun, W., Song, W., Ma, Y., Liu, X., Xie, G., Chen, S., and Gao, S. (2019) Enhancing Energetic Performance of Multinuclear Ag(I)-Cluster MOF-Based High-Energy-Density Materials by Thermal Dehydration. ACS Applied Materials & Interfaces, 11(9), 9233-9238.
8. Li, Y., Yu, T., Zhang, Y., Hu, J., Chen, T., Wang, Y., and Xu, K. (2019) Novel Energetic Coordination Polymers Based on 1,5-Di(nitramino)tetrazole With High Oxygen Content and Outstanding Properties: Syntheses, Crystal Structures, and Detonation Properties. Frontiers in Chemistry, 7(672), doi: 10.3389/fchem.2019.00672.
9. Li, S., Wang, Y., Qi, C.; Zhao, X., Zhang, J., Zhang, S., and Pang, S. (2013) 3D Energetic Metal–Organic Frameworks: Synthesis and Properties of High Energy Materials. Angewandte Chemie International Edition, 52(52), 14031-14035.
10. Tang, Y., He, C., Mitchell, L. A.; Parrish, D. A., and Shreeve, J. n. M. 2016, Potassium 4,4′-Bis(dinitromethyl)-3,3′-azofurazanate: A

Highly Energetic 3D Metal–Organic Framework as a Promising Primary Explosive. Angewandte Chemie International Edition, 55(18), 5565-5567.

11. Zhang, J., Zhang, J., Imler, G. H., Parrish, D. A., and Shreeve, J. n. M. 2019, Sodium and Potassium 3,5-Dinitro-4-hydropyrazolate: Three-Dimensional Metal–Organic Frameworks as Promising Super-heat-resistant Explosives. ACS Applied Energy Materials, 2, 7628-7634.

12. Zhang, J., Zhu, Z., Zhou, M., Zhang, J., Hooper, J. P., and Shreeve, J. n. M. (2020) Superior High-Energy-Density Biocidal Agent Achieved with a 3D Metal–Organic Framework. ACS Applied Materials & Interfaces, 12(36), 40541-40547.

13. Meyer, R; Köhler, J.; and Homburg, A. Explosives, 6th ed., Wiley-VCH Gmbh & Co. KGaA, 2007.

14. Gao, W., Liu, X., Su, Z., Zhang, S., Yang, Q., Wei, Q., Chen, S., Xie, G., Yang, X., and Gao, S. (2014) High-energy-density materials with remarkable thermostability and insensitivity: syntheses, structures and physicochemical properties of Pb(ii) compounds with 3-(tetrazol-5-yl) triazole. Journal of Materials Chemistry A, 2(30), 11958-11965.

15. Liu, X., Yang, Q., Su, Z., Chen, S., Xie, G., Wei, Q., and Gao, S. (2014) 3D high-energy-density and low sensitivity materials: synthesis, structure and physicochemical properties of an azide–Cu(ii) complex with 3,5-dinitrobenzoic acid. RSC Advances, 4(31), 16087-16093.

16. Biegańska, J. (2021) The Effect of the Reaction pH on Properties of Lead(II) Azide. Materials, 14(11), 2818.

17. Qi, C., Li, S.-H., Li, Y.-C., Wang, Y., Chen, X.-K., and Pang, S.-P. (2011) A novel stable high-nitrogen energetic material: 4,4'-azobis(1,2,4-triazole). Journal of Materials Chemistry, 21(9), 3221-3225.

18. Blair, L. H., Colakel, A., Vrcelj, R. M., Sinclair, I., and Coles, S. J. (2015) Metal–organic fireworks: MOFs as integrated structural scaffolds for pyrotechnic materials. Chemical Communications, 51(61), 12185-12188.

19. Larin, A. A., Muravyev, N. V., Pivkina, A. N., Suponitsky, K. Y., Ananyev, I. V., Khakimov, D. V., Fershtat, L. L., and Makhova, N. N. (2019) Assembly of Tetrazolylfuroxan Organic Salts: Multipurpose Green Energetic Materials with High Enthalpies of Formation and Excellent Detonation Performance. Chemistry – A European Journal, 25(16), 4225-4233.

20. Seth, S., and Matzger, A. J. (2017) Metal–Organic Frameworks: Examples, Counterexamples, and an Actionable Definition. Crystal Growth & Design, 17(8), 4043-4048.

21. Tang, Z., Zhang, J.-G., Liu, Z.-H., Zhang, T.-L., Yang, L., and Qiao, X.-J. (2011) Synthesis, structural characterization and thermal analysis of a high nitrogen-contented cadmium (II) coordination polymer

based on 1,5-diaminotetrazole. Journal of Molecular Structure, 1004(1), 8-12.

22. Wu, B.-D., Zhou, Z.-N., Li, F.-G., Yang, L., Zhang, T.-L., and Zhang, J.-G. (2013) Preparation, crystal structures, thermal decompositions and explosive properties of two new high-nitrogen azide ethylenediamine energetic compounds. New Journal of Chemistry, 37(3), 646-653.

23. Bushuyev, O. S., Peterson, G. R., Brown, P., Maiti, A., Gee, R. H., Weeks, B. L., and Hope-Weeks, L. J. (2013) Metal–Organic Frameworks (MOFs) as Safer, Structurally Reinforced Energetics. Chemistry – A European Journal, 19(5), 1706-1711.

24. Liu, X., Gao, W., Sun, P., Su, Z., Chen, S., Wei, Q., Xie, G., and Gao, S. (2015) Environmentally friendly high-energy MOFs: crystal structures, thermostability, insensitivity and remarkable detonation performances. Green Chemistry, 17(2), 831-836.

25. Feng, Y., Bi, Y., Zhao, W., and Zhang, T. (2016) Anionic metal–organic frameworks lead the way to eco-friendly high-energy-density materials. Journal of Materials Chemistry A, 4(20), 7596-7600.

26. Hao, W., Jin, B., Zhang, J., Li, X., Huang, T., Shen, J., and Peng, R. (2020) Novel energetic metal–organic frameworks assembled from the energetic combination of furazan and tetrazole. Dalton Transactions, 49(19), 6295-6301.

27. Zhang, J., Su, H., Dong, Y., Zhang, P., Du, Y., Li, S., Gozin, M., and Pang, S. (2017) Synthesis of Denser Energetic Metal–Organic Frameworks via a Tandem Anion–Ligand Exchange Strategy. Inorganic Chemistry, 56(17), 10281-10289.

28. Cohen, A., Yang, Y., Yan, Q.-L., Shlomovich, A., Petrutik, N., Burstein, L., Pang, S.-P., and Gozin, M. (2016) Highly Thermostable and Insensitive Energetic Hybrid Coordination Polymers Based on Graphene Oxide–Cu(II) Complex. Chemistry of Materials, 28(17), 6118-6126.

29. Chen, S., Zhang, B., Yang, L., Wang, L., and Zhang, T. (2016) Synthesis, structure and characterization of neutral coordination polymers of 5,5′-bistetrazole with copper(ii), zinc(ii) and cadmium(ii): a new route to reconcile oxygen balance and nitrogen content of high-energy MOFs. Dalton Transactions, 45(42), 16779-16783.

30. Shang, Y., Jin, B., Peng, R., Liu, Q., Tan, B., Guo, Z., Zhao, J., and Zhang, Q. (2016) A novel 3D energetic MOF of high energy content: synthesis and superior explosive performance of a Pb(ii) compound with 5,5′-bistetrazole-1,1′-diolate. Dalton Transactions, 45(35), 13881-13887.

31. Qin, J. S., Zhang, J. C., Zhang, M., Du, D. Y., Li, J., Su, Z. M., Wang, Y. Y., Pang, S. P., Li, S. H., and Lan, Y. Q. (2015) A Highly Energetic N-Rich Zeolite-Like Metal-Organic Framework with Excellent Air Stability and Insensitivity. Adv Sci (Weinh), 2(12), 1500150.

32. Zhang, M., Fu, W., Li, C., Gao, H., Tang, L., and Zhou, Z. (2017) (E)-1,2-Bis(3,5-dinitro-1H-pyrazol-4-yl)diazene – Its 3D Potassium Metal–Organic Framework and Organic Salts with Super-Heat-Resistant Properties. European Journal of Inorganic Chemistry, 2017(22), 2883-2891.

33. Zhang, J., Du, Y., Dong, K., Su, H., Zhang, S., Li, S., and Pang, S. (2016) Taming Dinitramide Anions within an Energetic Metal–Organic Framework: A New Strategy for Synthesis and Tunable Properties of High Energy Materials. Chemistry of Materials, 28(5), 1472-1480.

34. Qu, X., Zhang, S., Yang, Q., Su, Z., Wei, Q., Xie, G., and Chen, S. (2015) Silver(i)-based energetic coordination polymers: synthesis, structure and energy performance. New Journal of Chemistry, 39(10), 7849-7857.

35. Seth, S., McDonald, K. A., and Matzger, A. J. (2017) Metal Effects on the Sensitivity of Isostructural Metal–Organic Frameworks Based on 5-Amino-3-nitro-1H-1,2,4-triazole. Inorganic Chemistry, 56(17), 10151-10154.

36. Xu, J.-G.; Sun, C., Zhang, M.-J., Liu, B.-W., Li, X.-Z., Lu, J., Wang, S.-H., Zheng, F.-K., and Guo, G.-C. (2017) Coordination Polymerization of Metal Azides and Powerful Nitrogen-Rich Ligand toward Primary Explosives with Excellent Energetic Performances. Chemistry of Materials, 29(22), 9725-9733.

37. Nelson, A. P., and Trivedi, N. J. Host-guest complexes of solid oxidizers and metal-organic frameworks, U.S. Patent 8,197,619 B1.

38. McDonald, K. A., Bennion, J. C., Leone, A. K., and Matzger, A. J. (2016) Rendering non-energetic microporous coordination polymers explosive. Chemical Communications, 52(72), 10862-10865.

39. Kent, R. V., Vaid, T. P., Boissonnault, J. A., and Matzger, A. J. (2019) Adsorption of tetranitromethane in zeolitic imidazolate frameworks yields energetic materials. Dalton Transactions, 48(22), 7509-7513.

40. Seth, S., Vaid, T. P., and Matzger, A. J. (2019) Salt loading in MOFs: solvent-free and solvent-assisted loading of NH4NO3 and LiNO3 in UiO-66. Dalton Transactions, 48(35), 13483-13490.

41. Bambalaza, S. E., Langmi, H. W., Mokaya, R., Musyoka, N. M., Ren, J., and Khotseng, L. E. Compaction of a zirconium metal–organic framework (UiO-66) for high density hydrogen storage applications. Journal of Materials Chemistry A, 6(46), 23569-23577.

2

Solid-State [2+2] Photocycloaddition Reactions in Coordination Polymers and Metal-Organic Frameworks

Raghavender Medishetty, Akansha Ekka, Anshumika Mishra and Uma Kurakula*
Department of Chemistry, Indian Institute of Technology - Bhilai (IIT Bhilai), GEC Campus, Raipur, Chhattisgarh, India

1. Introduction

1.1 Solid-state [2+2] Cycloaddition Reaction

Solid-state photochemical reactions involving [2+2] cycloaddition have continued to be one of significant research interests for the past several decades. It was G. Schmidt and his fellow mates who paved way towards the advancement in the field of organic solid-state photo cycloaddition reactions and crystal engineering of cinnamic acid and its derivatives and their polymorphs.[1] Solid state [2+2] photochemical reaction has been recognized since 19th century as a green method to synthesize *regio-* and *stereo*-selective cyclobutane derivatives in quantitative yields, which are almost impossible to obtain through traditional solution based methods.[2-5] This is mainly due to the formation of strained cyclobutane and the occurrence of various side reactions like radical polymerization and cross-linking and so on. Hence, carrying out the [2+2] reactions with topochemical reactions becomes more interesting that involves initial arrangement of monomers which makes the reaction stereo-as well as regio-selective.

This topochemical reaction was systematically explored by Schmidt in 1970's who proposed the postulates for solid-state photo reactivity of olefins which are separated by 4.2 Å or below, rendering the possible movement of olefin groups in the solid, called as Schmidt's criteria.[1] Later this reaction was successfully utilized for the preparation of high-ly strained cyclobutane derivatives in quantitative yields by MacGillivray's group, through co-crystals by utilization of non-covalent interactions mostly.[2] This field of research further extended in the field of metal-complexes and coordination polymers (CPs) by several groups including, MacGillivray, Vittal, Biradha, Jian-Ping Lang, Barbour, Mir and so on.[2-7]

1.2 Photosalient Effect

Conversion of photo-energy into dynamic mechanical motions like *jumping, rolling, bending, splitting, crawling, curling, popping, wiggling and exploding etc,* is considered as *photosalient effect.*[8-9] These studies have been established in various photoreactions such as *cis-trans* isomerism, [4+4] cycloaddition, reversible cyclization of diarylethylenes to name a few.[10-11] Most of these studies have been established in organic crystals. Only, in recent years, solid-state [2+2] photoreactions have been utilized to observe similar photosalient effects. Similar to other reactions, photosalient effect driven by [2+2] cycloaddition has been observed in organic crystals. Naumov and Vittal's group has successfully presented the photosalient properties during the photo-polymerization of a metal complex (0D CP)[12] to one-dimensional CP (1D CP). Later various groups including, Biradha, Mir, and others have contributed significantly to this research field so far.[8-9] These photosalient motions in metal complexes and coordination polymers are of great interest due to the utilization of remotely controlled activation of mechanical motion by light with significant precision along with the incorporation of properties like magnetic, optical and others from meatl centres and dimensionality.

This photosalient effects are initially considered to be serendipitous, but currently they are at that stage of rational design in several systems through structure-property relationship. Such photomechanical motions triggered by [2+2] photocycloaddition reaction has proven to be great advantage towards the utility of light energy to develop stimuli responsive materials that can act as photo actuators and can be a potent material for remote sensors, artificial muscles, soft robotics etc.

This book chapter will highlight some of the recent studies in metal complexes and coordination polymers which has elevated this solid state [2+2] photoreaction mostly in crystalline hybrid materials and coordination polymers.

2. Photo-polymerization of Metal Complexes to Coordination Polymers

Photoreaction of monomers is the well-known strategy to obtain various kinds of polymers. This strategy has been successfully utilized by various researchers to obtain crystalline coordination polymers. Judicious choice of metal complexes/monomer and their photo-polymerization significantly utilized for the preparation of 1D CPs.[13] These photoreactions have also been demonstrated for the preparation of higher dimensional CPs through photoreaction of lower dimensional CPs using crystal engineering principles. Such design of coordination polymers may be difficult through direct or traditional synthetic procedures. In this section, only some of the interesting photo-polymerization of metal complexes will be covered.

First example of photo-polymerization of metal complex has been demonstrated by MacGillivray's group in Ag-complexes,[14] $[Ag_2(4spy)_4][CO_2CF_3]_2$ (**1**) where the photoreactive **4spy** linkers are arranged in head-to-head (HT) manner, with the alignment of olefin groups. Upon UV irradiation, these molecules underwent a cycloaddition reaction where the phenyl rings of the 4spy went far from the molecular plane, which brought these phenyl rings close to Ag(I). Thus there is a non-covalent interaction between the Ag(I) and phenyl carbon and lead in the formation of 1D CP. Thus, this work presented the photo-polymerization of a metal complex to CP.

Same group has presented a similar work by utilizing the Ag(I) metal complex, $[Ag_2(4spy)_4][CF_3SO_3]_2$ (**2**). Argentophilic interactions that are supported by the stacked π-π interactions between pyridyl and phenyl groups of neighboring olefins facilitated the alignment and the close proximity of the olefin group to a distance of 3.9 Å, following Schmidt's criteria. Photo-irradiation of this compound showed photo-polymerization of metal complex to 1D CP (Figure 1) established by SCSC (single-crystal to single-crystal) transformation of **2** by quantitative yield to, $[Ag_2(rctt\text{-}ppcb)_2][CF_3SO_3]$ (**3**) (Figure 1). Due to the presence of Argentophilic and stacked π-π interactions, the electrical conductivity of metal complex is found to be 20.8 ± 1.3 S·cm^{-1}. This electrical conductivity has been significantly enhanced to 37.0 ± 4.1 S·cm^{-1} (over 40% increase) upon photo-polymerization. Thus, this work highlights the importance of photo-polymerization of metal-complexes to the coordination polymer.[15]

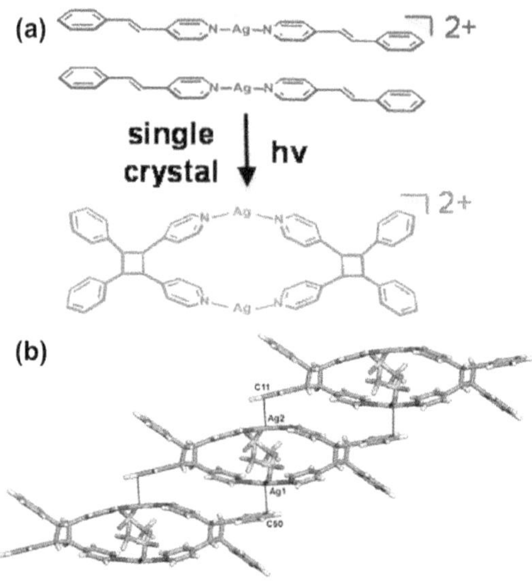

Figure 1. (a) Photo-cycloaddition of **2** and the formation of photo-dimer. (b) The interaction of Ag and phenyl-group, leading to the formation of 1D CP, **3**. Reproduced from Reference [15] with permission from American Chemical Society.

2.1 Mechanical Motion due to the Solid-state [2+2] Photoreaction

Vittal's group have extensively studied the photosalient effect of single crystals of metal complexes and coordination polymers through solid-state [2+2] photoreaction.[9, 16-17] The first photosalient behavior has been reported in three Zn(II) complexes, [Zn$_2$(benzoate)$_4$(L)$_2$] [L = **2F-4spy** in **4**, **3F-4spy** in **5** and **4spy** in **6**] during the solid-state photoreaction.[12] These mechanical motions are very similar to the bursting of popcorn over a hot surface. Similarly, the mechanism of such motions also follow rapid expansion of the material during the photoreaction and the formation of strain in the crystals through phase heterometry. Similar compound has been obtained using Mn(II) and the polymerization of metal complex as well as photomechanical motion of single crystal has been observed.[18]

Later nine Ag(I) metal-complexes have been prepared with the formula of AgL$_2$X$_2$ (L = **4spy**, **2F-4spy** and **3F-4spy**, X = BF$_4^-$, ClO$_4^-$ and

NO_3^-) having [AgL$_2$]$^+$ cations. The photoreactive 4spy and its derivative ligands are aligned in head-to-head (HH) and head-to-tail (HT) manner. The head-to-tail (HT) alignment of the linker in Ag (I) complexes resulted in the formation of 1D CP upon photoreaction. Thus, this work presented the transformation of metal-complex to 1D CP in four complexes, [Ag(4spy)$_2$]·BF$_4$ (**7**), [Ag(4spy)$_2$]·ClO$_4$ (**8**), [Ag(2F-4spy)$_2$]·BF$_4$ (**9**), [Ag(2F-4spy)$_2$]·ClO$_4$ (**10**). Among these, both **9, 10** have shown the polymerization along with photosalient behavior through [2+2] photoreaction (Figure 2).[19] Similar to the previous study, the primary driving force for the photosalient effect is considered to be the phase heterometry in the single crystal due to the formation of 1D CP as the photoproduct involving significant volume expansion during this photoreaction.

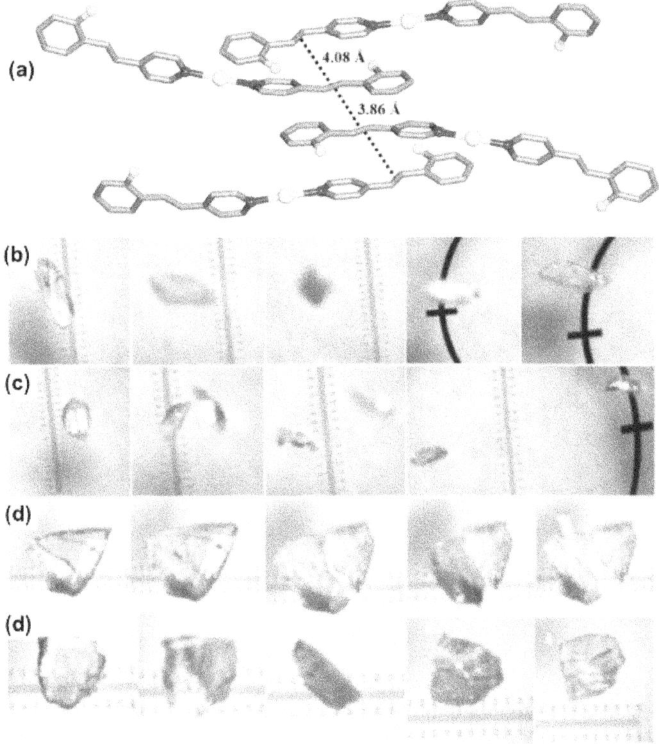

Figure 2. (a) Alignment of the olefins in **9**. Single crystal jumping in **9** (b, c) and **10** (d, e). Reproduced from Reference [19] with permission from American Chemical Society.

Effect of substituted fluorine on photochemical reaction: Mulijanto *et. al.*, have presented two polymorphs of Zn(II) complexes with substituted fluorine, both in the form of solvated, [Zn(NCS)$_2$(3F-4spy)$_2$]·DMF (**11·DMF**) and non-solvated, **11** system (Figure 3). The solvated system **11·DMF** underwent a rare curved morphology in dilute solution, when crystallized with different solvents. The binding of solvent with the Zn(II) caused the increase in the coordination number as well as alteration of the coordination geometry. Upon exposure of these solvated crystals to air, it splits into pieces as strain develops due to the loss of solvent molecules. The second polymorph without the guest solvent, **11** obtained as blocky crystals was photoreactive. This compound showed [2+2] photoreaction upon UV irradiation to form a 1D CP due to the quantitative photoreaction of 3F-4spy revealed by ^{1}H NMR spectroscopy. Photosalient effects were observed when the substitution of fluorine was different, **2F-4spy** instead of **3F-4spy** significant difference in mechanical motion was observed like popping, jumping and so on. This mechanical motion is due to the sudden anisotropic volume expansion during photoreaction, which leads to popping of the crystals under UV light. Hence, it has been found that substitution of different atoms like fluorine could significantly influence photosalient behavior and photoreactivity of structurally related Zn(II) complexes, while solvent effects explains curved morphology of single crystal.[20]

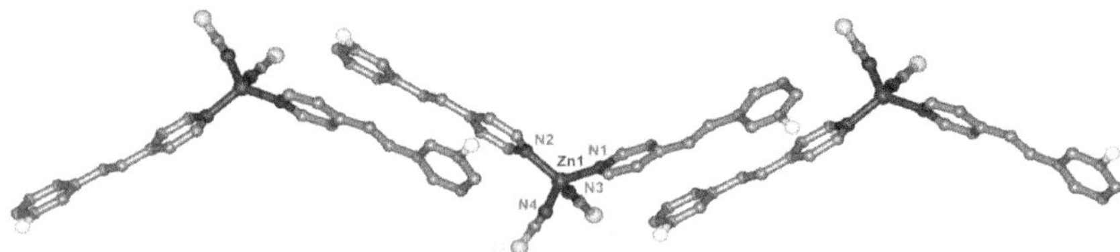

Figure 3. Alignment of olefins in **11·DMF** for the formation of 1D CP after the photoreaction. Reproduced from Reference [20] with permission from IUCr Journals.

Yadava et.al synthesized three Zn(II) complexes with 4spy and cinnamate anion, [Zn$_2$(cin)$_4$(2F-4spy)$_2$·Zn$_3$(cin)$_6$(2F-4spy)$_2$] (**12**), [Zn$_2$(cin)$_4$(4spy)$_2$] (**13**), and [Zn$_2$(cin)$_4$(3F-4spy)$_2$] (**14**). Of these, **13** and **14** are photoreactive with paddlewheel structure despite not obeying Schmidt's criteria, where the C=C bonds are separated by 4.35 Å in **13** and but 4.04 Å in **14**, satisfying Schmidt's criteria (Figure

4). In addition, these complexes also showed photosalient behavior driven by the photoreaction. Also, these mechanical motions are driven by significant expansion in the unit cell volume, *ca.*,11.93 % and 10.13 % for **13** and **14**, respectively, which is considered to be the main reason for this photosalient effect. Meanwhile, **12** showed photostability due to the far distance between the olefins. Thus, this work successfully presented the preparation of metal-complexes which could show significant mechanical motions due to solid-state photo-chemical reaction.[21]

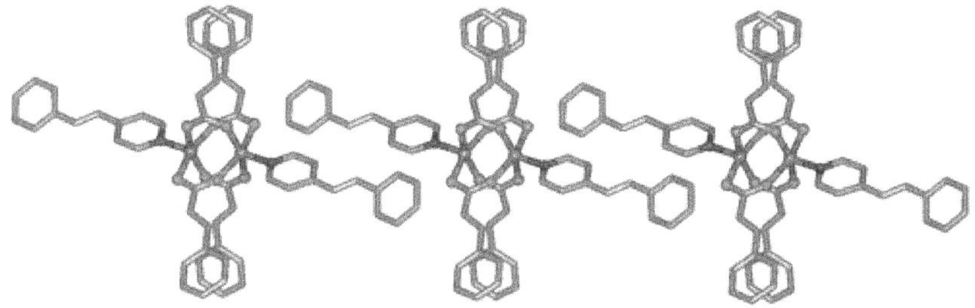

Figure 4. Arrangement of **4spy** linkers in **13**. Reproduced from Reference [21] with permission from American Chemical Society.

Vittal and co-workers studied the Zn(II) coordination complex, [Zn(benzoate)(2,5F-4spy)₂] (**15**). This compound underwent a [2+2] photoreaction despite not meeting Schmidt's topochemical criteria between olefinic bonds, 4.28 Å (Figure 5). Further, it displayed unpredicted photosalient effects like jumping, splitting and breaking of single crystals during UV irradiation. During the photo-irradiation, the cracks are developed in (010) plane which are associated with the weakest interactions. These cracks and photo-mechanical motion are driven by the strain generated through sudden anisotropic expansion of unit cell and the associated stress generated during photo-cycloaddition reaction. In addition, this compound exhibited multifunctional properties from being photoreactive and photosalient with unusual thermal expansion, and nonlinear optical properties such as two photon fluorescence and second-harmonic generation (SHG).[22]

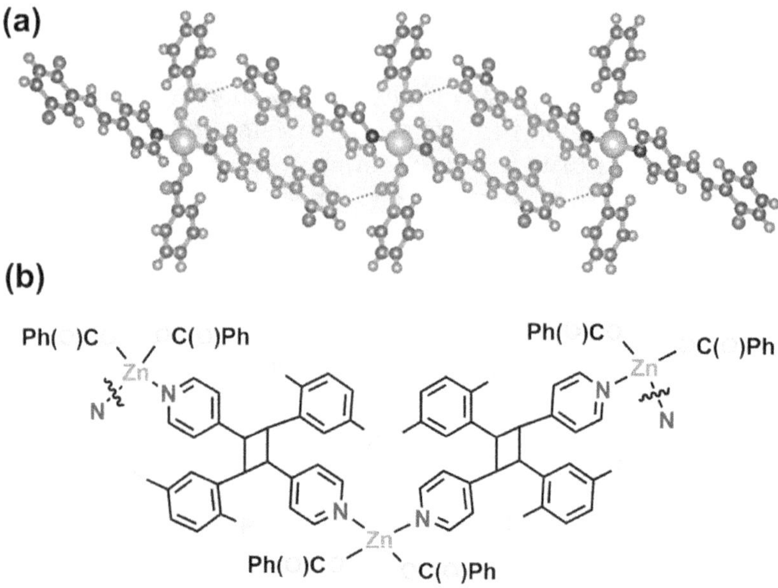

Figure 5. (a) The alignment of **2,5F-4spy** and, (b) the 1D CP formed after the photoreaction in **15**. Reproduced from Reference [22] with permission from American Chemical Society.

3. Photoreaction in Coordination Polymers and MOFs

CPs or MOFs are one of the famous and interesting hybrid crystalline materials that are used in various applications as discussed in other chapters.[23-24] Some of these applications have been discussed in other chapters of this book. As discussed in the previous sections, these photoreactions have been successfully employed for the synthesis of CPs of one-dimensional materials. In additionm these reactions have also been utilized for the conversion of lower dimensional materials to higher dimesional systems, which are almost impossible to obtain otherwise. Similarly, these reactions have also been successfully employed to alter the sorption, optical and other properties. This is mainly due to the change in polarity of the framework through the conversion of alkene with sp^2 hybridization to cyclobutane with sp^3 hybridization and alteration in the conjugation of the linker during the solid-state photoreaction.[23-24] Thus, the photoreactions have been successfully engaged both for the design and preparation of higher dimensional frameworks but also utilized for the tailoring of these materials for the gas sorption, selectivity, optical properties and other applications.[2-5, 25] Some of these studies are discussed in this section. For better description, this section is further divided into three sub-sections based on the dimensionality of the parent material.

3.1 Photoreaction in 1D Coordination Polymers

1D CPs are one of the basic MOF materials, where the metal nodes are connected by linkers in one direction mainly. The preparation of these materials can be acheived through polymerization of metal-complexes using photoreactions along with direct synthesis.[13] There are several groups including Vittal, Mir who have significantly contributed to this field which brought various advancements in the field.[26]

Photosalient nature of 1D CPs: Vittal's group has studied a photo-reactive 1D Pb(II) CP, [PbI$_2$(5-SPym)(DMF)] **(16)** which has demonstrated excellent photosalient effects during the solid-state [2+2] photoreaction under UV light. Unlike other examples, there are some crystals that maintained the single crystallinity even after 100% photoreaction, presenting one of the rare examples with SCSC transformation during the photosalient effect (Figure 6). This photosalient effect was mainly driven by the sudden anisotropic unit cell expansion during solid-state photoreaction which leads to macroscopic effects such as jumping, breaking, and splitting of the crystals under UV light. It is often observed in the literature that the rapid expansion of the compound resulted in the photosalient behavior. However, in this example the extent of volume expansion during the photoreaction is significantly low, but the crystals showed violent behavior. This is mainly due to the generation of stress during the formation of photoproduct and heterogeneous propagation of the reaction.[27]

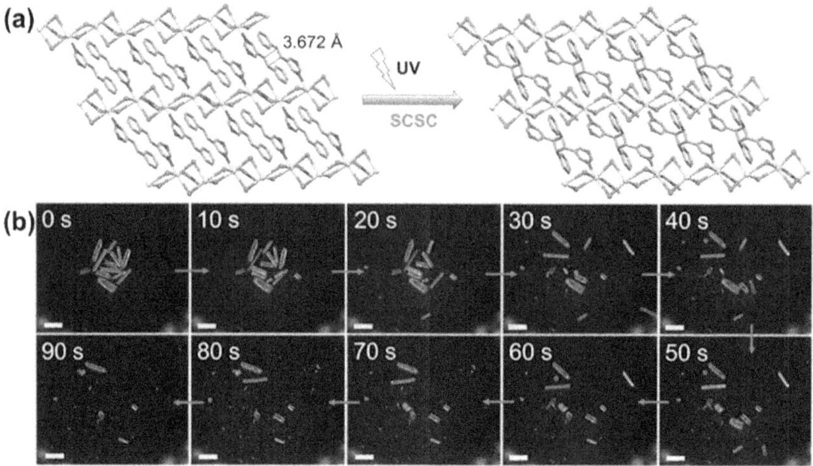

Figure 6. (a) The alignment of olefins in **16** and formation of olefins after the photoreaction is confirmed through SCSC measurements. (b) Microscopic images showing the photosalient nature upon UV irradiation. Reproduced from Reference [27] with permission from American Chemical Society.

Continuing their work, another 1D Pb(II) CP, $[PbBr_2(3F\text{-}4spy)_2]$ (**17**) showed great mechanical flexibility along with significant modulated photo-actuation properties. Herein, elastic bending of this slender crystal is observed on (001) face, without any fracture induced by mechanical force. Excellent flexibility of the CP was observed as crystals bend to an extent that can make a full circle and write few letters. This bending is facilitated by the anisotropic supramolecular interactions which cause the crystal bending over the (001) plane only. Photoinduced bending, twisting and curling has also been observed for the crystals with smaller width and thickness while the larger ones showed violent photosalient effects. The longer crystals (>3 mm) have also shown bending in backwards due to higher flexibility upon prolonged UV exposure (Figure 7). Meanwhile, the crystals showed splitting of the crystal when it was irradiated along (100) plane, at the tip due to the strain. Thus, showed influence of different mechanical motions based on the position of the UV irradiation.[28]

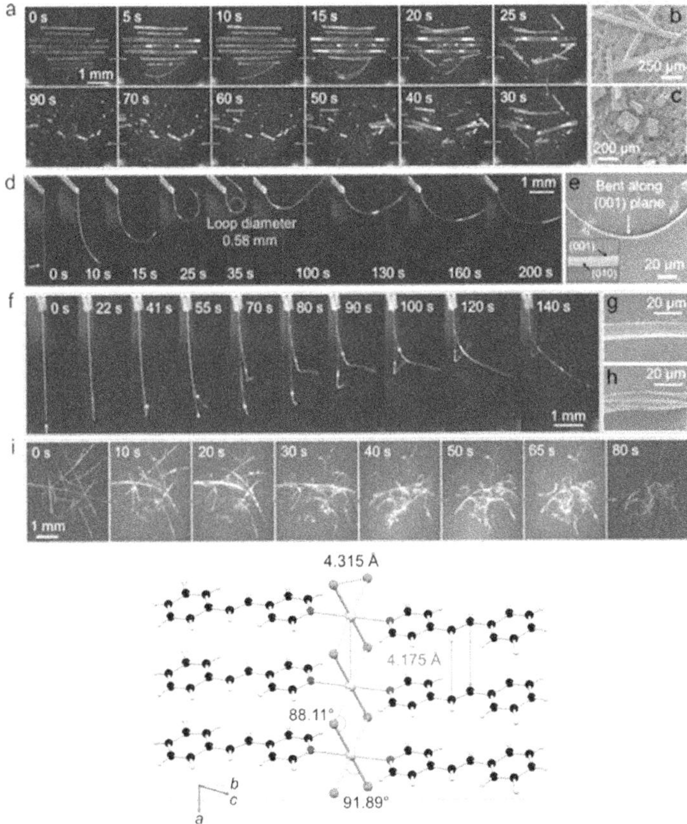

Figure 7. (top) Different mechanical motions shown by the single crystals upon UV irradiation and their SEM images. (bottom) Alignment of olefins in 1D CP of **17**. Reproduced from Reference [28] with permission from American Chemical Society.

Similar 1D Pb(II) CP, [Pb(SCN)$_2$(2F-4spy)$_2$] (**18**) has been synthesized by the same group which showed multi-responsive dynamic behaviors upon application of light, heat and mechanical force. The photoreactive **2F-4spy** ligands were aligned in HT manner, confirmed by SCXRD with C=C separation of 3.76 Å. This satisfies Schmidt's criteria for the [2+2] cycloaddition reaction. The obtained photoproduct has been characterized by ^{1}H NMR spectroscopy, confirming the formation of ***rctt*-2F-ppcb**. This photoreaction leads to the transformation of 1D CP to 2D structure along with significant photosalient effects such as hopping, jumping, breaking of the crystals into pieces

(Figure 8). In addition, this compound also showed thermosalient effects during heating and cooling, driven by reversible phase transitions, which made the crystals to bend and straighten upon heating and cooling respectively. Upon application of mechanical force on (001 plane), elastic deformation occurred due to the flexible $Pb(SCN)_2$ units and bending of thinner crystals were observed. Thus, these crystals not only proven to be photosalient, but considered as a multi-stimuli responsive material which is quite rare in CPs.[29]

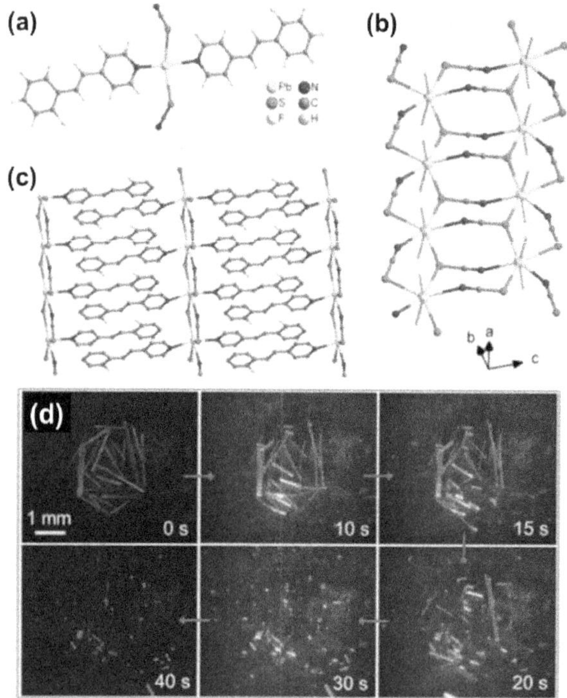

Figure 8. (a) Molecular structure of **18**, (b) formation of 1D CP through the bridging of Pb(II) through SCN, (c) alignment of olefins in head-to-tail fashion. (d) Photosalient nature of **18** under UV. Reproduced from Reference [29] with permission from American Chemical Society.

Mir and co-workers have studied 1D CP, $[Cd(adc)(nvp)_2(H_2O)]$ (**19**), obtained through slow diffusion methodology. The structure of this compound characterized through SCXRD which confirmed the photoreactive **nvp** ligands are aligned in HT fashion with C=C distance 3.67 Å which met the topochemical criteria. This compound showed solid-state [2+2] cycloaddition reaction under UV irradiation that led to the formation of dimerized product, [Cd(adc)(*rctt*-4-pncb)$_{1/2}$(nvp)(H$_2$O)] (**20**) as revealed by ¹H NMR analysis. Further, SCXRD of the UV irradiated product, **20** revealed through SCSC transformation and confirmed the formation of 1D ladderane structure to 2D layered structure which led to the reduction in adjacent distances and facilitate π-π stacking interactions and caused increase in electrical conductivity (Figure 9). Thus, this work showed the enhancement in electrical conductivity of the materials through photochemical reaction.[30]

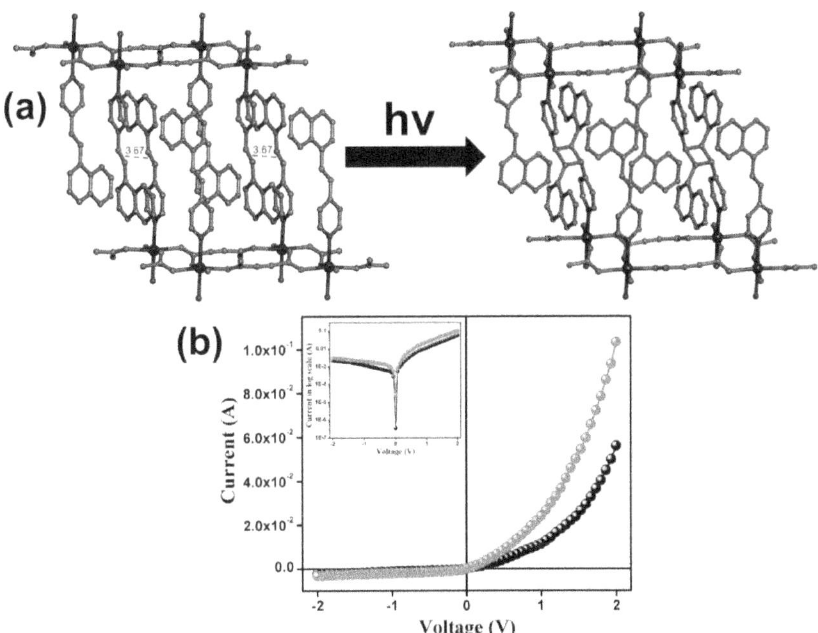

Figure 9. (a) Alignment of olefins and the formation of cyclobutanes under UV irradiation in **19**. The conductivity studies before and after the photoreaction (black-before and red-after photoreaction). Reproduced from Reference [30] with permission from American Chemical Society.

Same group reported another Cd(II) based 1D CP generated from the monomer [Cd (quin)$_2$(nvp)$_2$] (**21**) which underwent [2+2] cycloaddition under sunlight exposure as confirmed by ¹H NMR spectroscopy and confirmed the quantitative photoreaction. However, this photoreaction leads to decrease in the electrical conductivity. Further, SCXRD data confirms the formation of 100% photo-product [Cd(quin)$_2$(*rctt*-4-pncb)] (**22**) via crystal-to-crystal transformation

(Figure 10). The cyclobutane derivative photoproduct cleaved back upon heating at 200 °C to form monomer again. This structural transformation between a monomer and dimer accompanied with change in conductivity may lead to development of molecular switches.[31]

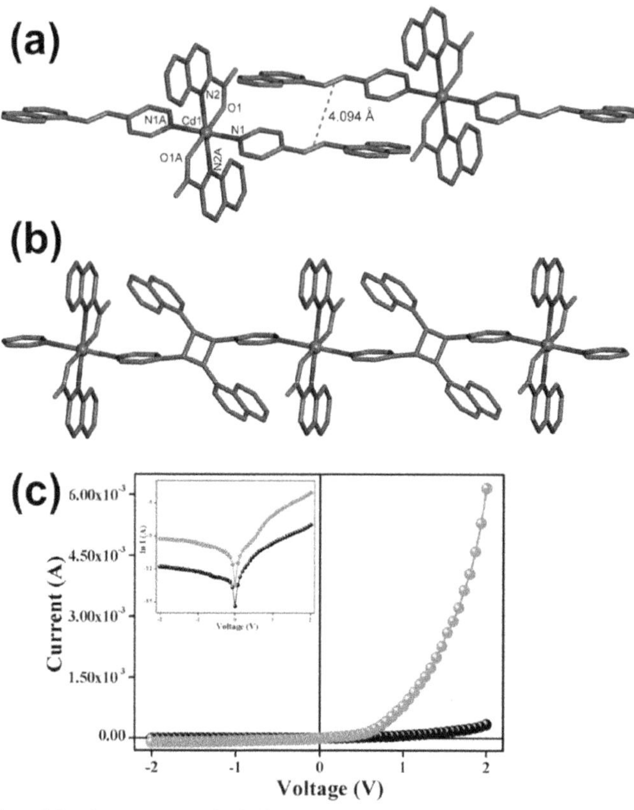

Figure 10. (a) Alignment of olefins in **21**. (b) Formation of cyclobutanes after the photoreaction. (c) Change in conductivity of the material after the photoreaction (black) compared to before photoreaction (red). Reproduced from Reference [31] with permission from American Chemical Society.

Another example was reported by the same group presenting a 1D CP, [Zn(nvp)(glu)] (**23**) that showed light driven mechanical motions not only in presence of UV light but also under sunlight. Elusive cyclobutane derivative has been synthesized during the solid state photoreaction and isolated the ligand via recrystallization. The olefins of **nvp** linkers from adjacent chains are aligned parallel with a separation distance of 3.79 Å, which meets Schmidt's criteria (<4.2 Å). During the photoreaction, olefin protons in ¹H NMR disappeared (δ 8.36

and 7.30) and new peaks appeared of cyclobutane (δ 5.19) which confirmed photodimerization. Due to the photosalient behavior, the crystals are shattered, thus, the dimerized product could not be measured, however, the photoproduct maintained the crystallinity, confirmed by the PXRD measurements. The photosalient behavior of this compound is mainly due to the strain created during the photo cycloaddition reaction. Meanwhile, the rate of mechanical motion in sun light is little slower compared to UV irradiation, might be due to the slower rate of photoreaction and strain creation.[32]

Effect of mechanical grinding on [2+2] photoreaction: Vittal's group reported a photostable helical coordination polymer [(Ph$_3$P)Ag(5-Spym)(CF$_3$SO$_3$)]·0.75MeOH (**24**), obtained through slow evaporation of a methanolic solution of AgCF$_3$SO$_3$, **5-Spym** and PPh$_3$. The connectivity of **5-spym** to the Ag(I) generated a helical CP along the *b*-axis of the unit cell. The unusual photoreaction took place between the olefin pairs after mechanical grinding with mortar and pestle. This brought the non-aligned olefin pairs in helical chains to get aligned with each other. Olefinic ligands with head-to-tail alignment from the adjacent CP underwent photochemical [2+2] cycloaddition with 83% photoconversion to obtain cyclobutane photoproduct as revealed by ^{1}H NMR analysis. Further, photoreaction caused the anisotropic volume expansion evident from the deccrease in density measured for the photoproduct.[33]

Cd(II) based 1D CP with diolefinic ligand, **bpvb** and auxiliary ligand **2·5-fdc²⁻**, [Cd(μ-OH2)(2,5-fdc)(bpvb)] (**25**) upon UV irradiation underwent regio-specific solid-state photoreaction to obtain a 2D network comprising cyclobutane derivative linkages in a SCSC manner.[34] This photoconversion demosntrated the photo-controlled fluorescence and this property of material could be employed in fluorescent devices. There are several other examples which has also reported the photoreaction in 1D CPs.[34-36] Meanwhile, double dimerization within a linker has also been successfully achieved in 1D MOFs.[37]

3.2 Photoreaction in 2D Networks

In this section also, significant contribution has been observed mainly from Vittal, Jian-Ping Lang, Barbour groups along with others.[38] This has brought various advancements in the field, especially towards the gas sorption, separation, luminescence properties, and recently mechanical motion of the crystals. Some of these work is being discussed in this section.

Vittal and coworkers successfully synthesized two 2D Zn(II) coordination polymers, $[Zn_2(2F-4spy)_4(Fumarate)_2(H_2O)_2]$ (**26**) and $[Zn_2(3F-4spy)_2(Fumarate)_2]$ (**27**) which was found to be photoreactive. Both **2F-4spy** and **3F-4spy** ligands of **26** and **27** are aligned in HT manner between the adjacent 2D layers that are interdigitated between the layers. Upon UV irradiation, 25% of **2F-4spy** dimerizes to form ***rctt*-2F-ppcb** in SCSC manner, whereas **3F-4spy** dimerized to give ***rctt*-3F-ppcb** by [2+2] photocycloaddition. Interestingly, CPs **27** gave another photoproduct consisting heterodimer **pfcb** by the cycloaddition of fumarate anions with **3F-4spy** along with dimer product (Figure 11), which is established by ^{1}H-NMR analysis through the formation of cyclobutane photo-products. Thus, based on the single crystal data of photo-reactant (2D CP), the product of this compound is supposed to be a non-interpenetrated 3D MOF.[39]

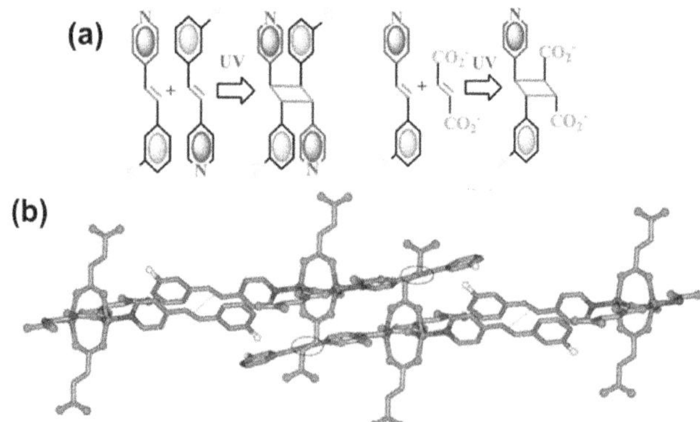

Figure 11. (a) ChemDraw images showing the photoreaction between both the **3F-4spy** linkers and **3F-4spy**-fumarate linkers. (b) Alignment of olefins between **3F-4spy** linkers and **3F-4spy** and fumarate linkers in **27**. Reproduced from Reference [39] with permission from American Chemical Society.

Same group have presented the transformation of 2D networks that could undergo photoreaction and form 3D interpenetrated networks.[40-41] Similar framework and the transformation has been utilized for the preparation of a hybrid device that could help to convert the light energy into mechanical force through photosalient effect.

Photomechanical motion in 2D MOF has been successfully demonstrated by the Lang's group. In this case, the authors have successfully blended **3F-4spy**, photoreactive linker into Zn(II) based 2D MOF, [Zn(bdc)(3F-4spy)] (**28**) existing in square pyramidal geometry where Zn(II) ions are coordinated by four oxygen of **bdc** to form 2D network structures and olefinic ligands attached from axial position, similar to the previous work.[40-41] The C=C separation is 3.76 Å between HT alignment of olefins that is feasible for [2+2] photocycloaddition reaction. Upon irradiation for 40 s with 365 nm UV light, complete photoconversion took place as confirmed by ^{1}H NMR analysis and corroborated by in situ PXRD measurements. The photoconversion was accompanied by different mechanical deformations depending upon the dimensions of crystals of **28** where thinner crystals bend towards the light source followed by twisting while thicker one's bend away from the light source followed by splitting upon prolonged exposure. Further, they studied photoreactivity of **28** with PVA resulting in bending of the crystal towards the light source. Moreover, IR stretching frequency suggests strong hydrogen interactions O-H⋯F between OH group of **PVA** and fluorine atom of **3F-4spy** whilst AFM measurements supported the mechanical deformations by a measure of roughness in the surface.[42]

Haynes and Barbour's group has presented similar regio-selective photoreaction in the bpeb linkers through solvent medicated alignment of olefins in the bpeb based 2D MOFs, [Cd(bpeb)(obc)]·2DMA (**29**) (**CdMOF-DMA**). This **DMA** solvated MOF double alignment (in-phase alignment) of olefins and the olefins are within Schmidt's criteria. This compound exhibited photoreaction upon UV irradiation at both the functional groups of the olefins of **bpeb**. Whereas upon exchange of solvent from **DMA** to **DMF** through soaking experiments, the alignment of olefins of **bpeb** has been significantly modified and upon UV irradiation, this compound showed slip-stacked alignment/out-of-phase alignment. Thus, this work has successfully pre-sented the solvent mediated alignment of olefins in MOFs where the MOF has been utilized as the template for the photoreactions.[43]

Later the role of temperature on the photoreaction of **CdMOF-DMA** has also been studied by the same groups, in its two isomers. Phase-1 of the MOF can be converted to another through temperature in-duced transformation. During this transformation, the alignment of olefins has been changed, and caused regio-selective photoreactivity of two olefins in bpeb linkers. **CdMOF-P1** (phase 1 of **CdMOF**) showed a reversible structural transformation to **CdMOF-P2** through thermal treatment, revealed during SCXRD analysis. During this str-uctural transformation, the confirmation of the bpeb ligands has been changed, which further altered alighment of olefins and their relative alignment of the bpeb in • • **MOF-P1** and **CdMOF-P2**. Irradiation of these two polymers results in formation of **isomer 1** and **isomer 2**. This allows temperature induced selective synthesis of both isomers in CdMOF, by carrying out the cycloaddition reaction (Figure 12).[44]

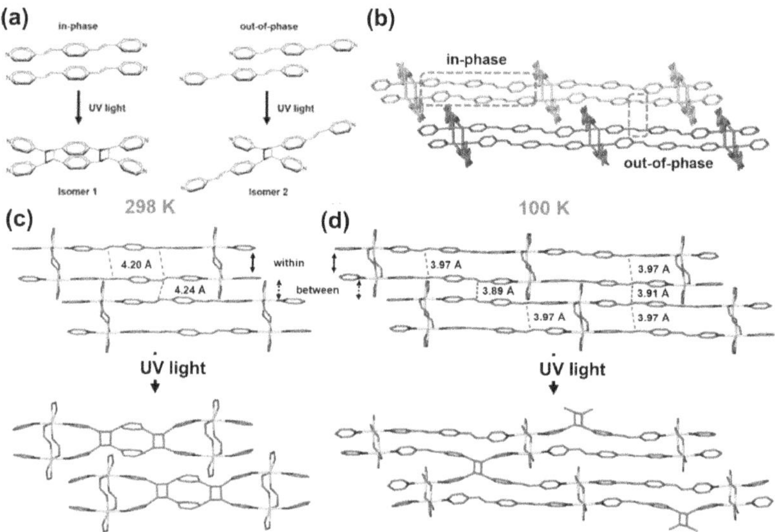

Figure 12. (a) In-phase and out-of-phase alignment of **bpeb** and their photo-products. (b) In-phase and out-of-phase alignment of **bpeb** in **CdMOF**. (c) and (d) In-phase and out-of-phase photoreaction of **bpeb** in **CdMOF**. Reproduced from Reference [44] with permission from American Chemical Society.

Jian Ping Lang and co-workers studied transformation of photo-inert diolefinic ligand, **bpbde** aligned *in-phase* and *out-of-phase* to obtain regio-selective photo products in metal-organic networks like, [Cd(adipate)(bpbde)] (**30**) and [Cd(pimelate)(bpbde)] (**31**). This regio selective reactions resulted in the formation of **tptco** and **bpbpvcb** due to the photoreaction of bpebde through *in-phase* and *out-of-phase* photoreaction respectively (Figure 13). Reaction and the structure of the compound has been characterized through SCSC transformation.[45] **30** is a 2D network where both the olefin groups of **bpbde** are parallelly aligned and underwent photoreaction into ladderane like linker, **tptco**. Meanwhile, the 3D network of **31** has resulted in the alignment of olefins in a slip stacked manner and formed **bpbpvcb** upon photoreaction of this compound. Such specific and differentiative alignment of olefins are possible only due to tailoring nature of metals and co-linkers, which are almost impossible to obtain similar linkers otherwise. Similar double dimerization of olefins in a linker also been reported by various other groups.[37]

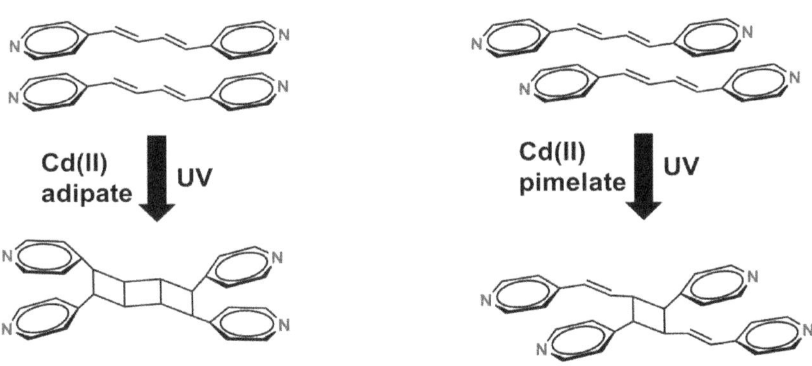

Figure 13. Photoreaction of bpbde in Cd-adipate and Cd-pimelate systems in **30** and **31**.

Polyrotaxane structures: Huiyeong Ju et.al constructed a 2D polyrotaxane-type MOF of Pb(II) containing [Pb$_2$(sdc)$_2$] units as wheels and **bpp** ligands as the axles, resulting in a doubly entangled 2D polyrotaxane structure with formula [Pb$_2$(bpp)(sdc)$_2$] (**32**). Pb(II) center is in penta coordinated structure with a hemispherical coordination geometry. The olefins of two **sdc** ligands are aligned parallel and are separated by 3.72 Å (follow Schmidt's criteria) and undergo quantitative [2+2] cycloaddition reaction upon UV irradiation, to form a 3D polyrotaxane framework through the linkage of the adjacent wheels by cyclobutane ring (Figure 14). Upon heating the obtained 3D polyrotaxane at 250 °C for 40 hours, partial thermal cleavage of the cyclobutane ring leads to a reverse reaction as confirmed by ^{1}H NMR and PXRD studies. The angular olefinic ligand (**sdc**) and noncollinear semirigid pillar ligands (**bpp**) at the Pb(II) centres in an orthogonal direction result in an interdigitated arrangement in 2D MOF, opening up new opportunities to obtain higher dimensional structures that may not be possible to obtain through traditional procedures.[46]

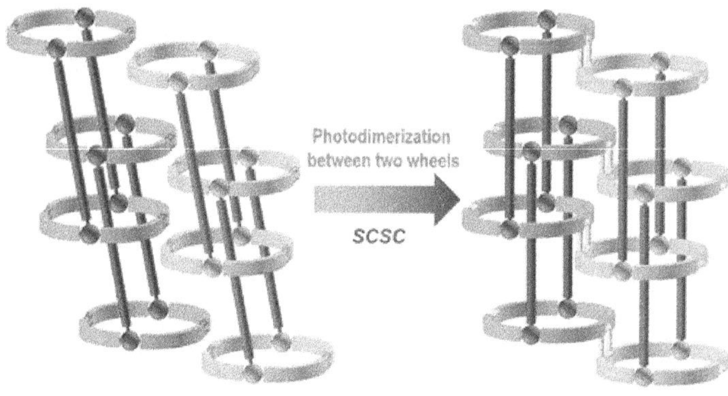

Figure 14. Schematic view showing the photoreaction of **32** polyrotaxane-like structures and its photoreaction. Reproduced from Reference [46] with permission from American Chemical Society.

Zn(II) based 2D MOF with (4,4) connectivity with **oba** linker has been successfully utilized to template the stereo-selective photoreaction of **4spy** linkers and their derivatives.[47] In most cases, the photoreaction between olefins leads to the formation of cyclobutane. This cyclobutane helps in the formation of new products with the tailored properties. By systematic alignment and activation groups, the aromatic groups can also be activated during the photoreaction. Such activation has been reported by Medishetty *et.al.*, during the polymerization of metal complex to coordination polymer where both the olefin-olefin reaction has been observed along with olefin-phenyl ring reaction.[48] Similar photoreaction of olefin-phenyl has been successfully reported by Matsuda's group during the transformation of interdigitated structure to 3D framework.[49] In this regard, an inter-digitated 2D MOF, [Zn(bpdc)(4spy)] (**33**) has been prepared where the olefins of **4spy** are well aligned and followed Schmidt's criteria. Upon inclusion of DMF solvent in this interdigitated framework resulted in the formation of alignment of phenyl-olefin instead of olefin-olefin due to the expansion of the 2D layers (Figure 15). Upon UV irradiation of this compound resulted in the photoreaction of phenyl-olefin cycloaddition reaction with high stereo-specificity which is almost impossible using regular solution procedures.

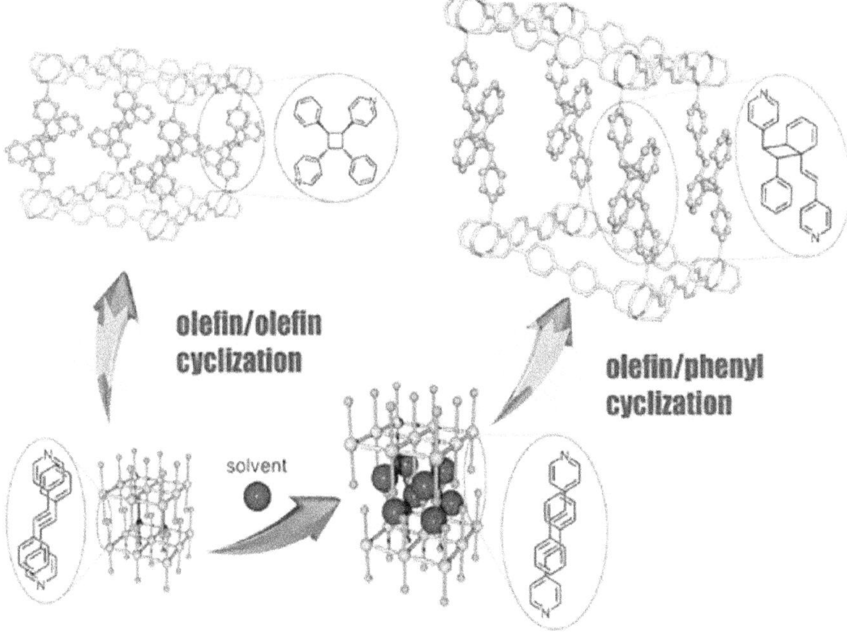

Figure 15. Schematic view showing the photoreaction of **4spy** in **33**. (left) Alignment of olefin-olefin groups and (right) alignment of phenyl-olefin groups driven by the solvent molecules and their photoreaction. Reproduced from Reference [49] with permission from American Chemical Society.

3.3 Photoreaction in 3D frameworks

3D MOFs are one of the well-studied hybrid materials in the literature. Although these materials were established in the 1990s, the photoreaction in these compounds was studied in 2010 by Vittal's group.[50] In the same year Jian Ping Lang also presented another interesting study of photoreaction in 3D MOFs.[51] Later several other groups have started working on the photoreactions in these 3D MOFs, considering their potentiality and applications. These studies have shown that the photochemical reactions in MOFs help to improve or incorporate the better properties compared to the pristine MOF and help in tailoring the properties of the framework through luminescence.[52] In addition, the photo-derivative MOFs are elusive or almost impossible to obtain through direct synthetic procedures. Other relevant studies were also reported.[53-54]

Recently Vittal's group have isolated two structural isomers of Zn(II) based pillared layer coordination polymer that arises from the different modes of bonding of carboxylates of bdc ligands to the Zn(II). One isomer with molecular formula $[Zn_2(bpeb)_2(bdc)_2]\cdot2DMF\cdot2H_2O\cdot0.5DMSO$ (**34**) obtained exclusively from solvothermal reaction while another isomer $[Zn_2(bpeb)_2(bdc)_2]\cdot DMA$ (**35**) was obtained as a minor product. Isomer **34** underwent double dimerization as confirmed by the ^{1}H-NMR data where quantitative cycloaddition reaction of *trans,cis,trans*-bpeb pairs occured and SCXRD revealed SCSC transformation to **tppcp** (Figure 16). Meanwhile, Isomer **35** underwent an incomplete photocycloaddition reaction. It is evident that photoreactivity is different in both the isomers as the respective void volume in **35** is less in which the compound doesn't have enough free volume for the double dimerization similar to **34**. Hence, the different bonding of carboxylates lead to formation of two isomeric CPs which differs in nature of interpenetration and photoreactivity.[55]

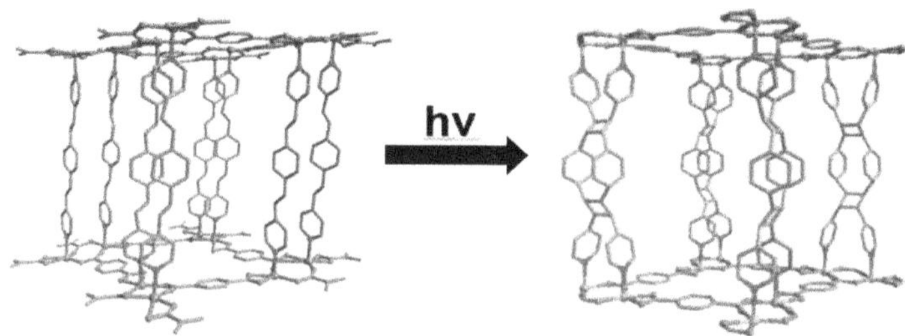

Figure 16: Double dimerization of bpeb in **34** and the product structure through SCSC transformation. Reproduced from Reference [55] with permission from Adapted with permission from IUCr Journals.

Park et al. synthesized three different Cd(II) based MOFs containing **sdb** ligands in a twisted manner to form a wavy ladder structure like coordination polymer, [Cd$_4$(bpe)$_4$(sdb)$_4$]·1.2DMF·DMSO (**36**), [Cd$_2$(bpe)$_2$(sdb)$_2$] (**37**), and [Cd$_4$(bpe)$_4$(sdb)$_4$]·1.7DMA·DMSO·H$_2$O (**38**) with **cds** topology. All the bpe pairs in **36** was found to be photoreactive and underwent [2+2] photocycloaddition to form cyclobutane under the UV light. This leads to increase in the stress in the crystal structure multiplied by the strain created by twisted **sdb** ligands. SCXRD revealed to obtain **39**, [Cd$_4$(*rctt*-tpcb)$_2$(sdb)$_4$(DMF)$_2$]·xSolvent as the photoproduct with broken ladder structure in a SCSC manner. The presence of **DMF** solvent prevented the disintegration of crystals. While partial photoreaction occurred in **37** and **38** to maintain the single crystallinity with ladder structure that remained fully intact. The partial photoreaction in desolvated **37** and with solvent **DMA** in **36** furnished unique patterning of unreacted **bpe** (**A**) and cyclobutane ring (**B**), ABAB and AABBAABB types. This arrangement/patterning is systematically modulated by the guest molecules absences, **37** and the **DMA** solvent molecules presence, **38**.[56]

Ghoshal, Reddy and co-workers have presented a doubly interpenetrated 3D Cd(II)- MOF, [Cd$_2$(bpe)$_2$(3,3-dmglu)$_2$] (**40**) with **bpe** ligand and **3,3-dmglu** with distinct C=C bonds undergo [2+2] cycloaddition in interconnected manner upon UV irradiation confirmed by SCXRD studies and thermal reversible transformation in a SCSC manner. The flexibility of the **bpe** ligands, as well as induced molecular mobility due to interpenetration, contribute to the photoinduced cycloaddition reaction (Figure 17). Crystals were softer and mechanical properties were comparable revealed by the nano-indentation technique which could be the reason for SCSC reversibility of the photoreaction.[25]

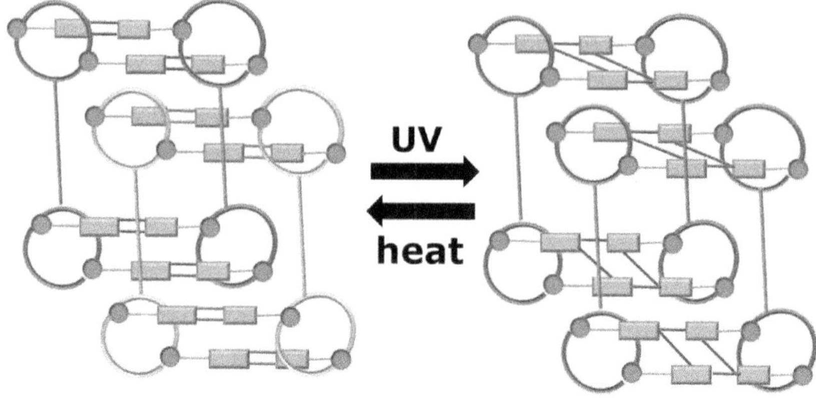

Figure 17. Schematic view showing the reversible dimerization of **40**.

Composites of organic polymer and coordination polymer have significant advantages, however, preparation of these composite materials in direct synthesis has significant challenges due to the poor solubility of the organic polymers, formation of crystalline system, characterization and so on. Solid-state photoreaction has been proven to be one of the best and simplest strategies to obtain these composite materials. Vittal's group is one of the first who presented the *in-situ* synthesis of organic polymers in MOFs through [2+2] cycloaddition reactions.[4, 57-58] This group and others have later presented various examples towards this concept of preparing the MOF with the impregnated organic polymers different polymeric structure. Recently they have studied the role of linker, **dhbdc** for the preparation and depolymerization of these composites. The SCSC photopolymerization of **bpeb** ligand in [Zn$_2$(bpeb)(dhbdc)(fa)$_2$] (**41**) via [2+2] cycloaddition reaction produced an organic polymer with cyclobutane ring assimilated framework, [Zn$_2$(dhbdc)(fa)$_2$] layers forming with the organic polymer-MOF composite material. The presence of hydrogen bonding between **dhbdc** and the carboxylate oxygen atoms caused not only photoreactivity of MOF in SCSC manner upon irradiation but also depolymerization of organic polymer **poly-bppcb** to **41** when heated in hot air oven at 250 °C (Figure 18). Due to this thermal depolymerization of the polymer to monomers, these composite materials can be utilized easily for various applications and are beneficiary to the environment.[58]

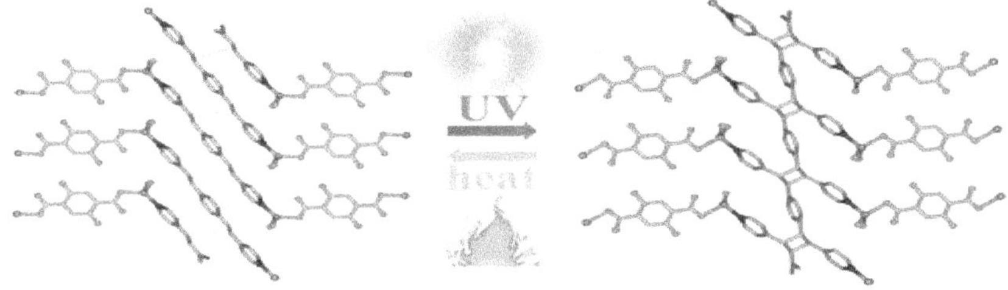

Figure 18. Reversible photo-polymerization of **41** through photore-action of **bpeb**. Reproduced from Reference [58] with permission from American Chemical Society.

Similar to the preparation of organic polymers, some photoreactive frameworks are very difficult to prepare through direct or traditional syntheses. Li-Hui Cao et.al successfully introduced a more flexible and photoreactive ligand, **bpe** in a pristine Cd-MOF via a SALE strategy. For this, **Cd-bpy-ipa** MOF has been used where the **bpy** linker has been successfully exchanged using SALE strategy. This resulted in the formation of **Cd-bpe-ipa** where the olefin linkers of this compound are well aligned and follow Schmidt's criteria. UV irradiation of this compound showed the photo reactivity of these olefins and presented novel methodology to obtain photoreactive MOF. Similar results were obtained by using **5NH$_2$-ipa**, where the **bpy** linker has been exchanged with **bpe** and formation of photoreactive compound. This synthetic strategy is important as these frameworks are not possible to obtain through traditional synthetic procedure as shown in Figure 19.[59] Thus, SALE strategy is proven to be one of the best strategies which provides desired photoreactive MOFs which are difficult to obtain through traditional routes.

Figure 19: Synthesis of Cd-IPA-BPY followed by the exchange of **bpy** with photoreactive **bpe** linker through SALE mechanism. Photoreaction of this ligand exchanged MOF through UV-irradiation and the formation of cyclobutane derived linker. This MOF with **bpe** is not possible to obtain through regular synthesis. Reproduced from Reference [59] with permission from American Chemical Society.

4. Conclusions and Perspectives

The solid-state photoreactions are proven to be one of the novel methods to obtain coordination polymers and MOFs towards various applications. It also has been proven to tailor the properties of these crystaline materials through the conversion of planar olefins to non-planar cyclobutane derivatives. During the preparation of higher dimensional CPs or MOFs through photoreaction, several of these MOFs have shown interesting applications including photo-mechanical motion which has significant applications in actuators, robotics, soft materials to name a few. Meanwhile, the photoreaction of CPs also helps in the change or improvement in the gas sorption, selectivity properties, electrical conductivity to name a few.

Since the role of photoreactions and the modification of the material property are not considered reaching the potentiality of the materials, considering the limited research groups working in this research area, there is a significant scope of improvement in the field is expected. As observed by the soft-crystalline and multi-responsive nature of materials in recent years, this field of research is expected to be dominating in the next few years considering the potentiality of applications in this direction along with tailoring of sorption and optical properties.

Abbreviations

CP = coordination polymer
MOF = metal-organic framework
SCXRD = single crystal X-ray diffraction
PXRD = powder X-ray diffraction
SCSC = single crystal-to-single crystal
1D CP = one-dimensional coordination polymer
2D CP = two-dimensional coordination polymer
3D CP = three-dimensional coordination polymer
bdc = benzene-1,4-dicarboxylate
dhbdc = 2,5-dihydroxy-benzene-1,4-dicarboxylate
bpdc = 4,4′-biphenyldicarboxylate
oba^{2-} = 4,4'-oxybis(benzoate)
3,3-dmglu = 3,3-dimethylglutarate
5NH$_2$-ipa = 5-aminoisophthalate
H$_2$obc = 4,4'-oxybisbenzoic acid
H$_2$IPA = isophthalic acid
H$_2$adc = acetylenedicarboxylic acid
2,5-fdcH$_2$ = 2,5-furan dicarboxylic acid
H$_2$glu = glutaric acid
Hquin = quinoline-2-carboxylic acid
H$_2$sdc = 3,3′-stilbenedicarboxylic acid
Hcin = cinnamic acid
sdb = 4,4'-sulfonyl dibenzoate
obc = 4,4'-oxybis(benzoic acid)
bpdc = 4,4′-biphenyldicarboxylate
bpe = trans-1,2- bis(4-pyridyl)ethylene
bpeb = 1,4-bis[2-(4-pyridyl)ethenyl]benzene
bpy = 4,4′-bipyridine
4spy = 4-styrylpyridine
2F-4spy = 2′-fluoro-4-styrylpyridine
3F-4spy = 3′-fluoro-4- styrylpyridine

nvp = 4-(1- naphthylvinyl)pyridine
25F-4spy = 2′,5′-difluoro-4-styrylpyridine
5-Spym = 5-styrylpyrimidine
rctt = regio-*cis, trans, trans[60]*
ppcb = 1,3-bis(4′-pyridyl)-2,4-bis(phenyl)cyclobutane
2F-ppcb = 1,3-bis(4′-pyridyl)-2,4-bis(2′-fluoro-phenyl)cyclobutane
3F-ppcb = 1,3-bis(4′-pyridyl)-2,4-bis(3′-fluoro phenyl)cyclobutane
4-pncb = 1,3-bis(4′-pyridyl)-2,4-bis(naphthyl)cyclobutane
bpvb = 1,3-bis[2-(3-pyridyl)vinyl]benzene
pfcb = 1-(4′-pyridyl)-2,3-bis(carboxylic acid)-4- (3′-fluoro benzene)cyclobutane
tppcp = tetrakis(4-pyridyl)-1,2,9,10- diethano[2.2]paracyclophane
bpp = 1,4-bis(4-pyridyl)piperazine
tpcb = tetrakis(4-pyridyl)cyclobutane
DMF = *N,N*-Dimethylformamide
DMA = *N,N*-Dimethylacetamide
DMSO = Dimethyl sulfoxide
SALE = Solvent-assisted linker exchange
PVA = Polyvinylalcohol

References

1. Schmidt, G. M. J. (1971) Photodimerization in the solid state. *Pure & Appl. Chem.*, **27**, 647-678.
2. MacGillivray, L. R., Papaefstathiou, G. S., Friščić, T., Hamilton, T. D., Bučar, D.-K., Chu, Q., Varshney, D. B., and Georgiev, I. G. (2008) Supramolecular Control of Reactivity in the Solid State: From Templates to Ladderanes to Metal–Organic Frameworks. *Acc. Chem. Res.*, **41**(2), 280-291.
3. Biradha, K., and Santra, R. (2013) Crystal engineering of topochemical solid state reactions. *Chem. Soc. Rev.*, **42**(3), 950-967.
4. Medishetty, R., Park, I.-H., Lee, S. S., and Vittal, J. J. (2016) Solid-state polymerisation via [2+2] cycloaddition reaction involving coordination polymers. *Chem. Commun.*, **52**(21), 3989-4001.
5. Kole, G. K., and Vittal, J. J. (2013) Solid-state reactivity and structural transformations involving coordination polymers. *Chem. Soc. Rev.*, **42**(4), 1755-1775.
6. Khan, S., Dutta, B., and Mir, M. H. (2020) Impact of solid-state photochemical [2+2] cycloaddition on coordination polymers for diverse applications. *Dalton Trans.*, **49**(28), 9556-9563.
7. Vittal, J. J., and Quah, H. S. (2017) Photochemical reactions of metal complexes in the solid state. *Dalton Trans.*, **46**(22), 7120-7140.

8. Khan, S., Akhtaruzzaman, Medishetty, R., Ekka, A., and Mir, M. H. (2021) Mechanical Motion in Crystals Triggered by Solid State Photochemical [2+2] Cycloaddition Reaction. *Chemistry – An Asian Journal*, **16**(19), 2806-2816.

9. Rath, B. B., and Vittal, J. J. (2021) Dynamic effects in crystalline coordination polymers. *CrystEngComm.*, **23**, 5738-5752.

10. Naumov, P., Karothu, D. P., Ahmed, E., Catalano, L., Commins, P., Mahmoud Halabi, J., Al-Handawi, M. B., and Li, L. (2020) The Rise of the Dynamic Crystals. *J. Am. Chem. Soc.*, **142**(31), 13256-13272.

11. Naumov, P., Chizhik, S., Panda, M. K., Nath, N. K., and Boldyreva, E. (2015) Mechanically Responsive Molecular Crystals. *Chem. Rev.*, **115**(22), 12440-12490.

12. Medishetty, R., Husain, A., Bai, Z., Runčevski, T., Dinnebier, R. E., Naumov, P., and Vittal, J. J. (2014) Single Crystals Popping Under UV Light: A Photosalient Effect Triggered by a [2+2] Cycloaddition Reaction. *Angew. Chem. Int. Ed.*, **53**(23), 5907-5911.

13. Leong, W. L., and Vittal, J. J. (2011) One-Dimensional Coordination Polymers: Complexity and Diversity in Structures, Properties, and Applications. *Chem. Rev.*, **111**(2), 688-764.

14. Chu, Q., Swenson, D. C., and MacGillivray, L. R. (2005) A Single-Crystal-to-Single-Crystal Transformation Mediated by Argentophilic Forces Converts a Finite Metal Complex into an Infinite Coordination Network. *Angew. Chem. Int. Ed.*, **44**(23), 3569-3572.

15. Hutchins, K. M., Rupasinghe, T. P., Ditzler, L. R., Swenson, D. C., Sander, J. R. G., Baltrusaitis, J., Tivanski, A. V., and MacGillivray, L. R. (2014) Nanocrystals of a Metal–Organic Complex Exhibit Remarkably High Conductivity that Increases in a Single-Crystal-to-Single-Crystal Transformation. *J. Am. Chem. Soc.*, **136**(19), 6778–6781.

16. Chen, C.-X., Yin, S.-Y., Wei, Z.-W., Qiu, Q.-F., Zhu, N.-X., Fan, Y.-N., Pan, M., and Su, C.-Y. (2019) Pressure-Induced Multiphoton Excited Fluorochromic Metal–Organic Frameworks for Improving MPEF Properties. *Angew. Chem. Int. Ed.*, **58**(40), 14379-14385.

17. Wojnarska, J., Gryl, M., Seidler, T., and Stadnicka, K. M. (2020) Effect of Substituent Exchange on Optical Anisotropy in Multicomponent Isostructural Materials Containing Sulfathiazole and 2-Aminopyridine Derivatives. *Cryst. Growth. Des.*, **20**(10), 6535-6544.

18. Yadava, K., and Vittal, J. J. (2019) Solid-State Photochemical [2+2] Cycloaddition Reaction of MnII Complexes. *Chem. Eur. J.*, **25**(44), 10394-10399.

19. Medishetty, R., Sahoo, S. C., Mulijanto, C. E., Naumov, P., and Vittal, J. J. (2015) Photosalient Behavior of Photoreactive Crystals. *Chem. Mater.*, **27**(5), 1821-1829.

20. Mulijanto, C. E., Quah, H. S., Tan, G. K., Donnadieu, B., and Vittal, J. J. (2017) Curved crystal morphology, photoreactivity and photosalient behaviour of mononuclear Zn(II) complexes. *IUCrJ*, **4**(1), 65-71.

21. Yadava, K., and Vittal, J. J. (2019) Photosalient Behavior of Photoreactive Zn(II) Complexes. *Cryst. Growth. Des.*, **19**(5), 2542-2547.

22. Bolla, G., Chen, Q., Gallo, G., Park, I.-H., Kwon, K. C., Wu, X., Xu, Q.-H., Loh, K. P., Chen, S., Dinnebier, R. E., Ji, W., and Vittal, J. J. (2021) Multifunctional Properties of a Zn(II) Coordination Complex. *Cryst. Growth. Des.*, **21**(6), 3401-3408.

23. Hazra, A., Bonakala, S., Adalikwu, S. A., Balasubramanian, S., and Maji, T. K. (2021) Fluorocarbon-Functionalized Superhydrophobic Metal–Organic Framework: Enhanced CO2 Uptake via Photoinduced Postsynthetic Modification. *Inorg. Chem.*, **60**(6), 3823–3833.

24. Sato, H., Matsuda, R., Mir, M. H., and Kitagawa, S. (2012) Photochemical cycloaddition on the pore surface of a porous coordination polymer impacts the sorption behavior. *Chem. Commun.*, **48**(64), 7919-7921.

25. Pahari, G., Bhattacharya, B., Reddy, C. M., and Ghoshal, D. (2019) A reversible photochemical solid-state transformation in an interpenetrated 3D metal–organic framework with mechanical softness. *Chem. Commun.*, **55**(83), 12515-12518.

26. Rath, B. B., Kole, G. K., and Vittal, J. J. (2018) Structural Transformation of Photoreactive Helical Coordination Polymers to Two-Dimensional Structures. *Cryst. Growth. Des.*, **18**(10), 6221-6226.

27. Rath, B. B., and Vittal, J. J. (2020) Single-Crystal-to-Single-Crystal [2 + 2] Photocycloaddition Reaction in a Photosalient One-Dimensional Coordination Polymer of Pb(II). *J. Am. Chem. Soc.*, **142**(47), 20117-20123.

28. Rath, B. B., and Vittal, J. J. (2021) Mechanical Bending and Modulation of Photoactuation Properties in a One-Dimensional Pb(II) Coordination Polymer. *Chem. Mater.*, **33**(12), 4621–4627.

29. Rath, B. B., Gupta, M., and Vittal, J. J. (2022) Multistimuli-Responsive Dynamic Effects in a One-Dimensional Coordination Polymer. *Chem. Mater.*, **34**(1), 178-185.

30. Dutta, B., Dey, A., Sinha, C., Ray, P. P., and Mir, M. H. (2018) Photochemical Structural Transformation of a Linear 1D Coordination Polymer Impacts the Electrical Conductivity. *Inorg. Chem.*, **57**(14), 8029–8032.

31. Dutta, B., Dey, A., Sinha, C., Ray, P. P., and Mir, M. H. (2019) Sunlight-Induced Topochemical Photodimerization and Switching of the Conductivity of a Metal–Organic Compound. *Inorg. Chem.*, **58**(9), 5419-5422.

32. Dutta, B., Sinha, C., and Mir, M. H. (2019) The sunlight-driven photosalient effect of a 1D coordination polymer and the release of an elusive cyclobutane derivative. *Chem. Commun.*, **55**, 11049-11051

33. Rath, B. B., Kole, G. K., Morris, S. A., and Vittal, J. J. (2020) Rotation of a helical coordination polymer by mechanical grinding. *Chem. Commun.*, **56**, 6289-6292.

34. Yang, F., Li, N. Y., Ge, Y., and Liu, D. (2021) Single-crystal to single-crystal transformation of a coordination chain to a two-dimensional coordination network through a photocycloaddition reaction. *CrystEngComm.*, **23**(15), 2783-2787.

35. Al-Mohsin, H. A., AlMousa, A., Oladepo, S. A., Jalilov, A. S., Fettouhi, M., P. and Peedikakkal, A. M. (2019) Single-Crystal-to-Single-Crystal Transformation of Hydrogen-Bonded Triple-Stranded Ladder Coordination Polymer via Photodimerization Reaction. *Inorg. Chem.*, **58**(15), 10167-10173.

36. Sawant, V. A., Wu, J., and Vittal, J. J. (2018) Competition Between Head-to-Head and Head-to-Tail Photocycloaddition Reaction in the Solid State: A Case Study. *Cryst. Growth. Des.*, **18**(6), 3661-3667.

37. Li, N.-Y., Liu, D., Ren, Z.-G., Lollar, C., Lang, J.-P., and Zhou, H.-C. (2018) Controllable Fluorescence Switching of a Coordination Chain Based on the Photoinduced Single-Crystal-to-Single-Crystal Reversible Transformation of a syn-[2.2]Metacyclophane. *Inorg. Chem.*, **57**(2), 849-856.

38. Li, W.-X., Gu, J.-H., Li, H.-X., Dai, M., Young, D. J., Li, H.-Y., and Lang, J.-P. (2018) Post-synthetic Modification of a Two-Dimensional Metal–Organic Framework via Photodimerization Enables Highly Selective Luminescent Sensing of Aluminum(III). *Inorg. Chem.*, **57**(21), 13453–13460.

39. Quah, H. S., Yap, J. L. X., Sambasivam, U., Chanthapally, A., Donnadieu, B., and Vittal, J. J. (2018) Contrast Solid-State Photoreactive Behavior of Two Two-Dimensional Zn(II) Coordination Polymers. *Cryst. Growth. Des.*, **18**(6), 3693-3696.

40. Medishetty, R., Tandiana, R., Koh, L. L., and Vittal, J. J. (2014) Assembly of 3D Coordination Polymers from 2D Sheets by [2+2] Cycloaddition Reaction. *Chem. Eur. J.*, **20**(5), 1231–1236.

41. Medishetty, R., Koh, L. L., Kole, G. K., and Vittal, J. J. (2011) Solid-State Structural Transformations from 2D Interdigitated Layers to 3D Interpenetrated Structures. *Angew. Chem. Int. Ed.*, **50**(46), 10949-10952.

42. Shi, Y.-X., Zhang, W.-H., Abrahams, B. F., Braunstein, P., and Lang, J.-P. (2019) Fabrication of Photoactuators: Macroscopic Photomechanical Responses of Metal–Organic Frameworks to Irradiation by UV Light. *Angew. Chem. Int. Ed.*, **58**(28), 9453-9458.

43. Claassens, I., Nikolayenko, V., Haynes, D., and Barbour, L. J. (2018) Solvent-mediated synthesis of cyclobutane isomers in a photoactive cadmium(II) porous coordination polymer. *Angew. Chem. Int. Ed.*, **57**(47), 15563-15566.

44. Claassens, I. E., Barbour, L. J., and Haynes, D. A. (2019) A Multistimulus Responsive Porous Coordination Polymer: Temperature-Mediated Control of Solid-State [2+2] Cycloaddition. *J. Am. Chem. Soc.*, **141**(29), 11425-11429.

45. Li, N.-Y., Liu, D., and Lang, J.-P. (2019) Regioselective Photochemical Cycloaddition Reactions of Diolefinic Ligands in Coordination Polymers. *Chemistry – An Asian Journal*, **14**(20), 3635-3641.

46. Ju, H., Shin, M., Park, I.-H., Jung, J. H., Vittal, J. J., and Lee, S. S. (2021) Construction of 2D Interdigitated Polyrotaxane Layers and their Transformation to a 3D Polyrotaxane by a Photocycloaddition Reaction between Wheels. *Inorg. Chem.*, **60**(11), 8285-8292.

47. Hu, F.-L., Mi, Y., Zhu, C., Abrahams, B. F., Braunstein, P., and Lang, J.-P. (2018) Stereoselective Solid-State Synthesis of Substituted Cyclobutanes Assisted by Pseudorotaxane-like MOFs. *Angew. Chem. Int. Ed.*, **57**(39), 12696-12701.

48. Medishetty, R., Bai, Z., Yang, H., Wong, M. W., and Vittal, J. J. (2015) Influence of Fluorine Substitution on the Unusual Solid-State [2 + 2] Photo-Cycloaddition Reaction between an Olefin and an Aromatic Ring. *Cryst. Growth. Des.*, **15**(8), 4055-4061.

49. Kusaka, S., Kiyose, A., Sato, H., Hijikata, Y., Hori, A., Ma, Y., and Matsuda, R. (2019) Dynamic Topochemical Reaction Tuned by Guest Molecules in the Nanospace of a Metal–Organic Framework. *J. Am. Chem. Soc.*, **141**(40), 15742-15746.

50. Mir, M. H., Koh, L. L., Tan, G. K., and Vittal, J. J. (2010) Single-Crystal to Single-Crystal Photochemical Structural Transformations of Interpenetrated 3 D Coordination Polymers by [2+2] Cycloaddition Reactions13. *Angew. Chem. Int. Ed.*, **49**(2), 390-393.

51. Liu, D., Ren, Z.-G., Li, H.-X., Lang, J.-P., Li, N.-Y., and Abrahams, B. F. (2010) Single-Crystal-to-Single-Crystal Transformations of Two Three-Dimensional Coordination Polymers through Regioselective [2+2] Photodimerization Reactions. *Angew. Chem. Int. Ed.*, **49**(28), 4767-4770.

52. Li, N.-Y., Liu, D., Abrahams, B. F., and Lang, J.-P. (2018) Covalent switching, involving divinylbenzene ligands within 3D coordination polymers, indicated by changes in fluorescence. *Chem. Commun.*, **54**, 5831-5834.

53. Hazra, A., Jain, A., Deenadayalan, M. S., Adalikwu, S. A., and Maji, T. K. (2020) Acetylene/Ethylene Separation and Solid-State Structural Transformation via [2 + 2] Cycloaddition Reactions in 3D Microporous ZnII Metal–Organic Frameworks. *Inorg. Chem.*, **59**(13), 9055-9064.

54. Hazra, A., Bonakala, S., Bejagam, K. K., Balasubramanian, S., and Maji, T. K. (2016) Host–Guest [2+2] Cycloaddition Reaction: Postsynthetic Modulation of CO2 Selectivity and Magnetic Properties in a Bimodal Metal–Organic Framework. *Chem. Eur. J.*, **22**(23), 7792-7799.

55. Park, I.-H., Ju, H., Kim, K., Lee, S. S., and Vittal, J. J. (2018) Isomerism in double-pillared-layer coordination polymers - structures and photoreactivity. *IUCrJ*, **5**(2), 182-189.

56. Park, I.-H., Lee, E., Lee, S. S., and Vittal, J. J. (2019) Chemical Patterning in Single Crystals of Metal–Organic Frameworks by [2+2] Cycloaddition Reaction. *Angew. Chem. Int. Ed.*, **58**(42), 14860-14864.
57. Park, I.-H., Chanthapally, A., Zhang, Z., Lee, S. S., Zaworotko, M. J., and Vittal, J. J. (2014) Metal–Organic Organopolymeric Hybrid Framework by Reversible [2+2] Cycloaddition Reaction. *Angew. Chem. Int. Ed.*, **53**(2), 414-419.
58. Park, I.-H., Sasaki, K., Quah, H. S., Lee, E., Ohba, M., Lee, S. S., and Vittal, J. J. (2019) Metal Organo-Polymeric Framework via [2 + 2] Cycloaddition Reaction: Influence of Hydrogen Bonding on Depolymerization. *Cryst. Growth. Des.*, **19**(3), 1996-2000.
59. Cao, L.-H., Xu, X.-Q., Tang, X.-H., Yang, Y., Liu, J., Yin, Z., Zang, S.-Q., and Ma, Y.-M. (2021) Controllable Strategy for Metal–Organic Framework Light-Driven [2 + 2] Cycloaddition Reactions via Solvent-Assisted Linker Exchange. *Inorg. Chem.*, **60**(4), 2117-2121.
60. Nagarathinam, M., Peedikakkal, A. M. P., and Vittal, J. J. (2008) Stacking of double bonds for photochemical [2+2] cycloaddition reactions in the solid state. *Chem. Commun.*, (42), 5277-5288.

3

Magneto-Structural Analysis of Metal-Organic Frameworks

*Rupesh Kumar, Archana Yadav and Kafeel Ahmad Siddiqui**
Department of Chemistry, National Institute of Technology
Raipur, Chhattisgarh, India

1. Introduction

Metal-organic frameworks have acquired considerable lot of interest in the field of inorganic chemistry due to their unique design and numerous advantages. MOFs have numerous advantages over other porous materials due to its – flexibility; tunable porosity; uniform pore structures; extensive variability's; and atomic-level structural uniformity in network topology, dimension, geometry, and chemical functionality. Aside from their technical importance, they also have the potential to provide strong skeleton for magnetic coupling along with their porous structure.

A class of materials called molecular-based magnets is capable of displaying complex magnetic phenomena. This class broadens the properties of materials commonly associated with magnets, such as low density, low-temperature fabrication, electrical insulation, and transparency. Due to its unique design, molecules can generate magnetic networks and have various unique properties. Some of these compounds have exhibited attractive properties - long-range magnetic order and electrical conductivity that are useful for various industrial applications.

In spite of the advantages of molecule-based magnets, their low operating temperatures make them less advantageous than solid-state inorganic ones. Long-range magnetic ordering involves strong interactions in multiple dimensions. MOFs may be further functionalized to enhance these properties or add new ones via organic linkers, resulting in long-range magnetic ordering.

Classes of Magnets

Unlike paramagnets, permanent magnets have a permanent magnetic field. They have unpaired electrons with strong long-range interactions, letting spontaneously organize their dipoles at low temperatures T_c. The order in which ferromagnetic, antiferromagnetic, or ferrimagnetic are generated is determined by the magnitude and relative alignment of spin centers. [1-4]

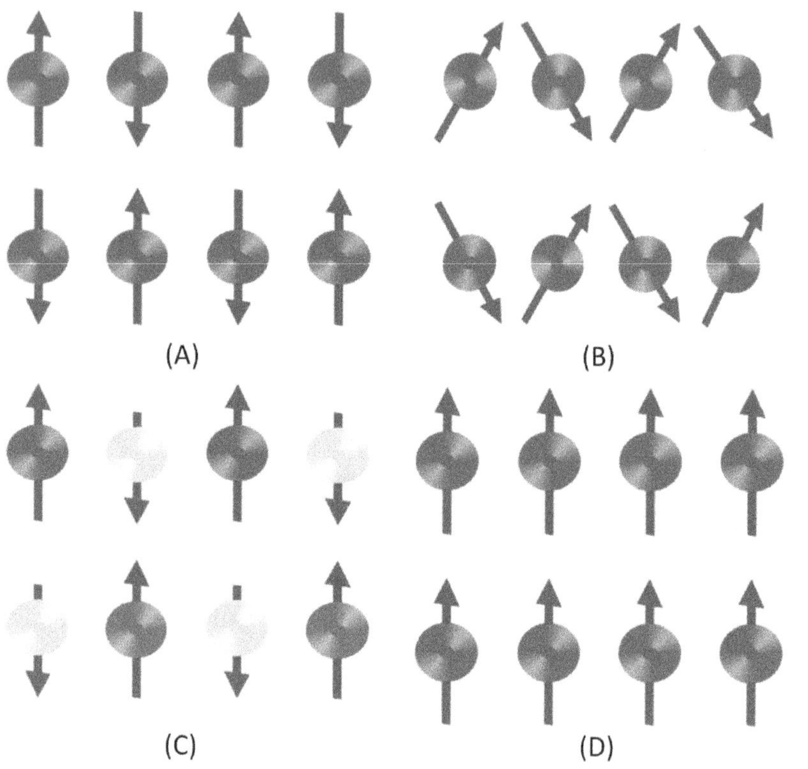

Figure 1. Demonstration of spin alignment (A) Ferromagnet (B) Spin-canted ferromagnetic (C) Ferrimagnet (D) Antiferromagnet.

The alignment of nearest neighbour spin leads to the formation of a magnetic state, when spins are aligned parallel it led to ferromagnetic state and if spins are aligned antiparallel it led to an antiferromagnetic state, with no residual spin in material. Spin canting is a type of arrangement that occurs when the spin arrangement of the objects changes in magnitude, magnetic moment is reduced. When the spin magnitudes are not equal, an uncompensated magnetic moment is generated in the ferrimagnetic state (Figure 1). Ferrimagnetism exhibits antiferromagnetism, although its magnetic behavior is closer to the properties of ferromagnet.

Meta-magnets, which exhibit a magnetic field-dependent change from antiferromagnetic to ferromagnetic behaviour, and spin glass, which exhibits short-range magnetic ordering are two other words used to describe magnets. The temperature-dependent magnetic susceptibility measurements (χ_{mT}) and the field dependence of the magnetization below Tc provide insight into the type of magnetic order (Figure 2). Various experimental methods, including direct and alternating magnetic susceptibility (χM), neutron diffraction, magnetic hysteresis, and heat capacity measurements, can be used to verify the existence of extended magnetic order. In a magnetic field, molar magnetic susceptibility (χm) of a substance is the sum of its diamagnetic and paramagnetic components. χ_{mT} increases continually as the temperature is dropped to Curie temperature (T_c) are ferromagnets. Ferrimagnet shows a minimum χ_{mT} values in the paramagnetic range above T_c, where the Neel temperature (T_N) is the critical temperature.

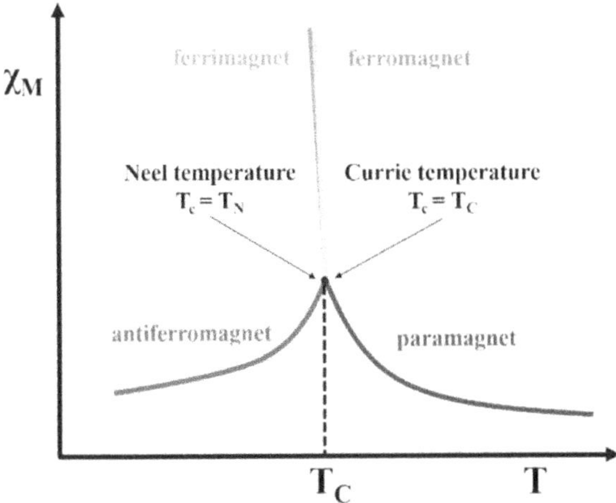

Figure 2. Plot between temperature (T) and dc magnetic susceptibility (χM) for different magnetism. The intense increase of χM below the Curie temperature (T_C) considered as ferromagnets and below the Neel temperature (T_N) considered as ferrimagnets while the antiferromagnets show rapid decrease in χM below T_N.

Other metrics can be used to evaluate the performance of permanent magnets. They exhibit magnetic ordering temperature and hysteretic behavior, which occurs when a field in the reverse way is, must to remove the material's magnetism. The hysteresis loop is a measurement that shows the width of the field that's needed to demagnetize of magnetized material. The hysteresis loop's maximum width is equivalent to double the coercive field's (Hc), known as a coercivity (Figure 3).

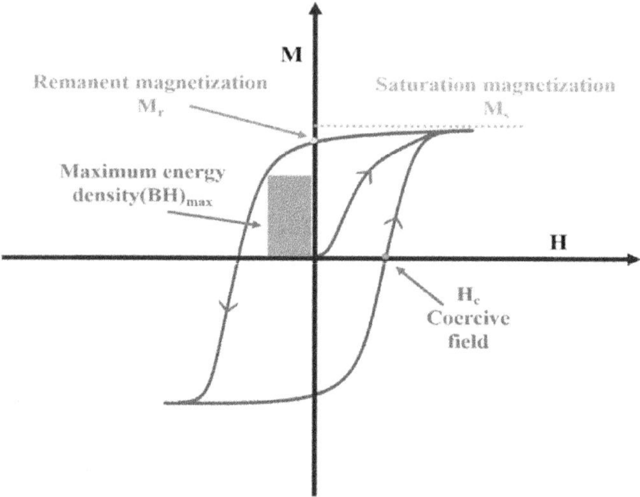

Figure 3. Illustrative diagram of a magnetic hysteresis loop has been discovered for, spin canted antiferromagnets ferrimagnets ferromagnets. Key parameters: remanent magnetization (Mr), maximum energy density, coercive field (Hc), and saturation magnetization (Ms).

The coercive field, shown above, is a measure of a magnet's capacity to endure an external field of magnetic material without getting demagnetized. The magnetization that persists after the magnetic field applied under the hysteresis measurement is withdrawn is referred as remanent magnetization (Mr), and it differs from saturation magnetization (Ms). The coercive field values are used to find a hard and soft magnet, the hard magnet has a high coercivity (Hc>100Oe) and can continue to maintain its magnetization without an external magnetic field, in opposition to those that are easily magnetized and demagnetized is termed as a soft magnet (Hc<100Oe).

The present chapter describes the origins and progress in optimizing molecular magnets on several areas of application: thermo electrics, energy storage materials, and catalysts. Solid-state spin-orbit coupling present in many molecule magnets is investigated as a possible explanation for their improved magnetic properties.

The magnetic characteristics of 3D MOFs, based on axial and equatorial axis orientation, are discussed in this book chapter.

Table 1. Distance of axial atom (X and Y), magnetic susceptibilities (χ_{mT}), Weiss constant (θ), Magnetic Exchange (ME), and geometry of 3D framework of Manganese complexes

S. No	RefCode	Axial X=Y/ X≠Y	Mn—X (A°)	Mn—Y (A°)	χ_{mT}	θ	ME	Geometry	Ref
1.	OQOSUK	O,O	2.168	2.187	9.15	-1.72	AF	DO	5
2.	DEYBUG	O,O	2.199	2.172	8.61	-7.60	AF	DO	6
3.	BUJHOG	O,O	2.216	2.211	4.02	-19.67	AF	DO	7
4.	WAJMAY	O,O	2.200	2.200	4.45	-25.05	AF	DO	8
5.	SFMOHO1	O,O	2.151	2.251	4.38	-0.74	AF	DO	9
6.	HUCJOG	O,O	2.173	2.173	4.23				10
7.	DEYCAN	O,O	2.188	2.155	4.38	-20.8	AF	DO	6
8.	DEYBOA	O,O	2.246	2.246	4.49	-0.40	AF	DO	6
9.	GITBUI	O,O	2.191	2.160	4.22	-30.70	AF	DO	11
10.	WOXKOL	O,O	2.246	2.152	9.04	-8.53	AF	DO	12
11.	DEZKAW	O,N	2.176	2.322	9.51	-5.5	AF	DO	13
12.	MIFQOJ	O,N	2.132	2.283	8.64	-24.76	AF	DO	14
13.	TIXLAP	O,N	2.176	2.329	4.10	-9.95	AF	DO	15
14.	SAMTIM	O,N	2.243	2.198	8.15	-9.56	AF	DO	16
15.	LAXMOP	O,O N,N	2.309 2.265	2.324 2.235	9.13	-2.84	AF	DO	17
16.	FIVQOT	O,O O,O	2.189 2.189	2.169 2.297	13.50	-10.91	AF	DO	18

17.	XULNOJ	O,O O,O	2.177 2.177	2.258 2.129	13.04	-13.71	AF	DO	19
18.	WEKKUU	O,O O,O	2.197 2.197	2.195 2.274	13.21	-16.93	AF	Pseudo O	20
19.	ACUWIH	O,O N,N	2.158 2.158	2.277 2.777	4.09	-17.5	AF	O	21
20.	TEDMAT	O,O N,O	2.163 2.251	2.339 2.121	10.22	-8.25	AF	DO	22
21.	WEKLAB	N,N N,O	2.249 2.241	2.185 2.053	13.18	-19.97	AF	DO	20
22.	DEZKEA	N,N O,N	2.308 2.268	2.213 2.331	9.04	-28.2	AF	DO	13

From the table above, we may deduce that all monometallic and bi-metallic complexes have axial Mn-X and Mn-Y bond lengths of 2.132 Å -2.309 Å and 2.053 Å -2.339 Å, respectively. The χ_{mT} value of all the complexes is between 4.02 - 13.50 cm³mol⁻¹ K and the θ value is between -30K and -0.40 K. The majority of the complexes are antiferromagnetic, although some of it is weak antiferromagnetic and canted antiferromagnetic. The majority of the complexes have distorted octahedral geometry, while others have octahedral and pseudo octahedral geometry.

Table 2. Distance of axial atom (X and Y), magnetic susceptibilities (χ_{mT}), Weiss constant (θ), Magnetic Exchange (ME), and geometry of 3D framework of Cobalt complexes

S. No	RefCode	Axial X=Y/X ≠ Y	Co—X (A°)	Co—Y (A°)	χ_{mT}	θ	ME	Geometry	Ref
1.	DIDHUV	O,O	2.061	2.062	2.92	−23.16	AF	DO	23
2.	DAHFUQ	O,O	2.048	2.146	5.38	7.67	F	DO	24
3.	WONYAC	O,O	2.201	2.115	4.37	-35.71	AF	DO	25
4.	KUWJIY	N,N	2.159	2.187	2.62	-25.72	AF	DO	26
5.	WONYOQ	N,N	2.200	2.200	3.16	-23.4	AF	DO	25
6.	YOZRUC	O,N	2.042	2.176	3.40	-21.20	AF	DO	26
7.	LAXMIJ	O,O N,N	2.253	2.310	5.38	-32.5	AF	DO	17

			2.229	2.320					
8.	MAZQAI	O,O O,O	2.087 2.216	2.143 2.126	6.36	-35.7	AF	DO	27
9.	MAZPUB	N,N N,N	2.216 2.169	2.156 2.064	4.17	-41.0	AF	DO	27
10.	NOCFAO	O,N O,O	2.186 2.157	2.099 2.076	6.52	-48.9	AF	DO	28
11.	GECKOR	O,N O,O	2.108 1.948	2.108 1.972	10.19	-78.16	AF	DO	29
12.	GECKUX	O,N O,O	2.098 2.062	2.107 2.079	8.95	-43.89	AF	DO	30
13.	QUVZEO	N,O O,O	2.064 2.036	2.186 2.036	8.26	-28.38	AF	DO	31
14.	EBADET	N,O O,O	2.187 2.153	2.291 2.153	8.36	-16.6	AF	DO	32

We can deduce from the preceding table that all monometallic and bimetallic cobalt complexes have axial Mn-X and Mn-Y bond lengths of 1.948 Å -2.253 Å and 1.172 Å -2.320 Å, respectively. All of the complexes have a χ_{mT} value is in the range of 2.92-10.19 cm³mol⁻¹ K, and θ value ranging from -78.16 K to +7.67K. The majority of the complexes are antiferromagnetic, with the exception of one that has ferromagnetic magnetic characteristics. All of the complexes have distorted octahedral geometry.

2. Magnetic Properties of Manganese Mononuclear Three Dimensional Metal Organic Frameworks

2.1 Both Axial Positions Occupied by Oxygen Atom from Ligand

2.1.1 Equatorial (4N)

Crystal structure of OQOSUK, the Mn(II) ion is bonded to four atoms of nitrogen (N1, N9, N8, N5) in equatorial position from ligand and two oxygen atoms (O1, O2) of two H_2O molecules in axial position forming a distorted octahedral geometry (Figure 4A). Magnetic susceptibilities depended on temperature measurement shows, χ_{mT} value, spin-only value, curie constant (C) and Weiss constant (θ) of the complex are 9.15, 8.76, 9.18 cm^3 mol^{-1} K & -1.72K respectively.

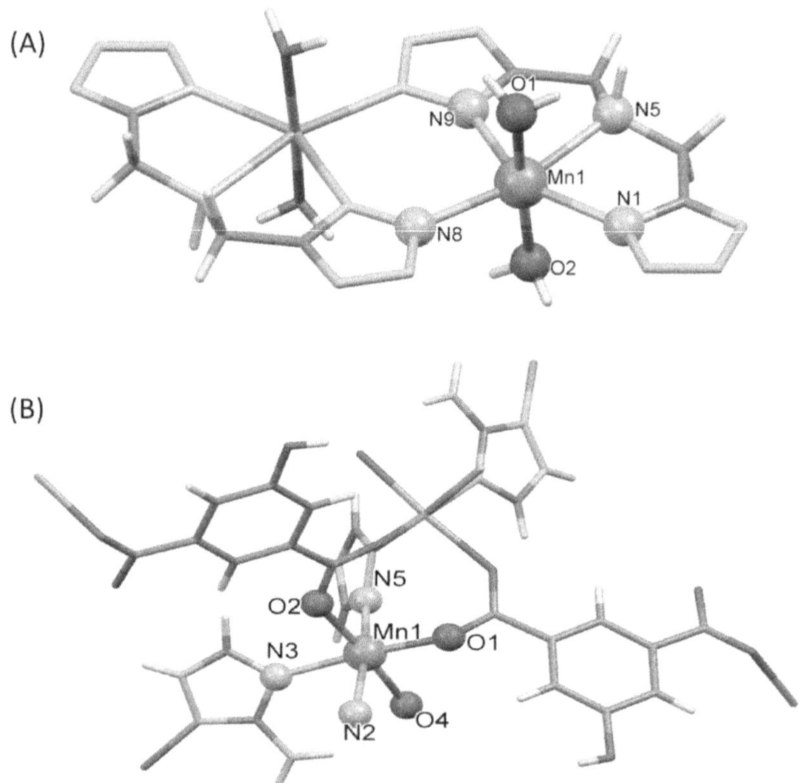

Figure 4. Illustration of coordination environment of Mn (II) in (A) OQOSUK (B) DEYBUG.

2.1.2 Equatorial (3N, 1O)

In DEYBUG, the equatorial position of Mn(II) ion is occupied by three atoms of nitrogens (N3, N5, N2A), one atom of oxygen (O1) of carboxylate group while two oxygen atoms (O2A, O4B) of carboxylate group of ligand is occupy axial position of coordination environment forming a distorted octahedral geometry (Figure 4B). The Axial bond lengths Mn-O are 2.199 Å and 2.17 Å. Temperature-dependent magnetic susceptibilies show, χ_{mT} value, spin only value, curie constant(C) and weiss constant (θ) of the complex are 8.61, 8.74, 8.91 $cm^3mol^{-1}K$ & -7.60 K respectively.

2.1.3 Equatorial (2N, 2O)

In the crystal structure of BUJHOG, two oxygen atom (O2A, O5B) and two nitrogen atoms (N3, N4) with two carboxy oxygen atoms (O1, O4) occupies axial and equatorial position of Mn(II) ion respectively forming an octahedral geometry. Bond lengths of axial are Mn1-O5B = 2.211Å Mn1- O2A = 2.216 Å, was observed (Figure 5A). The temperature-dependent magnetic susceptibilies, χ_{mT} value, spin only value, curie constant (C) and weiss constant (θ) of the complex is 4.02, 4.38, 4.32 $cm^3mol^{-1}K$ & - 19.67 K respectively.

In WAJMAY, the Mn (II) ion forms distorted octahedral geometry, in which two oxygen atoms (O2) are coordinated in axial position and two oxygen atoms (O1) with two nitrogen atoms (N2) are coordinated in an equatorial position (Figure 5B). The Mn–O bond distances range from 2.092 - 2.200 A°. The temperature-dependent magnetic susceptibilies, χ_{mT} value, spin only value, curie constant (C) and weiss constant (θ) of the complex are 4.45, 4.38, 4.82 cm^3mol^{-1} K & - 25.05K respectively.

In SEFMOH01 complex the Mn(II) ion form an octahedral geometry bonded by two O atoms (O2, O2c) and two N atoms (N1, N1) in equatorial position, and two O atoms (O3, O3) of waters in axial position (Figure 6A). The temperature-dependent magnetic susceptibilies, χ_{mT} value, spin only value, C and θ of complex are 4.386, 4.376, 4.4 cm^3 $mol^{-1}K$ & -0.74 K respectively.

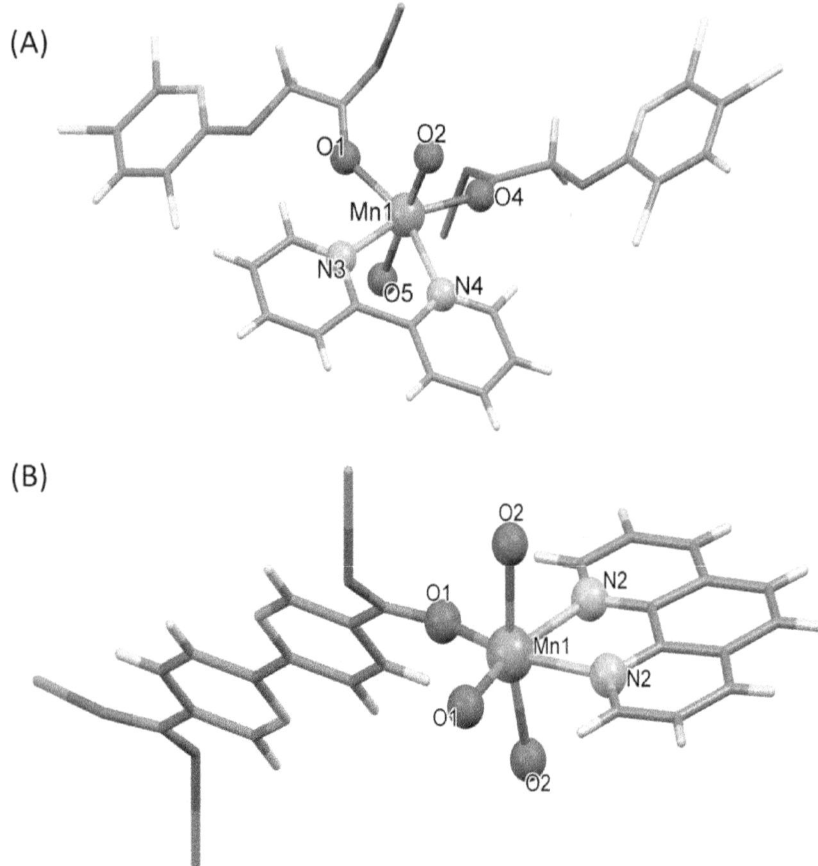

Figure 5. Illustration of coordination environment of Mn(II) in (A) BUJHOG (B) WAJMAY.

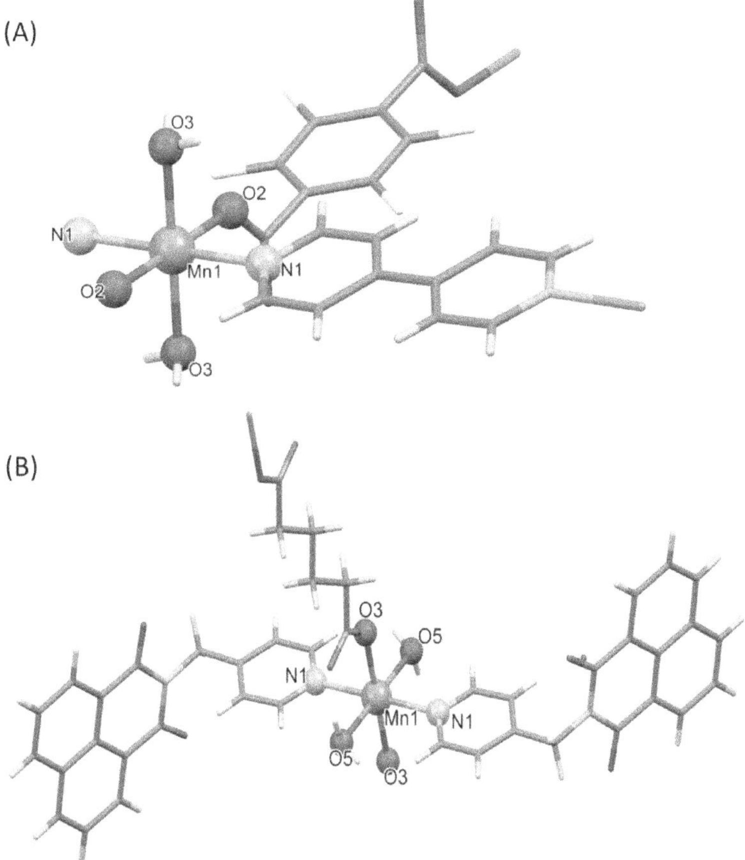

Figure 6. Illustration of coordination environment of Mn(II) in (A) SEFMOH01 (B) HUCJOG.

In an octahedral geometry of HUCJOG, two N atoms (N1) and, two O atoms (O5, O5) of water molecule occupies equatorial position, and two oxygen atoms (O3, O3) occupies axial position to form hexa-coordinate Mn(II) ion (Figure 6B). The temperature-dependent magnetic susceptibilities show, χ_{mT} value and spin only value of the complex are 4.23 & 4.37 cm^3 mol^{-1} K respectively.

In the crystal structure of DEYCAN, two oxygen atoms (O1, O4), and two nitrogen atoms (N3, N4) and two carboxy oxygen atoms (O1, O4) occupies axial and equatorial position of Mn(II) ion respectively forming an octahedral geometry. Axial bond lengths of Mn-O are 2.188 Å and 2.155 Å and bond lengths of equatorial Mn-N/Mn-O are observed in the ranges from 2.108Å.-2.268 Å. Temperature-dependent magnetic susceptibilities, χ_{mT} value, spin only value, C & θ of the complex are 4.38, 4.37, 4.68 cm³ mol⁻¹K & -20.8 K respectively (Figure 7).

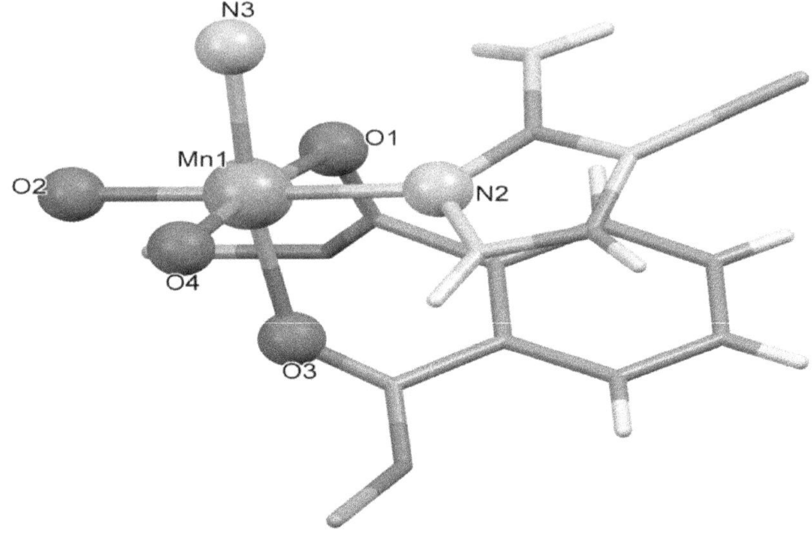

Figure 7. Illustration of coordination environment of Mn(II) in DEYCAN.

The Mn(II) ion in DEYBOA is surrounded by two axial water oxygen atoms (O5W, O5) and four equatorial 2N 2O (O1, O1, N3, N3), adopting an octahedral coordination arrangement. Bond lengths of Axial Mn-O are 2.251 Å and 2.251 Å and bond lengths of equatorial Mn-N/Mn-O are observed in the ranges from 2.144Å.-2.246 Å. The temperature-dependent magnetic susceptibilities, χ_{mT} value, spin only value, C, & θ of the complex are 4.49, 4.48, 4.50 cm³ mol⁻¹ K & -0.40 K respectively (Figure 8).

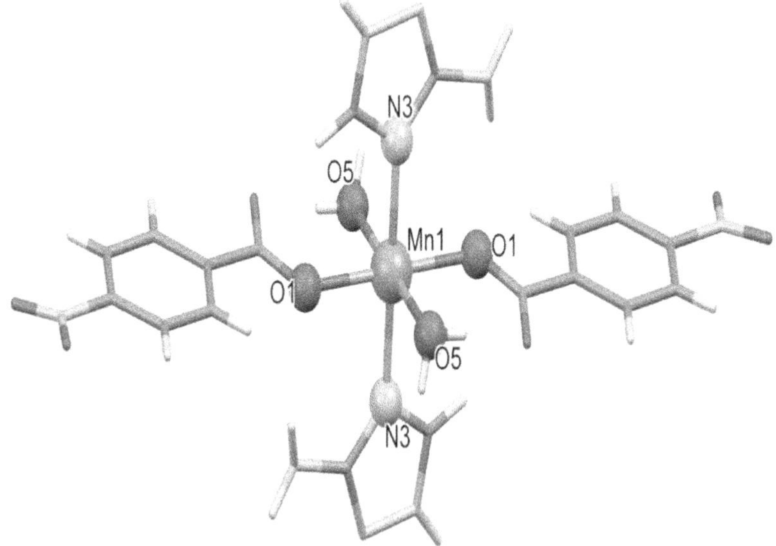

Figure 8. Illustration of coordination environment of Mn(II) in DEYBOA.

2.1.4 Equatorial (4O)

In GITBUI, Mn(II) ion is hexacoordinated with two oxygen atoms (O4, O2) from two carboxylate groups axially and four oxygen atoms (O1, O3, O4, O5) equatorially, and has a distorted octahedral geometry. Bond lengths of Axial Mn-O are Mn1-O4 = 2.191 Å and Mn1-O2 = 2.150Å and bond length of equatorial Mn-O are observed in the range of 2.122-2.188Å (Figure 9). The temperature-dependent magnetic susceptibilities, χ_{mT} value, spin only value, C & θ of the complex are 4.49, 4.22, 4.76 cm^3 mol^{-1} K -30.70K respectively.

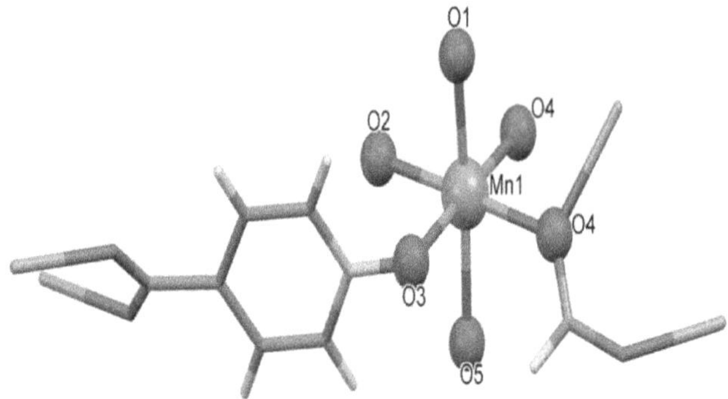

Figure 9. Illustration of coordination environment of Mn(II) in GITBUI.

2.1.5 Equatorial (3O, 1N)

In WOXKOL Mn(II) ion bonded, three oxygen atoms (O2W, O1D, O2E), one nitrogen atom (N2) in equatorial position and two coordinating oxygen atoms (O1W, O4), in axial position and has a distorted octahedral geometry. Bond lengths of axial Mn-O are Mn1-O1W = 2.246 Å and Mn1-O4 = 2.152 Å and bond length of equatorial Mn-N/Mn-O are observed in the scale of 2.119-2.246Å (Figure 10). The temperature-dependent magnetic susceptibilities, χ_{mT} value, spin only value, C & θ of the complex are 9.04, 8.75, 9.12 cm^3 mol^{-1} K & -8.53K respectively.

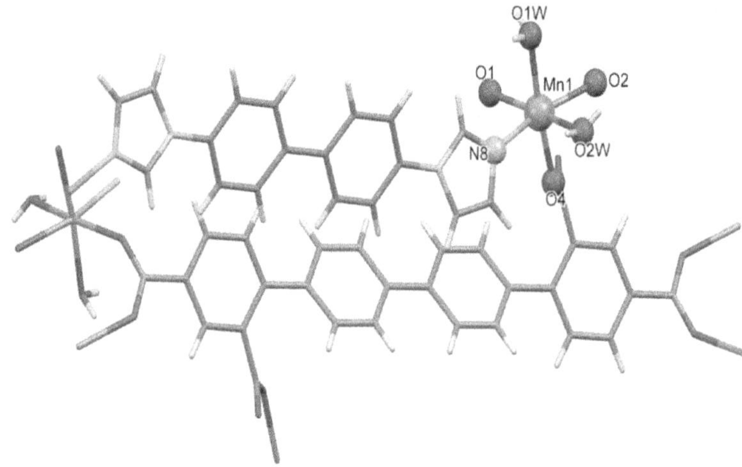

Figure 10. Illustration of coordination environment of Mn(II) in WOXKOL.

2.2 Axial Position Occupied by Oxygen and Nitrogen Atoms from Ligand

2.2.1 Equatorial (4N)

In DEZKAW, Mn(II) ion is hexacoordinated by four nitrogen atoms (N2, N3, N4, N6) equatorially, and one nitrogen (N7) and one oxygen atom (O1) from H_2O molecules, form not a regular octahedral geometry. Axial Mn-O bond distances are Mn1-N7 = 2.322Å and Mn1-O = 2.176Å and bond length of equatorial Mn-N are observed in the scale of 2.210–2.322 Å (Figure 11). The temperature-dependent magnetic susceptibilities, χ_{mT} value, spin only value, C & θ of the complex are 9.51, 8.75, 9.7 cm³ mol⁻¹ K & -5.5 K respectively.

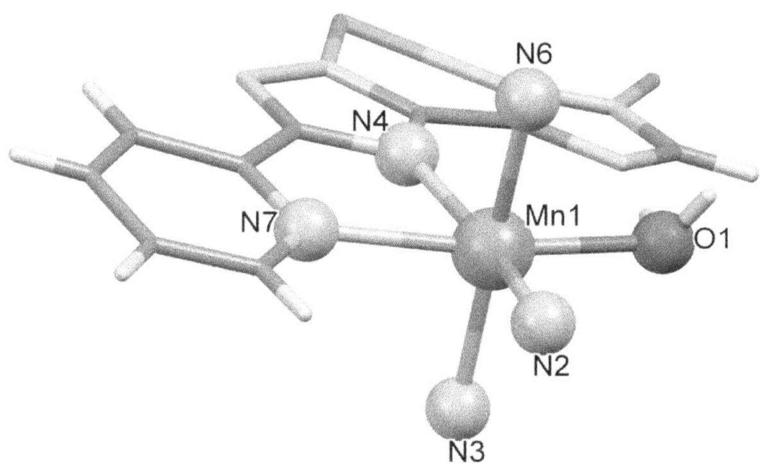

Figure 11. Illustration of coordination environment of Mn(II) in DEZKAW.

2.2.2 Equatorial (2N, 2O)

In octahedral distorted geometry of MIFQOJ, the Mn(II) centre is bonded by three nitrogen atoms (N2) and three oxygen atoms (O1). Bond lengths of axial are Mn1- O1 = 2.132 Å, Mn1-N2 = 2.283 Å and equatorial Mn-O/Mn-N are observed in the scale of 2.132 Å -2.283 Å (Figure 12A). The temperature-dependent magnetic susceptibilities,

χ_{mT} value, spin only value, C & θ of the complex are 8.64, 8.75, 9.38 cm^3 mol^{-1} K & -24.76K respectively.

2.2.3 Equatorial (3O, N)

In the complex TIXLAP, the oxygen and nitrogen atom (N1B, O2C) in axial position and nitrogen atom as well as three oxygen atoms (N2D, O3, O1, O4A) in equatorial position form a distorted structure of octahedral geometry for the Mn(II) atom. Bond lengths of axial are Mn1-O2 = 2.176 Å, Mn1-N1 = 239 Å and equatorial Mn-N/Mn-O are observed in the scale of 2.211 Å -2.287 Å (Figure 12B). The temperature-dependent magnetic susceptibilities, χ_{mT} value, spin only value, C & θ of the complex are 4.10, 4.38, 4.30 cm^3 mol^{-1} K & -9.95 K respectively.

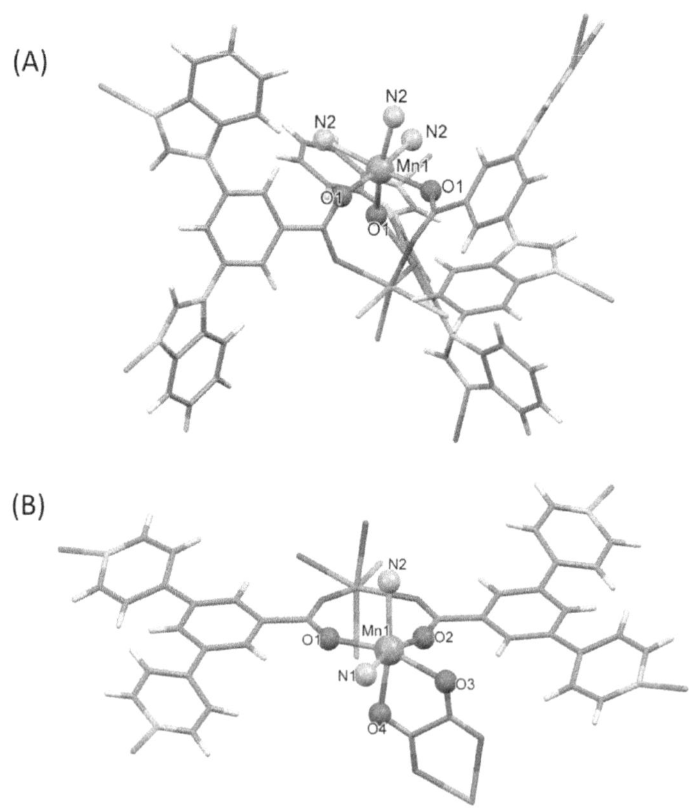

Figure 12. Illustration of coordination environment of Mn(II) in (A)MIFQOJ (B) TIXLAP.

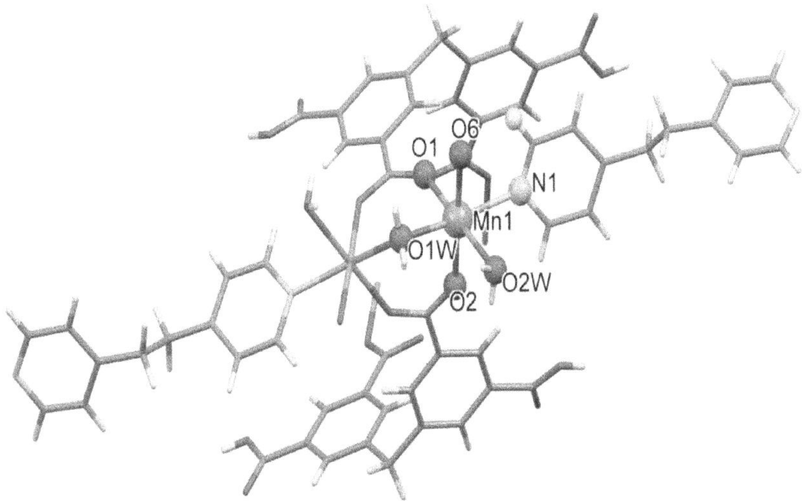

Figure 13. Illustration of coordination environment of Mn(II) in SAMTIM.

2.2.4 Equatorial (4O)

In SAMTIM, one nitrogen atom (N1) and five oxygen atoms (O1, O2, O1W, O2W, O6) form an octahedral with distorted geometry surrounding of Mn(II) atom. Bond lengths of axial are Mn1- O2 = 2.198Å, Mn1-N1 = 2.243Å and equatorial Mn-O/Mn-N are observed in the order of 2.119 Å -2.243 Å (Figure 13). The temperature-dependent magnetic susceptibilities, χ_{mT} value, spin only value, C & θ of the complex are 8.15, 8.75, 8.49 cm³ mol⁻¹ K & -9.56K respectively.

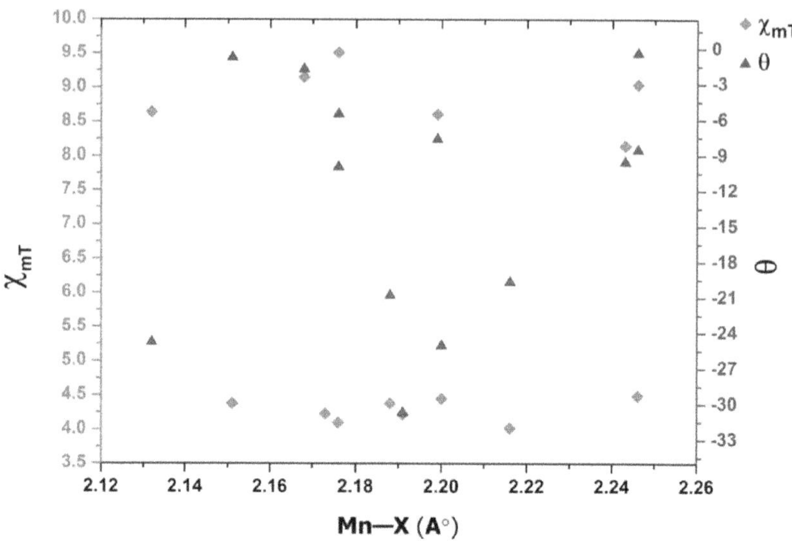

Figure 14. Scatter plot between χ_{mT}, θ, and Mn-X (Å) of monometallic Mn(II) complexes.

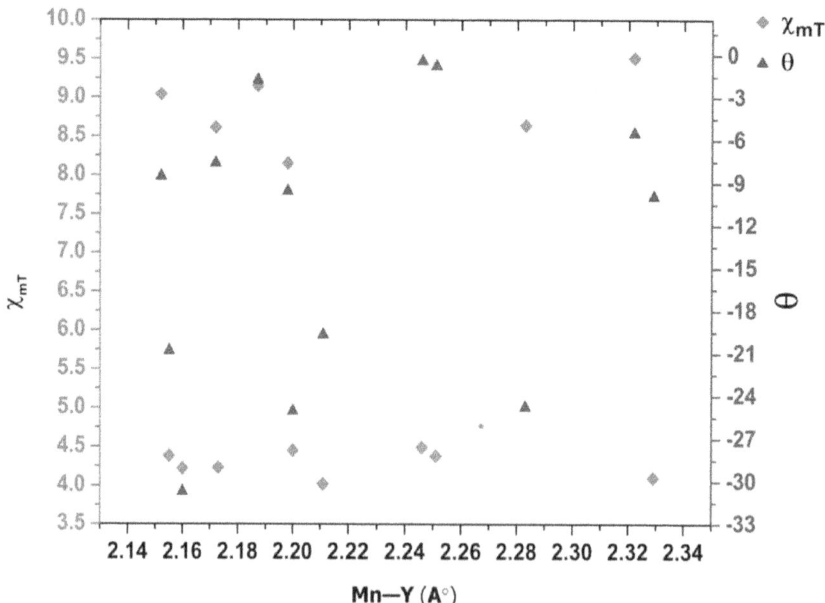

Figure 15. Scatter plot between χ_{mT}, θ, and Mn-Y (Å) of monometallic Mn(II) complexes.

The scatter plot between χ_{mT}, θ, Mn-X (Å) and Mn-Y (Å) indicate that the maximum numbers of the complexes are having Mn-X distance in the range 2.16 Å – 2.21 Å (Figure 14) and for Mn-Y it is in the range of 2.15 Å -2.21 Å (Figure 15). The χ_{mT} value for Mn-X and Mn-Y lies 4.0-9.5 & 4.0-9.1 cm^3 mol^{-1} K respectively and corresponding Weiss constant (θ) for Mn-X and Mn-Y lies -31 to +2 &-31 to +2K respectively.

3. Magnetic Properties of Bimetallic Mn Metal Organic Frameworks

3.3.1 Mn1 (Both Aaxial is Oxygen)

In the complex LAXMOP, Mn(II) ion is crystallized in two types, Mn1 and Mn2. The slightly distorted octahedral structure of Mn1 is hexacoordinated having four nitrogen atoms (N5, N5, N6, N6) in equatorial plane of bond length in the order of 2.259 Å-2.262 Å and two oxygen atoms (O1, O1) in axial position; (Mn1-O1 = 2.309 Å, O1B- Mn1 = 2.265 Å) from coordinated water molecules. Mn2 is form octahedral geometry, coordinated by six nitrogen atoms. Both axial and equatorial position is coordinated to nitrogen atom (axial- N7, N4 and equatorial- N3, N3, N4, N4) with the bond length of 2.225 Å - 2.235 Å range (Figure 16A). The temperature-dependent magnetic susceptibilities, χ_{mT} value, spin only value, C & θ of the complex are 9.73, 8.75, 2.86 cm^3 mol^{-1} K & -2.84 K respectively.

In the complex FIVQOT, Mn1 is bonded by two axial oxygen atoms (O1, O1A) with bond lengths are Mn-O1=2.189 Å and four equatorial carboxylate oxygen atoms (O5C, O5B, O8E, O8D) with bond distance in the scale of 2.127 Å -2.161Å and shows distorted octahedral geometry. The second manganese ion is hexacoordinated by two axial oxygen atoms (O9, O10) and equatorial position is occupied by three oxygens (O1, O2, O6) and one nitrogen atom (N2) with bond lengths of all Mn-O is in the order of 2.092 Å -2.297 Å (Figure 16B). The temperature-dependent magnetic susceptibilities, χ_{mT} value, spin only value, C & θ of the complex are 13.58, 13.2, 17.94 cm^{3}mol^{-1} K & -10.91 K respectively.

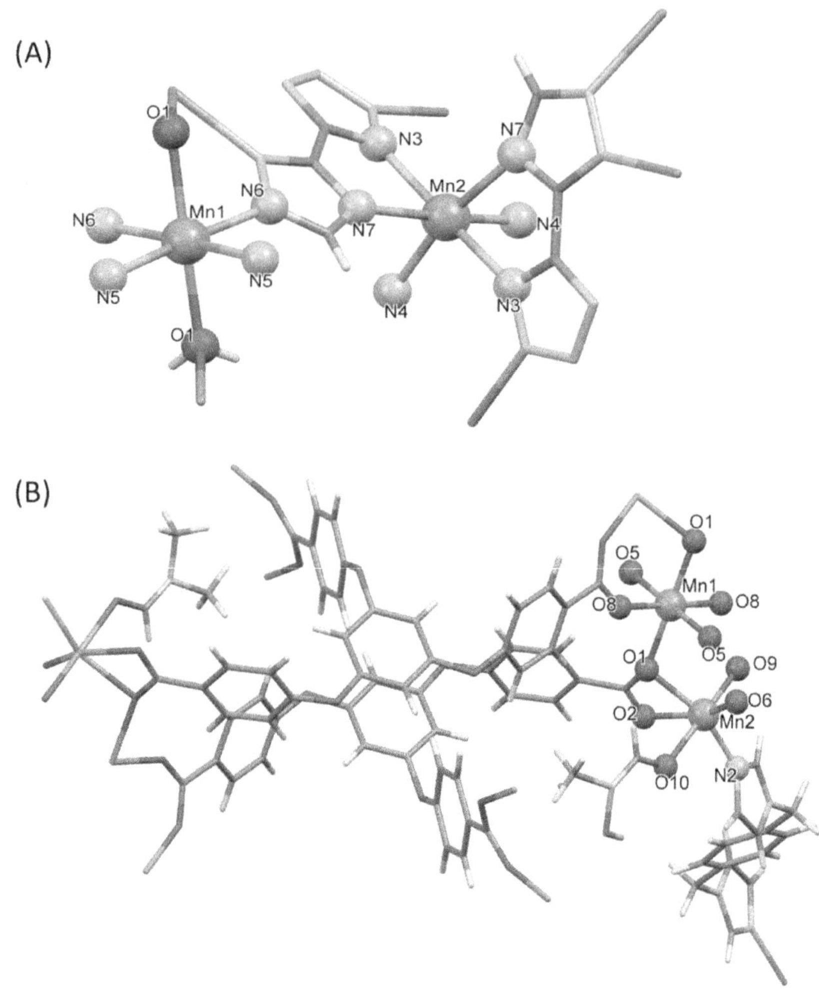

Figure 16. Illustration of coordination environment of Mn(II) in (A) LAX-MOP (B) FIVQOT.

Both Mn(II) ion show distorted octahedral geometry in XULNOJ. Mn1 coordinates to four oxygen atoms (O1, O1A, O8C, O8E) at equatorial and two molecules of waters (O2, O2A) at axial position, both bond lengths of axial Mn-O are 2.177 Å, bond lengths of equatorial are in the scale of 2.119 Å -2.200 Å. Mn2 is coordinated to six oxygen with bond lengths in the order of 2.147Å -2.258Å (Figure 17). The temperature-dependent magnetic susceptibilities, χ_{mT} value, spin only value, C & θ of the complex are 13.04, 13.13, 13.70 cm³mol⁻¹ K & -13.71 K respectively.

In the complex WEKKUU, pseudo-octahedral geometry of Mn(1) is ligated by four O atoms (O3, O3, O1, O1) [Mn–O= 2.102–2.274 Å] in the equatorial and two aqua oxygen atoms (O7, O7) [Mn–O=2.195 Å] in the axial positions. The Mn(2) ion is bonded by five oxygen atoms (O5, O4, O6, O1, O2) and one oxygen atom (O11) of ligand and coordinated respectively [Mn–O 2.061–2.308 Å]. The temperature-dependent magnetic susceptibilities, χ_{mT} value, spin only value, C & θ of the complex are 13.21, 13.125, 13.76 cm³mol⁻¹ K & -16.93 K respectively (Figure 18).

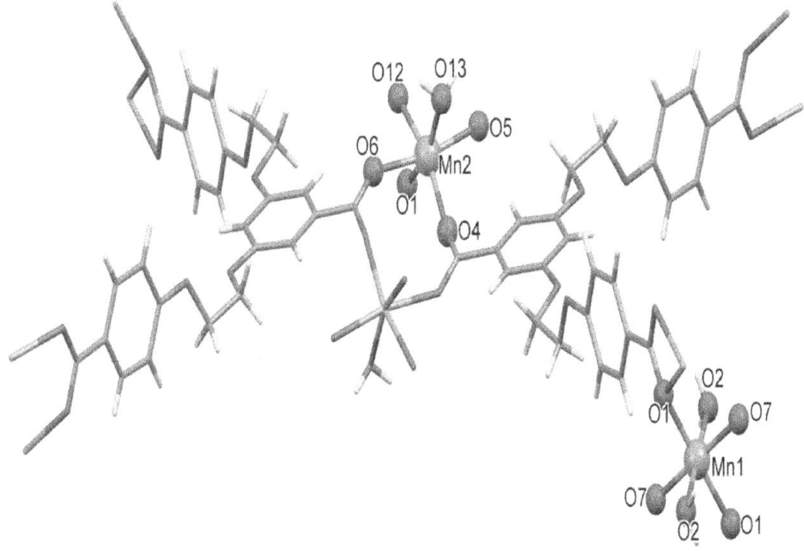

Figure 17. Illustration of coordination environment of Mn(II) in XULNOJ.

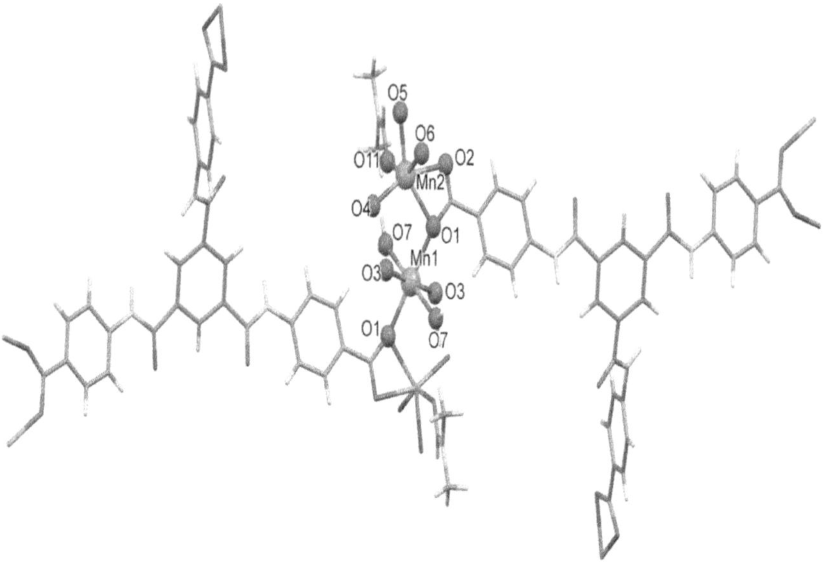

Figure 18. Illustration of coordination environment of Mn(II) in WEKKUU.

In the complex ACUWH, Mn1 is surrounded by four oxygens, two nitrogens in the coordination. Both axial (O4, O4) bond lengths Mn-O are 2.158 and equatorial (O2, O2, N2, N2) bond distance in the order of 2.153-2.300 Å, Mn2 octahedral structure coordinated by two carboxylate oxygens, two nitrogens, and two aqua oxygens. The axially coordinated Mn−N bond lengths [2.277 Å] are longer than equatorial coordinated Mn−O [2.181 Å] (Figure 19). The temperature-dependent magnetic susceptibilities, χ_{mT} value, spin only value, C & θ of the complex are 4.09, 4.375, 8.94 cm³mol⁻¹ K & -17.5K respectively.

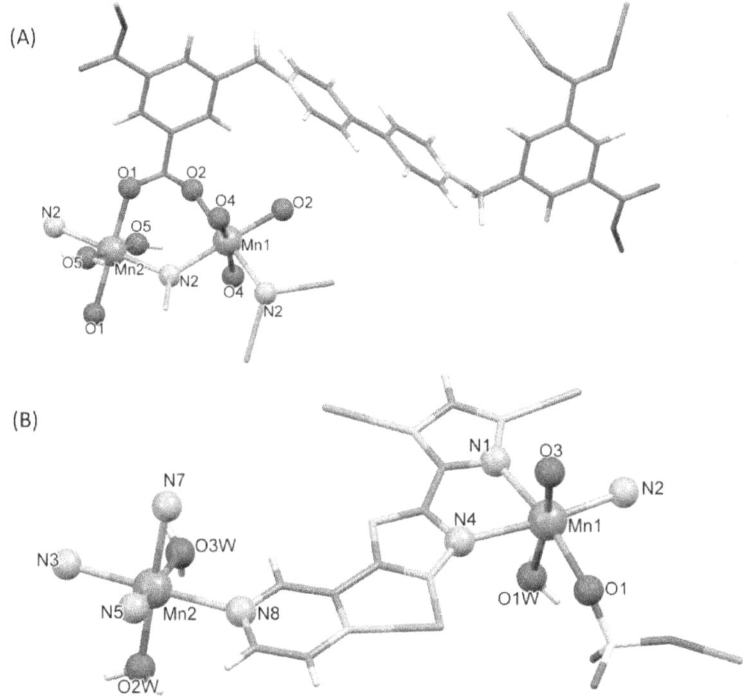

Figure 19. Illustration of coordination environment of Mn(II) in (A) ACUWH (B) TEDMAT.

Both Mn (II) ion show distorted octahedral geometry in TEDMAT. Mn1 is coordinated by two O atoms three N atoms, and one bonded aqua oxygen, axially occupied oxygen atom (O1W, O3) have bond length [Mn-O=2.163, Mn-O=2.195] and equatorially occupied (N1, N2, N4, O1) Mn-O/Mn-N with range of bond lengths is 2.195-2.317. Mn2 is coordinated by four N atoms (N3, N5, N7, N8) with range of bond lengths is 2.228-2.339 and two aqua oxygen atoms (O3w, O2w) with bond length of Mn-O=2.121-2.163. The temperature-dependent magnetic susceptibilities, χ_{mT} value, spin only value, C & θ of the complex are 10.22, 8.75, 10.28 cm³mol⁻¹ K & -8.25 K respectively.

3.2 Mn1 (Both Axial is Nitrogen)

In the complex WEKLAB, Mn1 exhibits not a regular octahedral coordination geometry, the four oxygen atoms are present in equatorial positions is occupied by four oxygen atoms [Mn–O = 2.109 – 2.341] and two nitrogen atoms occupy the axial positions [Mn–N= 2.242 Å]. The second Mn ion is bonded by oxygen atoms with the bond length of Mn–O are in the ranges from 2.060 –2.325 Å, and one nitrogen having the distance of Mn–N 2.185 Å which show distorted trigonal bipyramid (Figure 20A). The temperature-dependent magnetic susceptibilities, χ_{mT} value, spin only value, C & θ of the complex are 13.18, 13.125, 13.84 cm^3 mol^{-1} K & -19.97 K respectively.

Each Mn(II) ion in complex DEZKEA show distorted octahedral geometry, Mn1 is surrounded by five nitrogens and one oxygen atom from one aqua ligand. The axial position is occupied by two nitrogen atoms (N9, N4) and equatorial position is occupied by three nitrogens and one oxygen (N7, N10, N12, O1) All bond lengths of the Mn–N are in the ranges from 2.210Å -2.342Å and Mn-O bond length is 2.222 Å. Mn2 ion is also hexacoordinated with four nitrogen atoms and two oxygen atoms from ligand (Figure 20B). The temperature-dependent magnetic susceptibilities show, χ_{mT} value, spin only value, C & θ of the complex are 13.21, 13.125, 9.1 cm^3mol^{-1} K & -2.82K respectively.

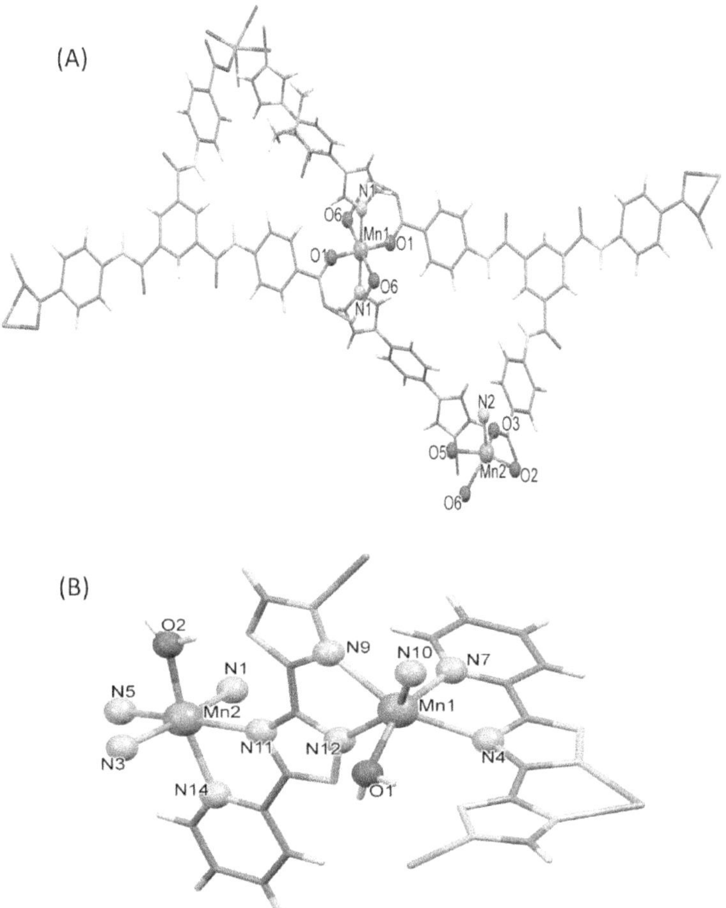

Figure 20. Illustration of coordination environment of Mn(II) in (A)WE-KLAB (B) DEZKEA.

Mn Bimetallic (M1)

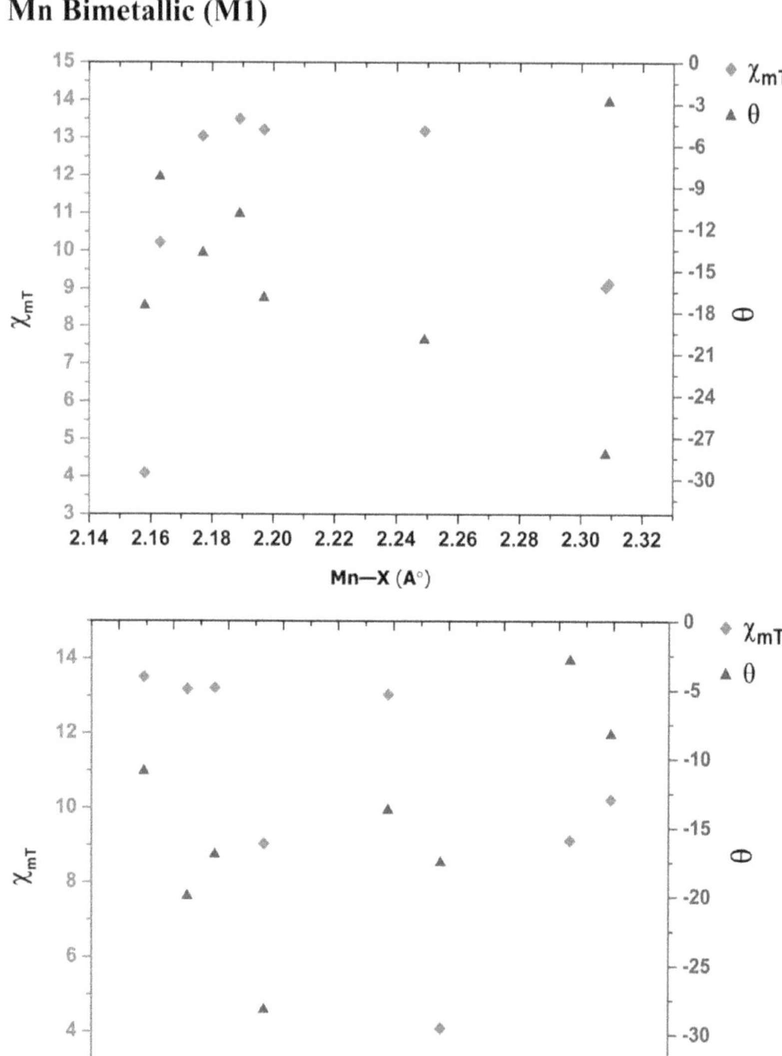

Figure 21. Scatter plot between χ_{mT}, θ, and **above-** Mn1-X (Å); **below-** Mn1-Y (Å) of bimetallic Mn(II) complexes.

The scatter plot between χ_{mT}, θ, Mn1-X (Å) and Mn1-Y (Å) indicate that the maximum numbers of the complexes are having Mn1-X distance in the range 2.16 Å – 2.20 Å (Figure 21 Above) and for Mn1-Y it is in the range of 2.17Å -2.20 Å (Figure 21 Below). The χ_{mT} value and corresponding weiss constant (θ) for Mn1-X lies 4.1.0-13.5 cm³ mol⁻¹K and -18 to -7K while for Mn1-Y it is 4.0-13.5cm³ mol⁻¹ K and -28 to -11K.

Mn Bimetallic (M2)

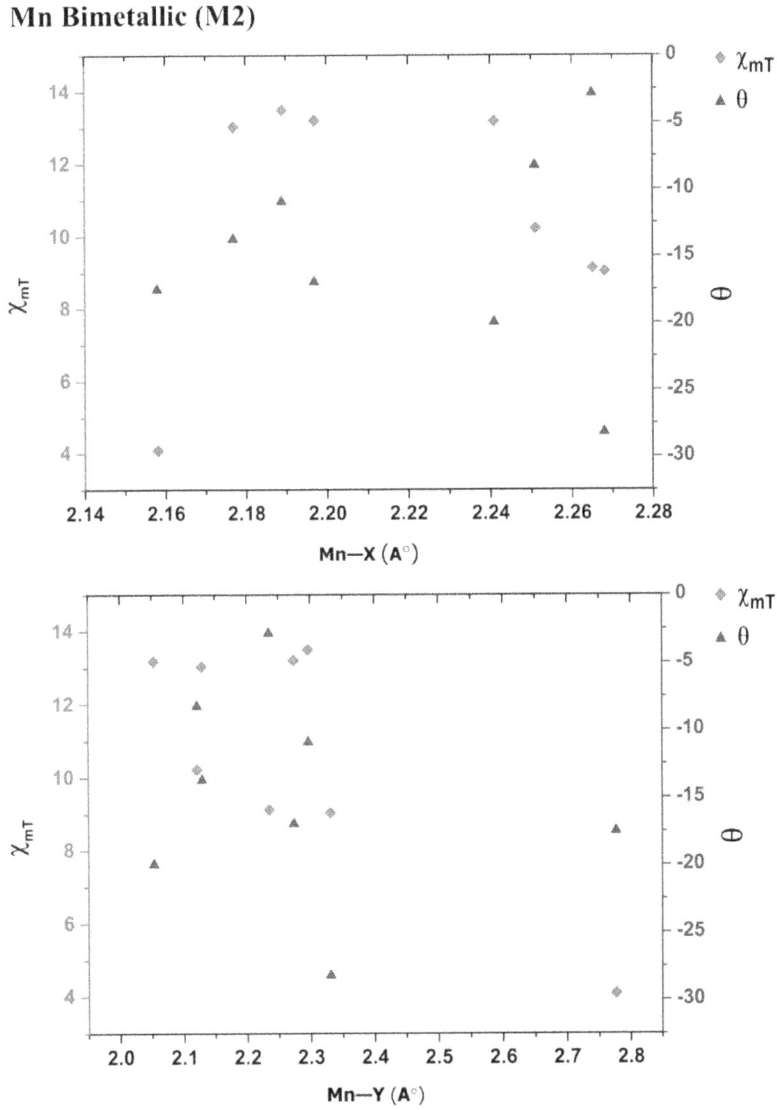

Figure 22. Scatter plot between χ_{mT}, θ, and **above-** Mn2-X (Å); **below-** Mn2-Y (Å) of bimetallic Mn(II) complexes.

The scatter plot between χ_{mT}, θ, Mn2-X (Å) and Mn2-Y (Å) indicate that the maximum numbers of the complexes are having Mn-X distance in the range 2.17Å – 2.20 Å and 2.24 Å -2.27 Å (Figure 22 Above) and for Mn-Y it   is in the range of 2.05 Å -2.53Å (Figure 22 Below). The χ_{mT} value for Mn2-X and Mn2-Y are lies 9.0-13.5 & 9.0-13.2 cm^3 mol^{-1} K respectively and corresponding Weiss constant (θ) for Mn2-X and Mn2-Y are -20 to +2.5K & -28 to +2.5K respectively.

4 Magnetic Properties of Cobalt Mononuclear Three Dimensional Metal Organic Frameworks

4.1 Both Axial are Oxygen Atoms

4.1.1 Equatorial (2N, 2O)

In the crystal structure of DIDHUV, Co(II) ion is hexacoordinated to two nitrogen atoms (N1, N2A) and two oxygen atoms (O1W, O4) equatorially, and axially two oxygen atoms (O3W, O1W) from aqua ligand, form a distorted octahedral geometry. Axial Co-O bond lengths are Co-O2W = 2.062Å and Co-O3W= 2.061Å and bond length of equatorial Co-N/Co-O are observed in the range of 2083-2.211 Å (Figure 23A). The temperature-dependent magnetic susceptibilities, χ_{mT} value, spin only value, C & θ of the complex are 2.92, 1.87, 3.08 cm mol^{-1} K & -23.16 K respectively.

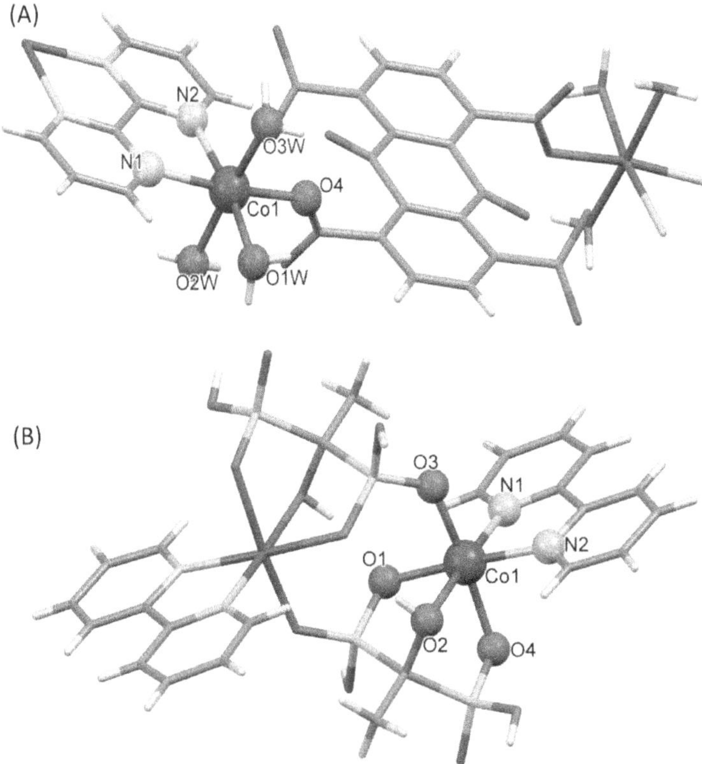

Figure 23. Illustration of coordination environment of Co(II) in (A) DIDHUV (B) DAHFUQ.

In DAHFUQ, Co(II) ion coordinated with two oxygen atoms (O3, O4) of two ligand groups axially and two oxygen atoms (O1, O2) and two nitrogen atoms (N1, N2) equatorially, and has a distorted octahedral geometry. Axial Co-O bond lengths are Co-O4 = 2.146 Å and Co-O3 = 2.048 Å and bond length of equatorial Mn-O/Mn-N are observed in the scale of 2.079-20234Å (Figure 23B). The temperature-dependent magnetic susceptibilities, χ_{mT} value, spin only value, C & θ of the complex are 5.38, 3.74, 5.27 cm³ mol⁻¹ K & 7.67 K respectively.

4.1.2 Equatorial (4O)

The Co (II) ion in WONYAC is surrounded by two axial oxygen atoms (O1, O2) and four equatorial oxygen atoms (O2, O3, O3, O4), adopting octahedral coordination arrangement. Axial Co-O bond lengths are 2.201 Å and 2.115 Å (Figure 24). The temperature-dependent magnetic susceptibilities, χ_{mT} value, spin only value, C & θ of the complex are 4.49, 1.875, 4.76 cm³ mol⁻¹ K & -35.71 K respectively.

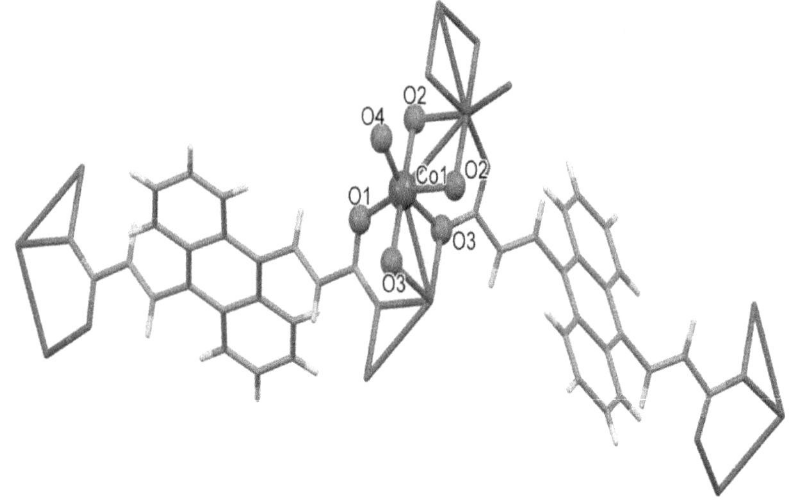

Figure 24. Illustration of coordination environment of Co(II) in WONYAC.

4.2 Both Axial Atoms are Nitrogen

4.2.1 Equatorial (4N)

The Co (II) ion in KUWJIY is surrounded by two axial nitrogen atoms (N13, N6) and four equatorial nitrogen atoms (N1, N8, N18, N10), adopting octahedral coordination arrangement. Axial bond lengths Co-O are 2.159 Å and 2.187 Å and bond lengths of equatorial Co-N are observed in the range of 2.105Å.-2.288 Å. The temperature-dependent magnetic susceptibilities show, χ_{mT} value, spin only value, C & θ of the complex are 2.62, 1.875, 2.87 cm³ mol⁻¹ K & -25.72 K respectively (Figure 25).

4.2.2 Equatorial (4O)

In the crystal structure of WONYOQ, Co(II) ion is hexacoordinated to, four carboxylate oxygen atoms (O1, O3, O1, O3) equatorially, and axially two nitrogen atoms (N1, N1) from ligand, form a distorted octahedral geometry. Axial Co-N bond lengths are Co-N1 = 2.048Å and Co-N1= 2.048Å and bond length of equatorial Co-O are observed in the range of 2.082-2.146 Å (Figure 26A). The temperature-dependent magnetic susceptibilities show, χ_{mT} value, spin only value, C & θ of the complex are 3.16, 3.75,3.33 cm³ mol⁻¹ K & -23.4 respectively.

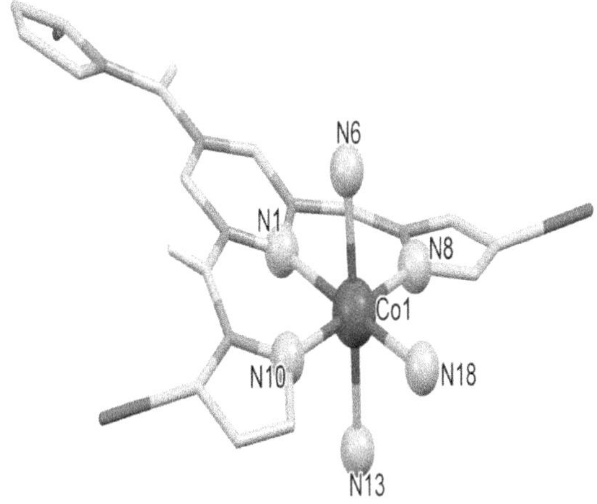

Figure 25. Illustration of coordination environment of Co(II) in KUWJIY.

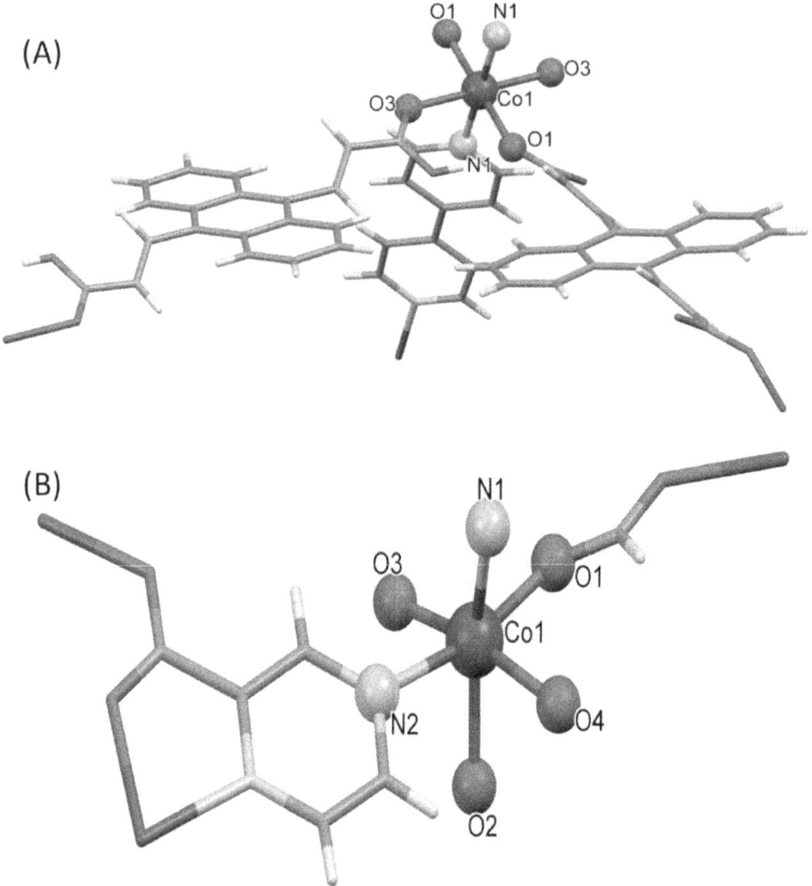

Figure 26. Illustration of coordination environment of Co(II) in (A) WONYO (B) YOZRUC.

4.3 Axial Position Occupied by Oxygen and Nitrogen Atoms

4.3.1 Equatorial (3N, 1O)

The Co(II) ion in YOZRUC is surrounded by axially one nitrogen atom (N1) and one oxygen atom (O2) and one nitrogen atom (N2), three oxygen atoms (O1, O3, O4) equatorially, adopting octahedral coordination arrangement. Axial bond lengths of Co-N= 2.187 Å and Co-O=2.081 Å and bond lengths of equatorial Co-N/Co-O are observed in the range of 2.042Å.-2.176 Å (Figure 26B). The temperature-dependent magnetic susceptibilities, χ_{mT} value of complex are 3.40 cm^3 mol^{-1} K. The Weiss constant θ = -21.20 K.

Co Monometallic

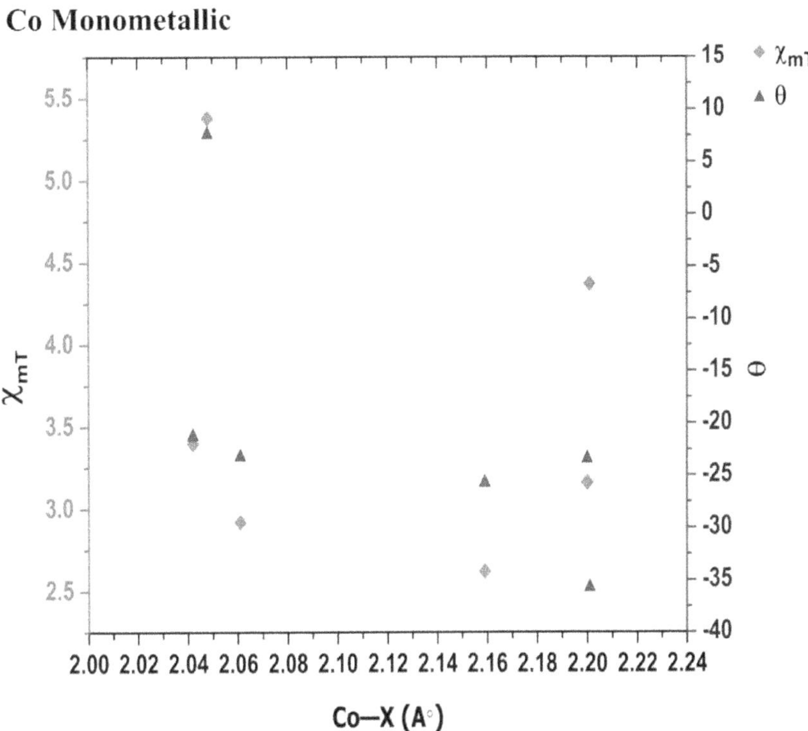

Figure 27. Scatter plot between χ_{mT}, θ, and Co-X (Å) of monometallic Co(II) complexes.

Co Monometallic

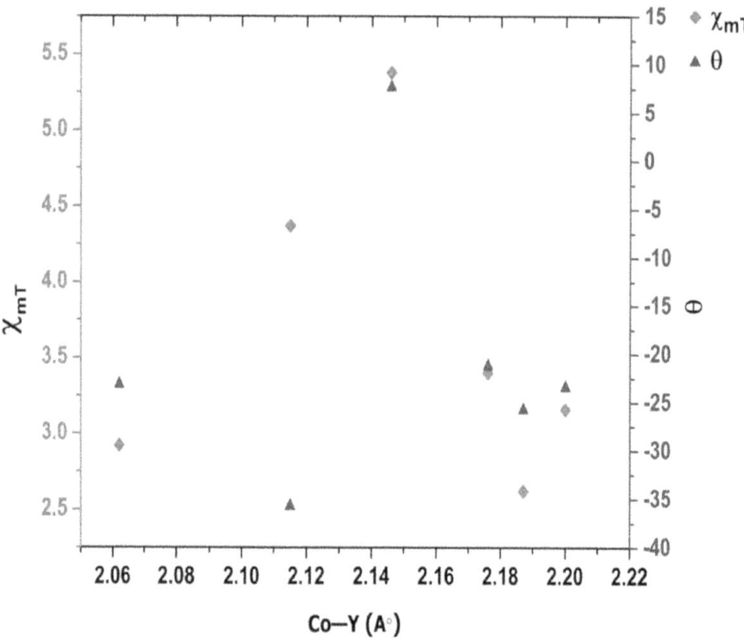

Figure 28. Scatter plot between χ_{mT}, θ, and Co-Y (Å) of monometallic Co(II) complexes.

The scatter plot between χ_{mT}, θ, Co-X (Å) and Co-Y (Å) indicate that the maximum numbers of the complexes are having Co-X distance in the range 2.04 Å – 2.06 Å and 2.16-2.20 (Figure 27) and for Co-Y it is in the range of 2.18 Å -2.20 Å (Figure 28). The χ_{mT} value and corresponding weiss constant (θ) for Co-X lies 2.5-3.5 cm³ mol⁻¹ K and -37 to -23K while for Co-Y it is 2.5-3.5 cm³ mol⁻¹ K and -37 to -20 K.

5. Magnetic Properties of Cobalt Bimetallic Complex

5.1 Co1 - Both Axial are Oxygen Atoms

In the complex LAXMIJ, Co(II) ion is crystallized in two types, Co1 and Co2, Co1 is distorted octahedral geometry. Co1 bonded by nitrogen atoms (N5, N5, N6, N6) having a bond length in the range of 2.253 Å-2.258Å in equatorial position and the axial position is filled by oxygen atoms (O1, O1); (Co1-O1 = 2.253 Å, Co1-O1 = 2.310 Å from coordinated water molecules. Co2 is formed octahedral coordination geometry, and connected by six nitrogen atoms from six ligands. Both axial and equatorial position is coordinated to nitrogen atoms (axial- N7, N2 and equatorial- N1, N7, N2, N1) with the bond lengths in the scale of 2.218 Å - 2.320 Å (**Figure 29A**). The temperature-dependent magnetic susceptibilities show, χ_{mT} value, spin only value & θ of the complex are 5.38, 3.76, 6.09 cm^3 mol^{-1} K & -32.5 K respectively.

In the complex MAZQAI, Co(II) ion is crystallized in two types, Co1 and Co2, Co1 is distorted octahedral geometry. Co1 consists of three nitrogen atoms (N5, N3, N4) and one oxygen atom(O5) having a bond length in the range of 2.074Å-2.275Å in equatorial position and the axial position is occupied by oxygen atoms (O1, O6); (Co1-O1 = 2.087 Å, Co1-O6 = 2.143 Å from coordinated water molecules. Co2 is formed octahedral coordination geometry, ligated by six oxygen atoms (O2, O2, O3, O3, O9, O9) from six ligands. Both axial and equatorial position is coordinated to oxygen atoms (O2, O2, O3, O3, O9, O9) with the bond lengths in the scale of 2.072 Å - 2.126 Å (**Figure 29B**). The temperature-dependent magnetic susceptibilities show, χ_{mT} value, spin only value, C & θ of the complex are 6.36, 5.625, 7.05 cm^3 mol^{-1} K & -35.7 K respectively.

5.2 Co1 - Both Axial are Nitrogen Atoms

In MAZPUB, Co(II) ion is crystallized in two types, Co1 and Co2, Co1 is distorted octahedral geometry. Co1 consists of four oxygen atoms (O2, O3, O7, O8) having a bond length in the range of 2.005Å-2.188Å in equatorial position and the axial position is occupied by nitrogen atoms (N5, N9); (Co1-N5 = 2.156 Å, Co1-N9 = 2.216 Å from coordinated water molecules. Co2 is formed octahedral coordination geometry. Axial position is occupied by two nitrogen atoms (N4, N10) having bond lengths Co-N4 = 2.169, Co-N10=2.064 and equatorial position is coordinated to four oxygen atoms (O1, O6, O7, O9) with the bond lengths in the scale of 2.072 Å - 2.142 Å (**Figure 30**). The temperature-dependent magnetic susceptibilities show, χ_{mT} value, spin only value, C & θ of the complex are 4.71, 3.75, 4.71 cm^3 mol^{-1} K & -41.0 K respectively.

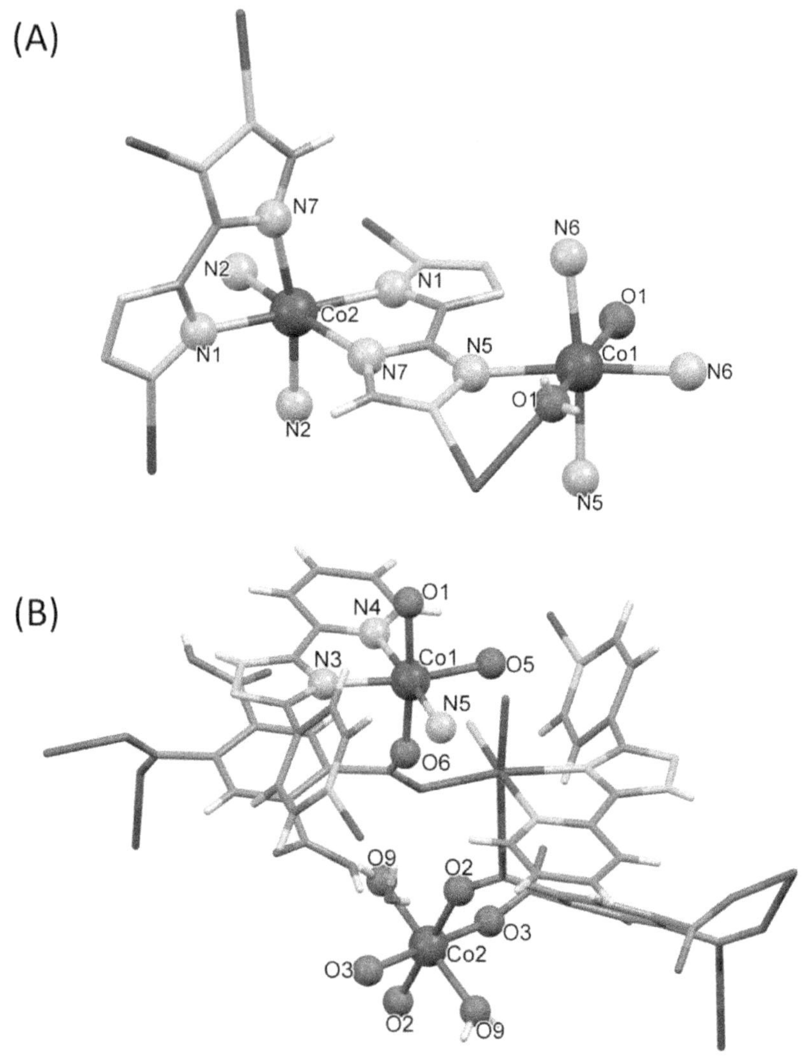

Figure 29. Illustration of coordination environment of Co(II) in (A) LAXMIJ (B) MAZQAI.

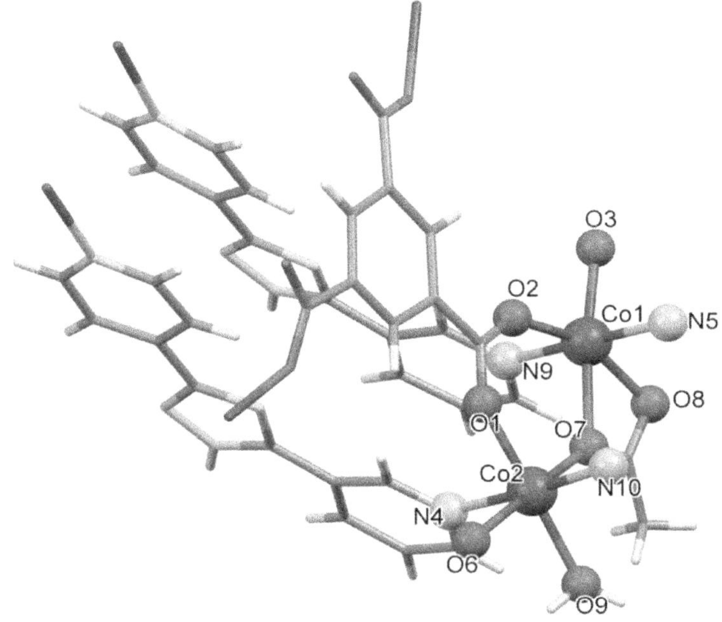

Figure 30. Illustration of coordination environment of Co(II) in MAZPUB.

5.3 Co1 - Axial Position Occupied by Oxygen and Nitrogen

In the complex NOCFAO, Co (II) ion is crystallized in two types, Co1 and Co2, Co1 is distorted octahedral geometry. Co1 consists of two nitrogen atoms (N5, N3) and two oxygen atoms (O2, O4) in equatorial plane, having a bond length in the range of 2.075Å-2.144Å and one oxygen atom (O3) and one nitrogen atoms (N6) in axial position; (Co1-O3 =2.186 Å, Co1-N6 = 2.099 Å. Co2 is formed octahedral coordination geometry. Axial position occupied by a two oxygen atoms (O5, O6) having bond lengths are 2.157 Å and 2.076Å, equatorial position is coordinated to one oxygen atoms (O1) and three nitrogen atoms (N1, N4, N2) with the bond lengths in the range of 2.107Å-2.154 Å (Figure 31A). The temperature-dependent magnetic susceptibilities shows, χ_{mT} value, spin only value, C & θ of the complex are 6.52, 3.76, 7.62 cm^3 mol^{-1} K & -48.9 K respectively.

In the complex GECKOR, Co (II) ion is crystallized in two types, Co1 and Co2, Co1 is distorted octahedral geometry. Co1 consists of four oxygen atoms (O1, O3, O5, O7) in equatorial plane having a bond length range- 2.073Å-2.155Å and the axial position have oxygen atom (O7) and nitrogen atoms (N1); (Co1-O7 =2.108 Å, Co1-N1 = 2.108Å). Co2 is formed tetrahedral geometry. Co2 is formed octahedral coordination geometry. Axial position occupied by a two oxygen atoms (O2, O4) having bond lengths are 2.062 Å and 2.079Å, equatorial position is coordinated to two oxygen atoms (O5, O6) and two nitrogen atoms (N4, N5) with the bond lengths in the range of 2.076Å-2.211Å (Figure 31B). The temperature-dependent magnetic susceptibilities show, χ_{mT} value, spin only value & θ of the complex are 8.95, 7.50, 13.18 cm³mol⁻¹ K & -78.16 respectively.

In the complex GECKUX, Co (II) ion is crystallized in two types, Co1 and Co2, Co1 is distorted octahedral geometry. Co1 consists of four oxygen atoms (O1, O3, O5, O6) having a bond length in the range of 2.075Å-2.167Å in equatorial and the axial position is occupied by one oxygen atom (O5) and one nitrogen atoms (N1); (Co1-O5 =2.107Å, Co1-N1 = 2.098Å. Co2 is formed octahedral coordination geometry. Axial position occupied by a two oxygen atoms (O2, O4) having bond lengths are 2.062 Å and 2.079Å, equatorial position is coordinated to two oxygen atoms (O5, O6) and two nitrogen atoms (N4, N5) with the bond lengths in the range of 2.076Å-2.211 Å (Figure 32A). The temperature-dependent magnetic susceptibilities shows, χ_{mT} value, spin only value, C & θ of the complex are 8.95, 7.50,10.49 cm³mol⁻¹ K & -43.89 K respectively.

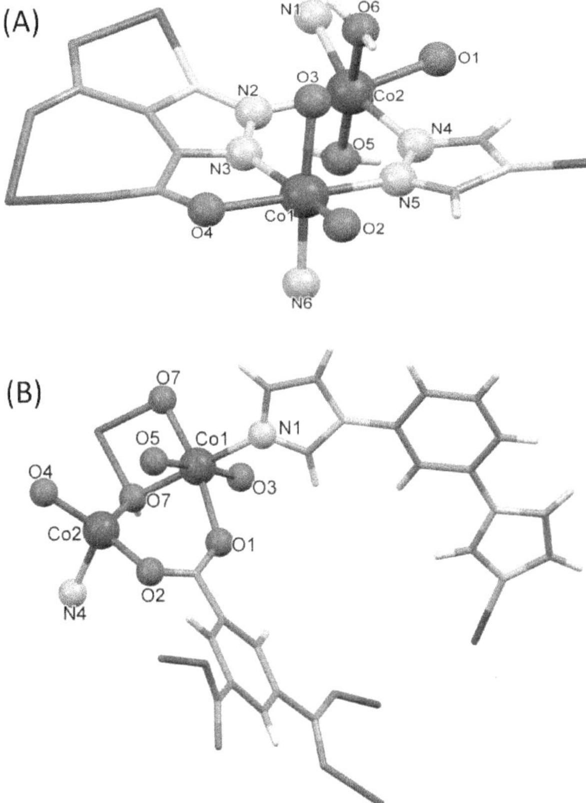

Figure 31. Illustration of coordination environment of Co(II) in (A) NOCFAO (B) GECKOR.

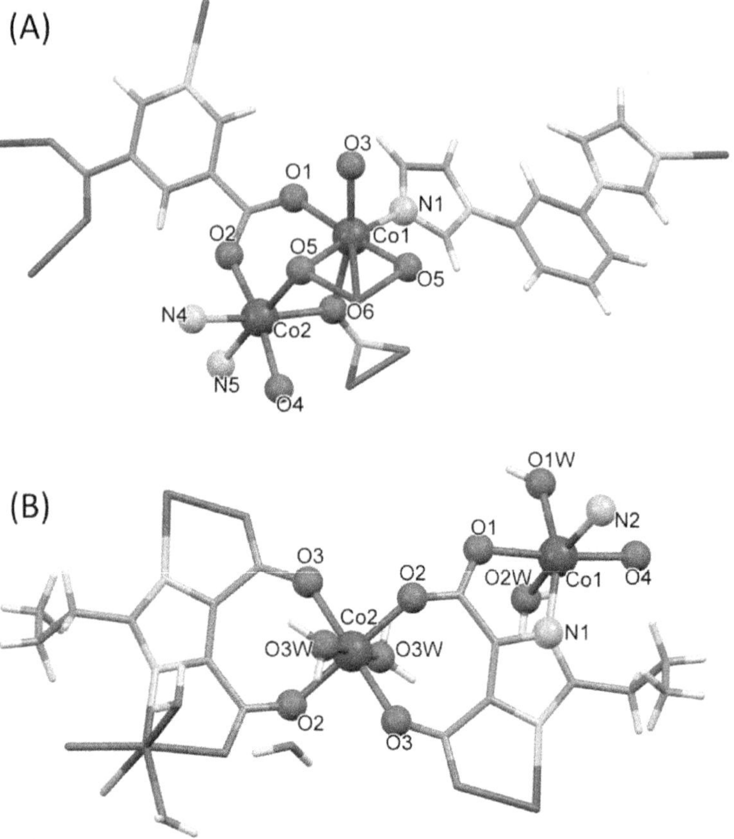

Figure 32. Illustration of coordination environment of Co(II) in (A) GECKUX (B) QUVZEO.

In the complex QUVZEO, Co (II) ion is crystallized in two types, Co1 and Co2, Co1 is distorted octahedral geometry. Co1 consists of one nitrogen atom (N1) and three oxygen atoms (O1, O1W, O4) in equatorial having a bond length in the range of 2.065Å-2.165Å and the axial position is occupied by one oxygen atom (O2W) and one nitrogen atoms (N2); (Co1-O2W = 2.186 Å, Co1-N2 = 2.046 Å. Co2 formed an octahedral coordination geometry where axial position is occupied by a two oxygen atoms (O3W, O3W) having bond length of 2.036Å. Equatorial position is coordinated to four oxygen atoms (O2, O2, O3, and O3) with the bond lengths in the order of 2.078Å-2.148Å (**Figure 32B**). The temperature-dependent magnetic susceptibilities show, χ_{mT} value, spin only value, C & θ of complex are 8.26, 5.625, 6.006 cm^3 mol^{-1} K & -28.383 K respectively.

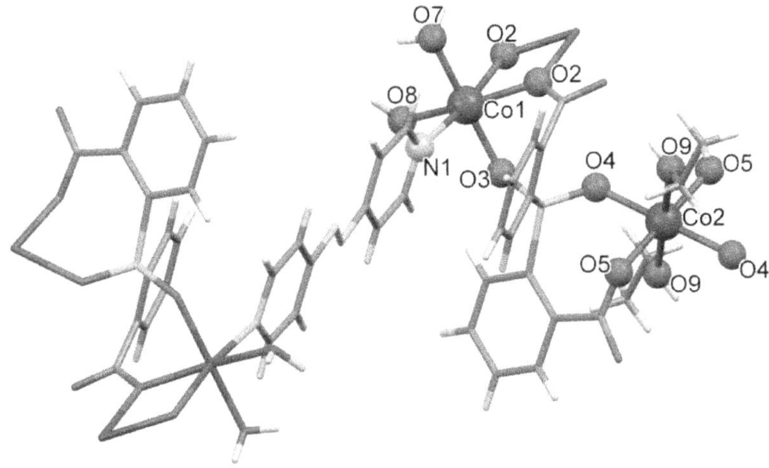

Figure 33. Illustration of coordination environment of Co(II) in ERADET.

In ERADET, Co (II) ion is crystallized in two types, Co1 and Co2, Co1 is distorted octahedral geometry. Co1 consists of four oxygen atoms (O2, O3, O7, O8) in equatorial plane having a bond length in the range of 2.019Å-2.135Å and the axial position is occupied by one oxygen atom (O2) and one nitrogen atom (N1); (Co1-O2 = 2.291Å, Co1-N1 = 2.187 Å. Co2 is formed octahedral coordination geometry. All position occupied by a six oxygen atom (O4, O9, O5, O4, O9, O5) having both bond length is in the scale of 2.0187-2.2908Å (Figure 33). The temperature-dependent magnetic susceptibilities show, χ_{mT} value, spin only value, C & θ of the complex are 8.36, 5.625, 8.81, cm³ mol⁻¹K & -16.6 K respectively.

Co Bimetallic (M1)

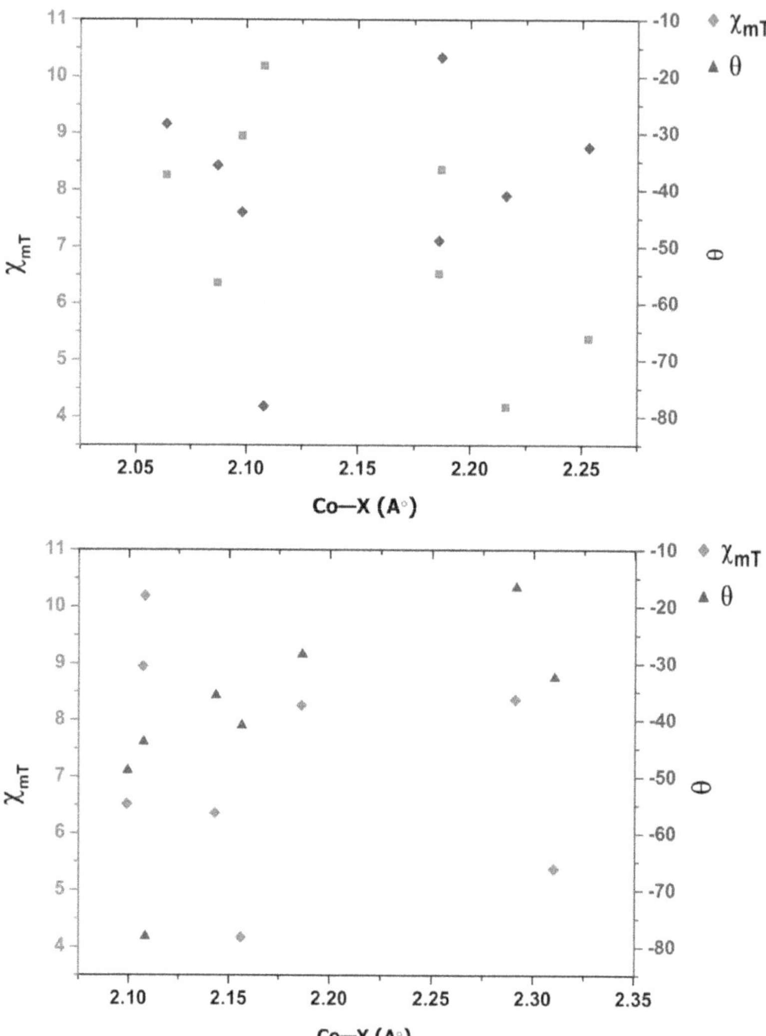

Figure 34. Scatter plot between χ_{mT}, θ, and **above-** Co1-X (Å); **below-** Co1-Y (Å) of bimetallic Co(II) complexes.

The scatter plot between χ_{mT}, θ, Co1-X (Å) and Co1-Y (Å) indicate that the maximum numbers of the complexes are having Co-X distance in the range 2.06 Å – 2.12 Å and 2.18-2.22 (Figure 34 Above) and for Co-Y it is in the range of 2.09Å -2.19 Å (Figure 34 Below). The χ_{mT} value for Co-X and Co-Y are lies 6.2.0-10.1 & 6.3 -10.2 cm³ mol⁻¹ K respectively and corresponding θ for Co-X and Co-Y are -50 to -15 K & -51 to -26 K respectively.

Co Bimetallic (M2)

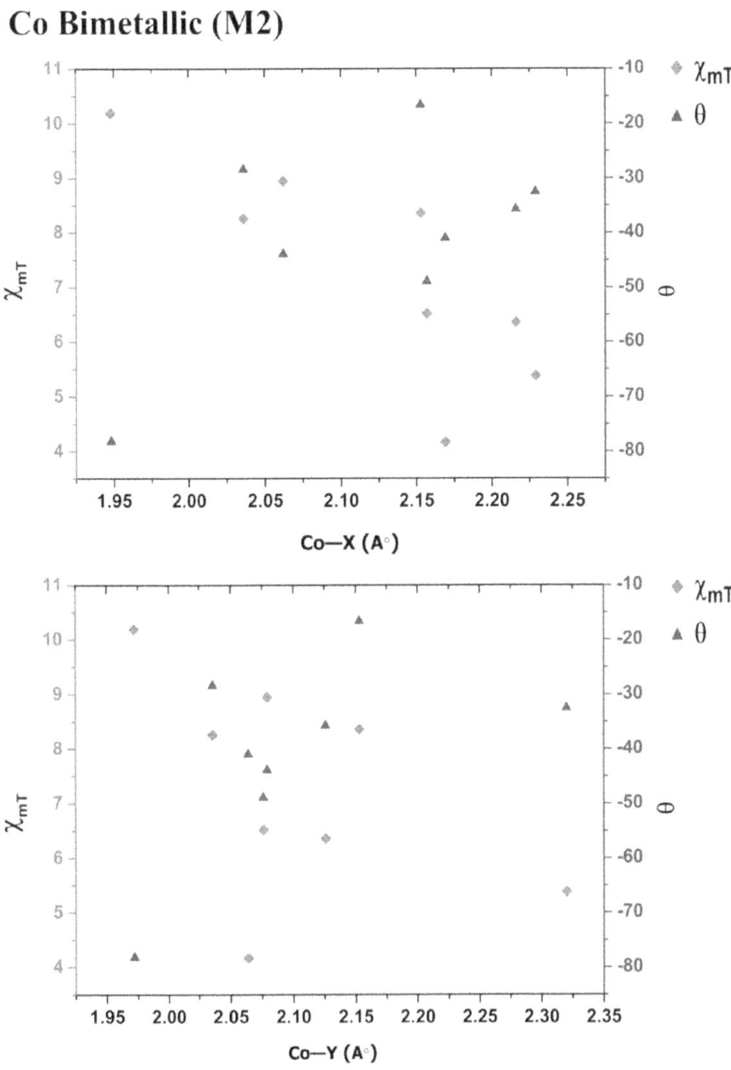

Figure 35. Scatter plot between χ_{mT}, θ, and **above-** Co2-X (Å); **below-** Co2-Y (Å) of bimetallic Co(II) complexes.

The scatter plot between χ_{mT}, θ, Co2-X (Å) and Co2-Y (Å) indicate that the maximum numbers of the complexes are having Co2-X distance in the range 2.15 Å – 2.24 Å (Figure 35 Above) and for Co2-Y it is in the range of 2.023Å -2.17Å (Figure 35 Below). The χ_{mT} value for Co2-X and Co2-Y lies 4.2.0-8.4 & 6.1 -8.9 cm^3 mol^{-1} K respectively and corresponding θ for Co2-X and Co2-Y are -50 to -29 K & -50 to -25 K respectively.

6. Conclusion and Future Outlook

By the analysis of data retrieved from CSD, number of complexes of both Cobalt and manganese, magneto-structural properties were analyzed on the basis of geometrical orientations and concluded in this chapter. The 3D MOFs of monometallic and bimetallic Co(II) and Mn(II) architectures reveal that they exhibit octahedral geometry connected by through ligand having O and N atoms. By these bridging atoms magnetic exchange phenomenon observed which is very important and main reason for magnetic properties of Cobalt and manganese complexes. We observed that in maximum complexes, metal generally exhibit +II oxidation state in which axial and equatorial bond length of octahedral geometry around metal ions play pivotal role for the structure and magnetic properties relationship in discussed 3D MOF.

Magnetic properties of monometallic and bimetallic of Mn(II) and Co(II) complexes were studied. Study shows that χ_{mT} value of Mn(II) complexes are observed in the range of 4 -9.5 cm^3 mol^{-1} K and 4.0 - 13.5 cm^3 mol^{-1} K of monometallic and bimetallic complexes respectively. Weiss constant (θ) of Mn (II) complexes are observed in the range of -31K to +2K and -28K to +2.5K of monometallic and bimetallic complexes respectively. χ_{mT} value of Co(II) complexes are observed in the range of 2.5-3.5 cm^3 mol^{-1} K and 4.2-10.2 cm^3 mol^{-1} K. Weiss constant (θ) are observed in the range of -37K to -20K and -51K to -15 K of monometallic and bimetallic Co(II) complexes respectively.

This book chapter mainly focuses on magneto-structural analysis of 3D MOFs based on the role of axial and equatorial atoms around metal (II) ion geometry. The study will enrich the area of MOFs as in designing of such materials with desired magnetic properties.

Acknowledgement

Dr. Kafeel Ahmad Siddiqui is thankful to Council of Scientific and Industrial Research, New Delhi, India for the award of research grant (01(2962)/18/EMR-II dated 01/05/2018) to support his "Crystal Engineering" Research.

References

1. Kahn, O., (1993) *Molecular Magnetism*; Wiley-VCH: New York.
2. Heck, C., (1974) *Magnetic Materials and Their Applications*; Butterworth Heinemann: Oxford.U.K.
3. Krishnan, K. M., (2016) *Fundamentals and Applications of Magnetic Materials*; Oxford University Press: Oxford, U.K.
4. Spaldin, N. A., (2011) *Magnetic Materials: Fundamentals and Applications*, 2nd ed.; Cambridge University Press: Cambridge, U.K.
5. Yan, W. H., Bao, S. S., Huang, J., Ren, M., Sheng, X. L., Cai, Z. S., and Zheng, L. M. (2013) Co(II) and Ni(II) complexes based on anthraquinone-1, 4, 5, 8-tetracarboxylic acid (H 4 AQTC): canted antiferromagnetism and slow magnetization relaxation in $\{[Co_2(AQTC)(H2O)_6]\cdot 6H_2O\}$. *Dalton Trans.*, **42**(23), 8241-8248.
6. Su, H. Q., Ma, K. R., Kan, Y. H., Qian, X. D., Hu, H. Y., and Li, R. Q. (2017) Synthesis, structure, and properties of a new Co (II) diphosphonate based on auxiliary ligand 2, 2'-bipyridine. *Inorg. Nano-Met. Chem.*, **47**(4), 608-613.
7. Fan, L., Liu, X., Zhang, L., Kong, X., Xiao, Z., Fan, W., Wang, R., and Sun, D. (2019) Four novel Co (II) metal-organic frameworks based on semi-rigid ligand and their secondary building units transformation. *J. Mol. Struct.*, **1197**, 87-95.
8. Wu, S., Li, M., Yang, Z., Xia, Z., Liu, B., Yang, Q., Wei, Q., Xie, G., Chen, S., Gao, S., and Lu, J. Y. (2020) Synthesis and characterization of a new energetic metal–organic framework for use in potential propellant compositions. *Green Chem.*, **22**(15), 5050-5058.
9. Liu, T., Gao, S. M., Xu, L. Y., Zhao, J. P., Liu, F. C., Hu, H. L., and Kang, Z. H. (2014) Design and synthesis of stable cobalt-based weak ferromagnetic framework with large spin canting angle. *Inorg. Chem.*, **53**(24), 13042-13048.
10. Wu, H., Li, M., Wei, Q., Xie, G., and Chen, S. (2017) CoII/MnII MOFs containing the characteristic double metal chains: Synthesis, structure and magnetic property. *Inorg. Chem. Commun.*, **80**, 23-26.

11. Wu, Y., Zhang, C., Xu, Z., Cai, C., Song, W., Zhang, X., and Liu, X. (2017) Two new cobalt-organic frameworks constructed from polycarboxylates and pyridine-triazole ligands: Synthesis, crystal structures and magnetic properties. *Inorg. Chem. Commun.*, **84**, 190-194.

12. Zou, J. Y., Shi, W., Gao, H. L., Cui, J. Z., and Cheng, P. (2014) Spin canting and metamagnetism in 3D pillared-layer homospin cobalt (II) molecular magnetic materials constructed via a mixed ligands approach. *Inorg. Chem. Front.*, **1**(3), 242-248.

13. Cao, C., Liu, S. J., Yao, S. L., Zheng, T. F., Chen, Y. Q., Chen, J. L., and Wen, H. R. (2017) Spin-Canted Antiferromagnetic Ordering in Transition Metal–Organic Frameworks Based on Tetranuclear Clusters with Mixed V-and Y-Shaped Ligands. *Cryst. Growth Des.*, **17**(9), 4757-4765.

14. Yao, R. X., Qiao, X. H., Cui, X., Jia, X. X., and Zhang, X. M. (2016) Hexagonal Co 6 and zigzag Co 4 cluster based magnetic MOFs with a pcu net for selective catalysis. *Inorg. Chem. Front.*, **3**(1), 78-85.

15. Wang, Z. X., Wu, L. F., Zhang, X., Xing, F., and Li, M. X. (2016) Structural diversity and magnetic properties of six cobalt coordination polymers based on 2, 2′-phosphinico-dibenzoate ligand. *Dalton Trans.*, **45**(48), 19500-19510.

16. Fan, J. Z., Du, C. C., and Wang, D. Z. (2016) Copper and manganese complexes based on bis (tetrazole) ligands bearing flexible spacers: Syntheses, crystal structures, and magnetic properties. *Polyhedron*, **117**, 487-495.

17. Zhang, X. Y., Liu, Z. Y., Liu, Z. Y., Yang, E. C., and Zhao, X. J. (2013) Three 3-Amino-1, 2, 4-Triazole-Based Magnetic Complexes Incorporated with Different Carboxylate-Containing Collagens: Synthesis, Structures, and Magnetic Properties. *Z. Anorg. Allg. Chem*, **639**(6), 974-981.

18. Li, J. X., Du, Z. X., Pan, Q. Y., Zhang, L. L., and Liu, D. L. (2020) The first 3, 5, 6-trichloropyridine-2-oxyacetate bridged manganese coordination polymer with features of $\pi\cdots\pi$ stacking and halogen$\cdots$ halogen interactions: Synthesis, crystal analysis and magnetic properties. *Inorganica Chim. Acta*, **509**, 119677.

19. Zhang, G. M., Li, Y., Zou, X. Z., Zhang, J. A., Gu, J. Z., and Kirillov, A. M. (2016) Nickel (II) and manganese (II) metal–organic networks driven by 2, 2′-bipyridine-5, 5′-dicarboxylate blocks: synthesis, structural features, and magnetic properties. *Transit. Met. Chem.*, **41**(2), 153-160.

20. Liu, Q., Yu, L., Wang, Y., Ji, Y., Horvat, J., Cheng, M. L., and Wang, G. (2013) Manganese-based layered coordination polymer: synthesis, structural characterization, magnetic property, and electrochemical performance in lithium-ion batteries. *Inorg. Chem.*, **52**(6), 2817-2822.

21. Nath, J. K., Mondal, A., Powell, A. K., and Baruah, J. B. (2014) Structures, magnetic properties, and photoluminescence of dicarboxylate coordination polymers of Mn, Co, Ni, Cu having N-(4-pyridylmethyl)-1, 8-naphthalimide. *Cryst. Growth Des.*, **14**(9), 4735-4748.

22. Wang, L., Zhao, R., Xu, L. Y., Liu, T., Zhao, J. P., Wang, S. M., and Liu, F. C. (2014) The synthesis, structure, and magnetic properties of two novel manganese (II) azido/formate coordination polymers with isonicotinic acid N-oxide as a coligand. *CrystEngComm*, **16**(10), 2070-2077.

23. Hu, T., Zheng, B., Wang, X., and Hao, X. (2015) Syntheses, structures and magnetic properties of four manganese (II) and cobalt (II) complexes. *CrystEngComm*, **17**(48), 9348-9356.

24. Dong, W. W., Li, D. S., Zhao, J., Ma, L. F., Wu, Y. P., and Duan, Y. P. (2013) Two solvent-dependent manganese (II) supramolecular isomers: solid-state transformation and magnetic properties. *CrystEngComm*, **15**(27), 5412-5416.

25. Agarwal, R. A., Aijaz, A., Sañudo, C., Xu, Q., and Bharadwaj, P. K. (2013) Gas adsorption and magnetic properties in isostructural Ni (II), Mn (II), and Co (II) coordination polymers. *Cryst. Growth Des.*, **13**(3), 1238-1245.

26. Hou, C., Zhao, Y., and Sun, W. Y. (2014) Synthesis, structure and magnetic property of cobalt (II) and manganese (II) complexes with mixed organic ligands. *Inorg. Chem. Commun.*, **39**, 126-130.

27. (a) Duan, X. Y., and Wei, M. L. (2017) From one-dimensional, two-dimensional to three-dimensional entangled architectures with polythreading feature: synthesis, structures, and properties. *Cryst. Growth Des.*, **17**(3), 1197-1207. (b) Wu, H., Li, M., Wei, Q., Xie, G., and Chen, S. (2017) CoII/MnII MOFs containing the characteristic double metal chains: Synthesis, structure and magnetic property. *Inorg. Chem. Commun.*, **80**, 23-26.

28. Zhang, J., Gao, L., Wang, Y., Zhai, L., Wang, X., Fan, L., and Hu, T. (2019) Two trinuclear cluster-based 3D interpenetrated metal-organic frameworks with selective adsorption and antiferromagnetic properties. *J. Solid State Chem.*, **271**, 303-308.

29. Zhang, Z. Y., Shi, P. F., Hu, H. C., Wu, Z. L., Cui, J. Z., and Zhao, B. (2015) Robust metal–organic framework with [Mn3] clusters: synthesis, structure and magnetic property. *Inorg. Chem. Commun.*, **53**, 76-79.

30. Chen, Y. Q., Liu, S. J., Li, Y. W., Li, G. R., He, K. H., Chang, Z., and Bu, X. H. (2013) Mn (ii) metal–organic frameworks based on Mn3 clusters: from 2D layer to 3D framework by the "pillaring" approach. *CrystEngComm*, **15**(8), 1613-1617.

31. Gong, T., Yang, X., Fang, J. J., Sui, Q., Xi, F. G., and Gao, E. Q. (2017) Distinct Chromic and Magnetic Properties of Metal–Organic Frameworks with a Redox Ligand. *ACS Appl. Mater. Interfaces*, **9**(6), 5503-5512.

32. Shi, W. J., Du, L. Y., Yang, H. Y., Zhang, K., Hou, L., and Wang, Y. Y. (2017) Ligand Configuration-Induced Manganese (II) Coordination Polymers: Syntheses, Crystal Structures, Sorption, and Magnetic Properties. *Inorg. Chem.*, **56**(16), 10090-10098.F

4

Thermal Expansion in Metal-Organic Frameworks

Prem Lama, Sonu Dhakla and Ashutosh Rawat*
*CSIR-Indian Institute of Petroleum, DHOP Division, Haridwar
Road, Mohkampur, Dehradun, India*

1. Introduction

Metal organic frameworks (MOFs) are the distinctive kind of porous material that are made using metal ions or metal cluster with organic linkers having multidentate coordinating sites. MOFs are rigid as well as flexible, thermally stable and due to its versatility, it can show various properties such as, luminescence, thermal expansion, gas sorption, magnetism etc. In addition, MOFs have shown various application in the fields like, catalysis, separation of liquid and gases, sensor etc. Thermal expansion is a common occurrence in materials such as zeolites, metal oxides and magnetic materials. In MOFs and organic molecular compounds, it occurs less frequently. Expansion/contraction in MOFs can occur mainly because of two important factors i.e temperature and pressure. However, we have seen the effect of guest solvent molecules on the expansion and contraction behaviour of MOFs. The word thermal is related to heat and hence in this condition, the thermal expansion is considered as the change in shape, area, volume, and density with temperature. In case of MOF, it is related to expansion or contraction in crystallographic/principal axes (lengths and angles) and volume with respect to the change in temperature. In this chapter, we mainly focused on those MOFs that show such unique property with variation in temperature.

2. MOFs Design and Characterization

In order to construct MOFs that display thermal expansion property, it is very important that MOFs as a system should have flexibility. This flexible nature can arise from the choice of metal ion or the linkers. In addition, solvent molecules that are present within the pores of the framework can also induce flexibility in MOF. If the metal ions have the tendency to form inert complexes, then flexibility must come mainly from the linker that has been used to construct the MOF.

The thermal expansion in crystalline MOF materials can generally be investigated by two methods. In first case, a good quality single crystal is used for variable temperature single crystal X-ray diffraction (VTSCXRD) while for polycrystalline materials variable temperature powder X-ray diffraction (VTPXRD) is employed. In both cases, data is collected at certain temperature interval and unit cell parameters are determined at each temperature. In addition, neuron scattering study has been performed along with theoretical studies to confirm the thermal expansion behaviour of MOF material.

3. Calculation of Thermal Expansion in MOFs

In case of MOFs, the calculation of thermal expansion depends on the type of crystal system that is associated with the crystalline material. For those MOF materials, where the crystallographic axes are orthogonal to each other i.e ($\alpha = \beta = \gamma = 90°$), the thermal expansion coefficients (α) is usually calculated by measuring the variation in length with the change in temperature (ΔT) at constant pressure (P) using equation 1. Similarly, the change in volume with temperature is calculated from equation 2.

$$\alpha_L = \frac{L_f - L_i}{L_i\, \Delta T} = \frac{1}{L}\left(\frac{\delta L}{\delta T}\right)_P \quad \text{(equation 1)}$$

$$\alpha_V = \frac{V_f - V_i}{V_i\, \Delta T} = \frac{1}{V}\left(\frac{\delta V}{\delta T}\right)_P \quad \text{(equation 2)}$$

where L is the length of the material along a particular orthogonal vector or unit cell parameter, V is the volume of MOF that crystallises in a particular crystal system, T is the temperature, subscript *f* and *i* indicates the final and initial Length/Volume.

However, the tricky situation arises in those crystal system of MOF materials where the crystallographic angles are not orthogonal to each other. In such case, new principal axes (orthogonal vectors) using cartesian coordinate system have to be constructed that are perpendicular to each other and hence the thermal expansion coefficient can be calculated by measuring the variation of length with temperature and applying the tensorial method.[1]

In this perspective, two important strain calculator softwares have been designed to calculate the thermal expansion of crystalline materials; 1) PASCal program (http://pascal.chem.ox.ac.uk/) and 2) Win_Strain program.[1] These two programs have been found to be very useful to calculate the thermal expansion of crystalline MOF materials along the principal axes where the crystallographic angle/angles are not perpendicular to each other. However, it is important to note that at least three or more unit cell parameters are required in order to get the best possible thermal expansion result for the crystalline compounds such as MOF.

4. Types of Thermal Expansion and Mechanism Associated with the MOF

In general, the phenomenon of thermal expansion can be classified into two main types: 1) isotropic thermal expansion (ITE) and 2) anisotropic thermal expansion (ATE). Materials that undergo isotropic thermal expansion (ITE) display an equivalent thermal response in all three orthogonal directions with a change in temperature. This has been reported for materials that show cubic crystal systems.[2,3] In case of anisotropic thermal expansion (ATE) materials undergo different thermal expansion, such as positive, negative and zero etc., to different extents along the three different perpendicular directions. It has thus far been possible for non-cubic crystal systems.

Mainly ITE is classified into three types and ATE into nine types. ITE can be divided into: Isotropic positive thermal expansion (PPP), Isotropic Negative thermal expansion (NNN) and Isotropic zero thermal expansion (ZZZ), where P = positive thermal expansion, N = negative thermal expansion and Z = zero thermal expansion along the three perpendicular directions. ATE, on the other hand, can be divided into: PPP, PPN, PPZ, PNZ, PZZ, NNN, NNP, NNZ and ZZN. The thermal expansion behaviour in these cases can be elucidated more efficiently in terms of an individual linear expansion coefficient.

4.1 Isotropic Thermal Expansion (ITE)

Isotropic thermal expansion such as with PPP, NNN and ZZZ involves equal thermal responses along all three orthogonal directions. When positive expansion takes place along all perpendicular directions we categorise it as PPP. PPP comprises of longitudinal/parallel thermal vibrations and especially in case of MOF materials these vibration occurs along the bond axis that results in the elongation of interatomic distances (**Figure 1**).

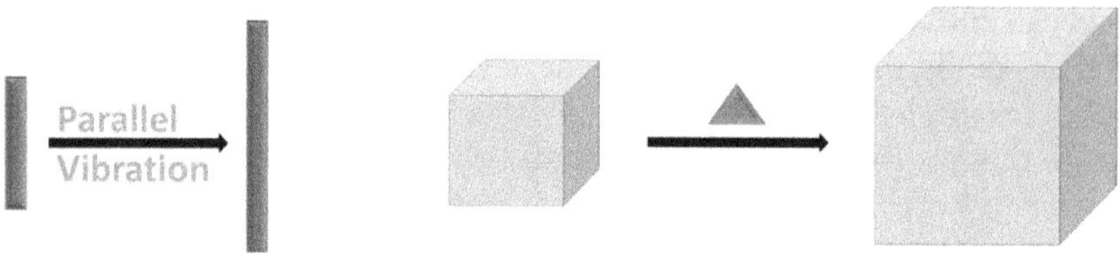

Figure 1. Schematic representation of parallel/longitudinal vibrations.

Material that has low frequency transverse vibrational modes, termed as 'rigid unit modes' (RUM), can experience reduction of the interatomic distances resulting in isotropic negative thermal expansion with the rise in temperature.[3] Negative thermal expansion along all three perpendicular directions is indicated by NNN (**Figure 2**) and is not as common as PPP.

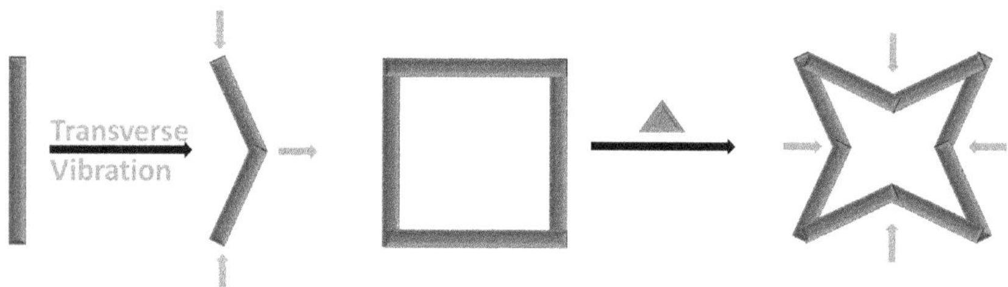

Figure 2. Representation of transverse vibrational modes present in NNN type.

Of the three types of isotropic thermal expansion, ZZZ type is very rare and it originates from the simultaneous effects of both PPP and NNN. This effect can be seen via three types of mechanisms: i) in case of ferromagnetic materials where PPP that results from longitudinal vibrations, is cancelled by NNN due to a transition from high magnetic moment to low magnetic moment, ii) composite materials combining both PPP and NNN ingredients and 3) materials containing both low frequency longitudinal and transverse vibrational modes.

4.2 Anisotropic Thermal Expansion (ATE)

Anisotropic thermal expansion is characterised by different magnitudes of expansion along the three perpendicular directions as well as combinations of positive and negative thermal expansion within the same material. As mentioned earlier ATE can be sub-divided into nine different categories namely: PPP, PPN, PPZ, PNZ, PZZ, NNN, NNP, NNZ and ZZN. Here we will discuss PPP, PPN, PPZ, NNN, NNP, and PNZ. The other types of ATE have still not been observed. In case of PPP or NNN in ATE, the expansion or contraction occurs with unequal change along the perpendicular direction.

For anisotropic thermal expansion such as PPN, PPZ, NNP and PNZ, materials undergo expansion/contraction along certain axis with a related compensation along the perpendicular direction due to the geometrical flexibility of the materials. These materials often have scissor- or hinge-like motion that gives rise to thermal expansion in grid-like networks, zigzag chains or helical chains (**Figure 3**).

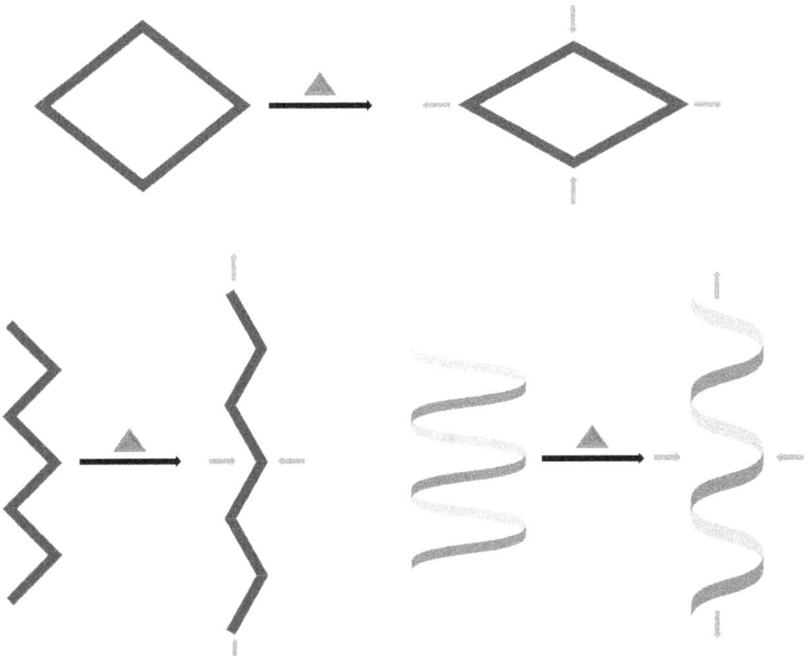

Figure 3. Graphical illustration of hinge like motions in grids, zigzag chains and helices.

5. Examples of MOFs with Mechanism

Barbour and co-workers [4] reported the synthesis of [Zn(L)$_2$(OH)$_2$]$_n$.nCH$_3$OH, (**1**$_{MeOH}$) { L= 4-(1H- naphtha [2,3-d]imidazole-1-yl)benzoic acid, a three dimensional (3D) metal organic framework, incorporating 1D square-shaped channels that depicts the anomalous thermal expansion within 100-370 K temperature. In addition the, framework also displayed an anomalous thermal expansion under the influence of guest solvent molecules like MeOH, EtOH, n-PrOH and i-PrOH. The apohost framework (**1**$_{apohost}$) of the MOF (**1**$_{MeOH}$) under stagnant vacuum demonstrates the anomalous PTE along *c* axis and NTE along *a* and *b* axes, within the temperature range of 100-370 K as characterized by SCXRD. The root cause of this anomalous thermal expansion is the increase of the intermolecular interactions (H- bonding) within the host framework and distortion of the flexible coordination angles nearby the metal nodes (**Figure 4**).

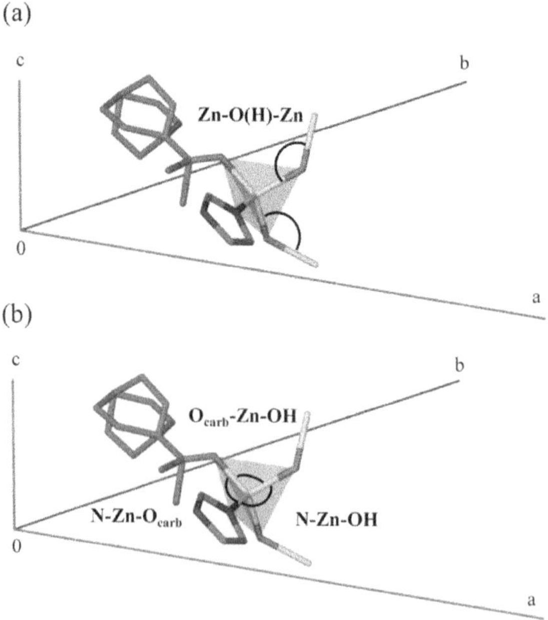

Figure 4. Coordination mode Zn(II) metal node with respect to unit cell axes. (a) Upon heating, the bond angle Zn-O(H)-Zn enlarge along *c* axis. (b) The bond angles (O$_{carb}$-Zn-OH and N-Zn-OH) undergoes contraction with increasing temperature that lies in the plane containing *a* and *b* axes. Reproduced from Reference [4] with permission from American chemical society.

In the temperature range 100-370 K, due to the enlargement in the bond angle (Zn-O(H)-Zn) by 4° along c axis, PTE was observed (α_c = 127 * 10^{-6} K^{-1}) that results in the simultaneous increase of the bond distance between the two Zn atoms from 3.431 to 3.482 Å across the hydroxyl bridge. Here, as a result of distortion of the bond angles (O_{carb}-Zn-OH, N-Zn-OH and N-Zn-O_{carb}) along a and b axes, NTE ($\alpha_{a,b}$ = -21 * 10^{-6} K^{-1}) was found.

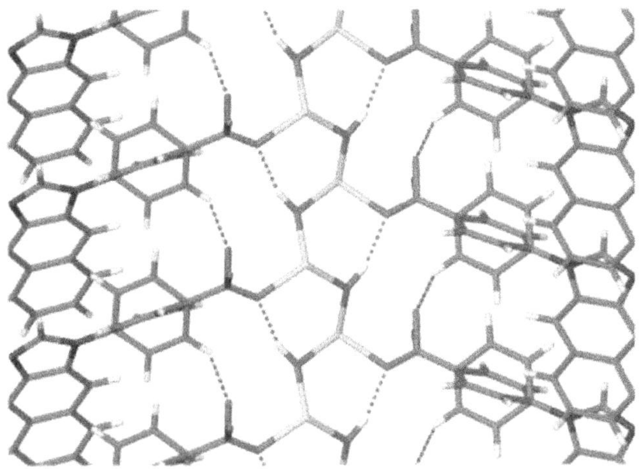

Figure 5. View showing the chain comprising of [Zn–OH–]$_n$ coordination orthogonal to the c axis (the dashed red lines represents the hydrogen bonds that grows closely parallel to the c axis). Reproduced from Reference [4] with permission from American Chemical Society.

The H-bond distance (**Figure 5**) between the coordinated oxygen of carboxyl and hydroxyl group increases from 2.875Å at 100 K to 3.065 Å at 370 K. **Table 1** gives an insight into the various changes in the **1**$_{apohost}$ framework that occurs with the change in the temperature.

Table 1. Variation in the angles containing Zn(II) metal ion of **1**$_{apohost}$ framework

Temp. (K)	∠Zn-O(H)-Zn (°)	∠O_{carb}-Zn-OH (°)	∠ N-Zn-OH (°)	∠ N-Zn-O_{carb} (°)
100	126.7(1)	105.9(1)	115.4(1)	111.6(1)
190	128.5(1)	105.0(1)	108.7(1)	112.0(1)
280	129.6(2)	104.6(2)	109.5(1)	112.4(1)
370	130.7(2)	104.2(2)	109.6(2)	113.5(3)

The porous nature (1D channels) of this synthesized MOF allows the incorporation of various guest molecules inside the pores that influences its thermal expansion character. The PTE coefficients for different guest molecules (MeOH, EtOH, n-PrOH and i-PrOH) runs parallel with their size/vander waals volume. This is due to the variation in the Zn-O(H)-Zn bond angle with different sized guest molecules. **Table 2** gives an insight into the variation in the coefficient thermal expansion with the change in guest molecules.

Table 2. Variation in the thermal expansion coefficient, guest volume and hydrogen bond with different guest molecules

Solvate	Guest volume (Å^3)[a]	α_c $(* 10^{-6}\ K^{-1})$[b]	$\alpha_{a,b}$ $(* 10^{-6}\ K^{-1})$[b]	$\angle$Zn-O(H)-Zn $(°)$	Hydrogen bond (Å)[c]
1_{MeOH}	38.8	166(4)	-41(3)	123.7(1)	2.808(4)
1_{EtOH}	54.0	144(5)	-17(5)	123.8(2)	2.812(5)
$1_{n\text{-}PrOH}$	70.7	138(4)	-38(4)	124.6(1)	2.832(8)
$1_{apohost}$	-	137(5)	-26(5)	126.7(1)	2.875(3)
$1_{i\text{-}PrOH}$	69.1	76(6)	23(6)	127.4(1)	2.915(3)
[a]The Vander waals surface is used to calculate guest volume. [b]The expansion value was calculated in the temperature range from 100-295 K. [c]Hydrogen bonding distance between coordinated oxygen of carboxyl group and hydroxyl group in intra-framework at 100 K.					

Due to the anomalous thermal expansion behaviour there occurs an increase in the crystal's length of 1_{MeOH} framework having needle-shaped (**Figure 6**) (1.10 mm to 1.16 mm) on heating (100 K to 370 K), which corresponds well with the NTE along *a* and *b* axes and PTE along *c* axis.

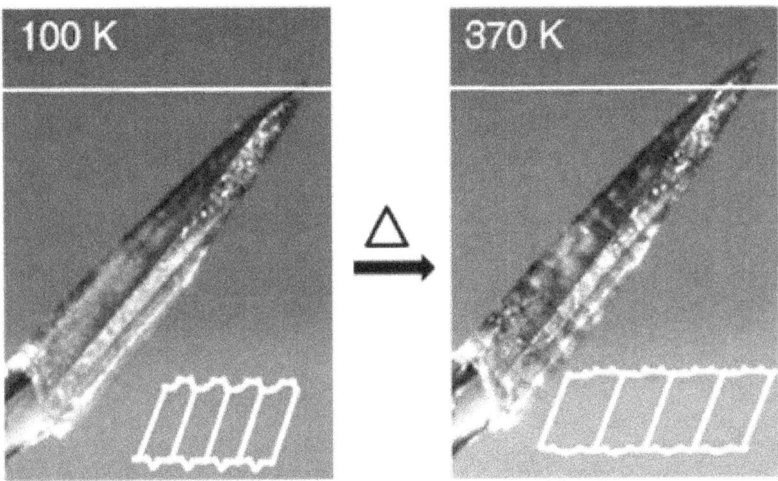

Figure 6. Picture showing the single crystal of 1_{MeOH} analysed at 100 and 370 K. Reproduced from Reference [4] with permission from American chemical society.

It is evident (**Table 2**) that α_c follows the pattern 1_{MeOH} > 1_{EtOH} > $1_{n\text{-}PrOH}$ > $1_{apohost}$ > $1_{i\text{-}PrOH}$. The thermal expansion (PTE, NTE) for solvate 1_{MeOH}, 1_{EtOH}, $1_{n\text{-}PrOH}$ was found to be greater than that for $1_{apohost}$, as these MOFs are in a "contracted" state i.e. Zn-O(H)-Zn bond angle is smaller at 100 K relative to empty framework which facilitates them to exhibit more thermal expansion. However, due to the existence of bulky i-PrOH molecules in solvate $1_{i\text{-}PrOH}$, the framework was already in the "expanded" state i.e. Zn-O(H)-Zn bond angle was already increased, hence there occurs less thermal expansion in solvate $1_{i\text{-}PrOH}$ framework. Overall, three important factors such as 1) Guest size and shape, 2) H-bonding interactions and 3) Temperature played a significant role and influenced the anomalous thermal expansion in this MOF material. Variation in any of the above mentioned three factors would bring obvious change in the coordination environment around the metal node that affect the host-guest interactions in the framework leading to thermal expansion behaviour.[5]

Lama *et. al* [6] presents a competent and interesting work towards the synthesis of new semi-flexible two dimensional (2D) MOF {[Cd(HBTC)(BPP)].1.5DMF.2H$_2$O}$_n$ (DMF = *N,N'*-dimethyl formamide) that crystallizes in triclinic system having space *P*-1 group and exhibits all three kinds of thermal expansions (PTE, NTE, ZTE) from

100 K to 260 K, owing to the rigid 1,3,5-benzenetricarboxylate (HBTC^{2-}) and 1,3-bis-(4-pyridyl)propane (BPP) as the flexible entity thus facilitating the combined stretching-tilting mechanism. VTSCXRD reveals that the MOF undergoes PTE {α = 133(2) MK^{-1}} along the principal axis X3 with coordinate vectors (0.0575, 0.1605, 0.9854), NTE {α = - 43.4(7) MK^{-1}, MK^{-1} = * 10^{-6} K^{-1}} along the principal axis X1 with coordinate vectors (0.4920, 0.7734, - 0.3997) and ZTE {α = 2(1) MK^{-1}} along the principal axis X2 with coordinate vectors (0.9134, - 0.4041, - 0.0480) (**Figure 7**).

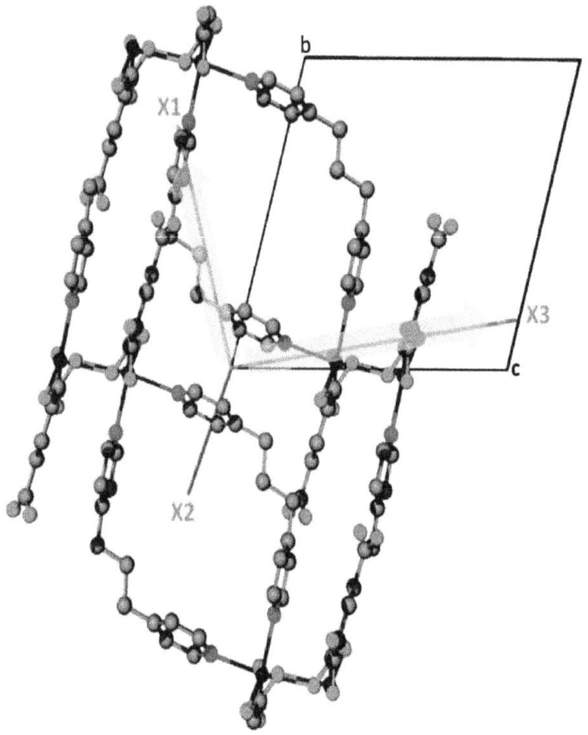

Figure 7. Schematic representation showing X1, X2 and X3 principal axes with the projection of X1 and X3 axes onto the crystallographic *bc* plane. Reproduced from Reference [6] with permission from Royal society of chemistry.

The nature of the ligands (flexible and rigid) play a very important role, which enables the MOF to show thermal expansion nature. In this MOF, the rigid nature of HBTC^{2-} i.e. the aromatic ring which was oriented parallel to the crystallographic *ab* plane remains largely unaffected in spite of temperature deviation, thus ZTE was observed along X2 principal axis. Whereas the flexible aliphatic unit of the BPP ligand (aligned parallel to X1) bends towards *c* axis resulting in the decrease of crystallographic angle α with increase in the temperature which simultaneously causes the macrocyclic unit (Cd_2BPP_2) to tilt in the same direction (*c* axis), thus leading to NTE along X1 and PTE along X3. However, macrocycle (Cd_2BPP_2) stretching unidirectionally leads to PTE along both principal axes X1 and X3. This stretching and tilting of macrocyclic unit (Cd_2BPP_2) simultaneously, causes overall PTE along X3. This stretching-tilting mechanism was supported by the facts that the ∠C15...C16...C17 get increased and the ∠Cd1...C16...Cd1' was decreased with the rise in the temperature (**Figure 8**, **Table 3** and **Table 4**). This thermal expansion phenomena of the MOF also holds its reversible nature with the temperature variation range.

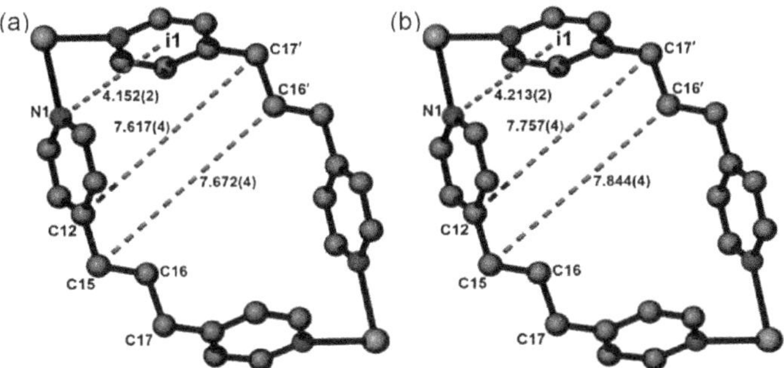

Figure 8. A macrocyclic unit (a) 100 K and (b) 260 K showing selected nonbonding distances. Reproduced from Reference [6] with permission from Royal society of chemistry.

Table 3. Variation in the non bonding distances of the macrocyclic unit with temperatures

Temp. (K)	C15...C16' (Å)	C12...C17' (Å)	N1...i1[a] (Å)
100	7.672(4)	7.617(4)	4.152(2)
140	7.703(4)	7.640(4)	4.157(2)
180	7.747(4)	7.678(4)	4.174(2)
220	7.797(4)	7.713(4)	4.192(2)
260	7.844(4)	7.757(4)	4.213(2)
i1[a] = centroid of C18' C19' C20' N2' C21' C22'.			

Table 4. Change of angles in MOF with variation in the temperature

Temp. (K)	θ^a (°)	ω^b (°)	τ^c (°)
100	113.0(2)	125.1(3)	2.3(3)
140	113.3(2)	124.8(3)	1.9(3)
180	113.6(2)	124.3(3)	1.4(3)
220	114.1(2)	123.8(3)	0.6(3)
260	114.5(2)	123.3(3)	0.5(3)

$^a\theta$ = $\angle$C15...C16...C17, $^b\omega$ = $\angle$Cd1...C16...Cd1' and $^c\tau$ = $\angle$N1...C15...C17...N2.

Choe and co-workers [7] reported the synthesis of HMOF-1 (3D hinged metal-organic framework) [$C_{40}H_{26}$ CdI_2N_8] that exhibit thermal expansion profile from 160-320 K. HMOF-1 holds cds-type framework that exhibits the hinged motion around the metal nodes. HMOF-1 crystallizes in the octahedral geometry, in which the cadmium centre is coordinated by four pyridyl groups in the equatorial position of the neighbouring porphyrins and two iodide anions occupied the axial site. Synchrotron PXRD [8] data exhibited that upon heating HMOF-1 undergoes PTE (α = + 177 * 10^{-6} K^{-1}) along *a* direction and NTE (α = -21 * 10^{-6} K^{-1}) along *b* direction from 160-320 K (**Table 5**).

Table 5. Thermal expansion coefficients along *a* and *b* directions

Temp. (K)	α_a (* 10^{-6} K^{-1})	α_b (* 10^{-6} K^{-1})
100	138.2	9.74
120	144.0	5.38
140	149.8	1.04
160	155.5	-3.30
180	161.3	-7.64
200	166.8	-11.98
220	172.3	-16.33
240	177.3	-20.69
260	182.8	-25.04
280	187.8	-29.42
295	192.1	-32.72
320	197.5	-38.19
340	203.1	-42.54
360	208.6	-46.94
380	241.1	-51.33

HMOF-1 also holds its reversible nature of anisotropic thermal expansion phenomena. The flexible nature of HMOF-1 was largely governed by the hinge angle θ (**Figure 9**) from 100-380 K, θ varies by 1.1° in HMOF-1. Since HMOF-1 belongs to the orthorhombic crystal system having *Pnnm* space group, so its cell parameters can be expressed in mathematical equation form as $a = 2d \sin\theta$ and $b = 2d \cos\theta$, where d = distance between two four connected nodes and θ = hinge angle. Following the equations, as hinge angle θ increases, a and b parameters exhibits PTE and NTE respectively.

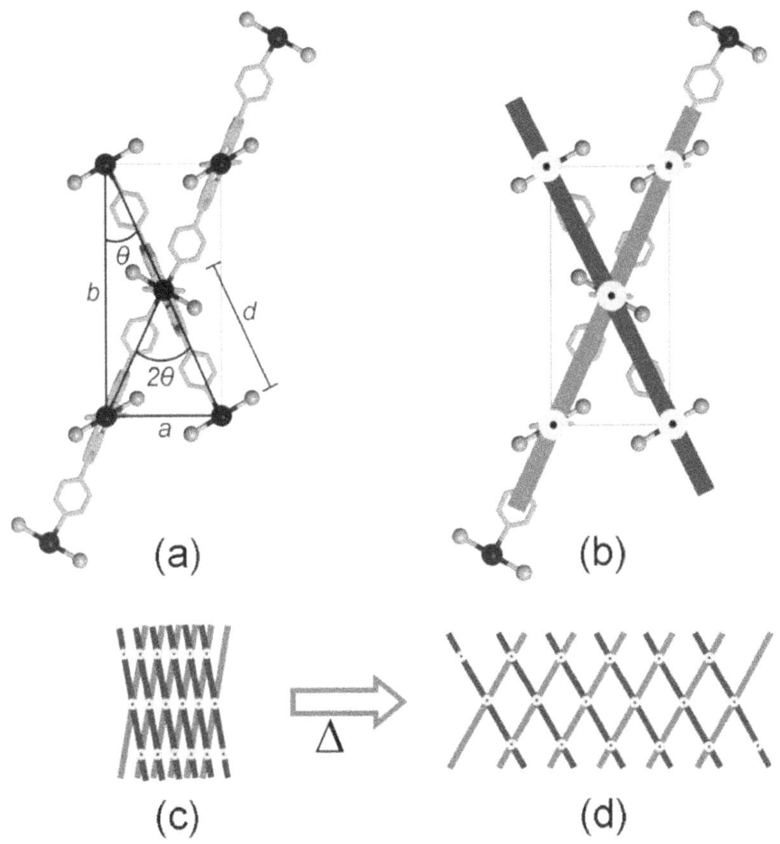

Figure 9. (a,b) Figure showing the structure of HMOF-1. (a) Diagram presenting the hinge angle θ along [001] direction and (b) more clarified figure to display porphyrin bars (red and blue) of a "lattice fence" with pivoting points (yellow) represented by cadmium ions. (c,d) Presentation of the "lattice fence" that undergoes expansion due to hinge motion on heating. Reproduced from Reference [7] with permission from American chemical society.

Usually, the thermal expansion occurs due to stretching or bending of bond around the metal centre, however in case of HMOF-1 this explanation was ruled out by the hinged type motion around the cadmium centre i.e. expansion and shrinkage of a "lattice fence" where the cadmium centre serve as the pivoting points (**Figure 9**). SCXRD data also revealed that there was a noticeable change in the distance between the two axially coordinated iodides to cadmium ion, which fluctuated from 4.43 Å (100 K) to 4.50 Å (200 K) and 4.69 Å at 297 K. The rise in the temperature facilitated the increase in I...I distance which favoured Cd-I bonds to rotate obligingly along the *c* axis to intensify such mechanical response.

Xing and co-workers [9] reported the synthesis of MIL-68(In) metal organic framework ($C_{24}H_{12}In_3O_{15}$) with orthorhombic space group Cmcm [10] that exhibits anisotropic 3D NTE in the temperature range from 125-600 K. SCXRD data revealed that on rising the temperature, MIL-68(In) shows lattice diminution (NTE), with thermal expansion coefficients α_a = - 5.6 * 10^{-6} K^{-1} (*a* axis), α_b = - 2.7 * 10^{-6} K^{-1} (*b* axis) and α_c = - 4.0 * 10^{-6} K^{-1} (*c* axis) in the temperature range from 125-600 K. In this temperature range, the unit cell axes *a*, *b* and *c* changes from 21.8796(5) Å, 37.3959(8) Å, 7.2352(1) Å to 21.8207(12) Å, 37.3479(20) Å and 7.2214(3) Å respectively.

The crystal lattice of MIL-68(In) was composed of rigid octahedrons of indium linked by 1,4-benzenedicarboxylate (BDC) along the *c* axis, thus constructing a kagome lattice in the *ab* plane with four triagonal and one hexagonal channels parallel to *c* axis (**Figure 10**).

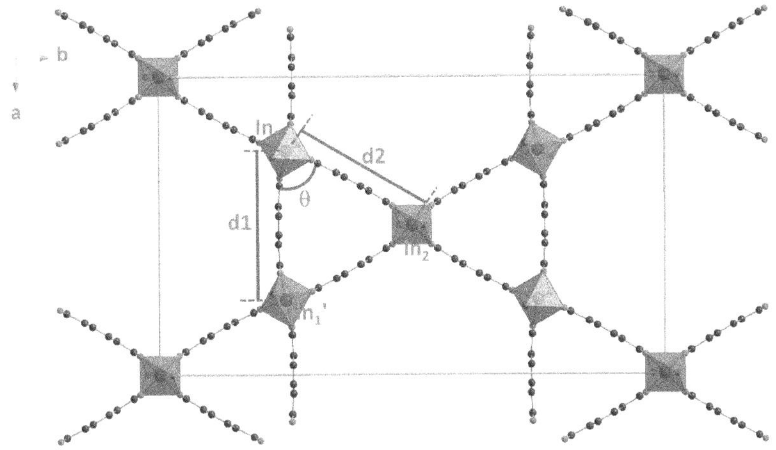

Figure 10. View of crystal structure along *c* axis. d1 indicates the distance between two bridged In_1 atoms, d2 indicates the distance between bridged In1 and In2, and θ indicates the angle $\angle In_1'...In_1...In_2$.

SCXRD analysis revealed that the rotation of rigid indium octahedrons with increasing temperature causes NTE along the *c* axis, whereas transverse vibration of phenyl ring of BDC ligands causes NTE in the *ab* plane. The variation in the distances of d1, d2 with temperature is shown in **Table 6**.

Table 6. Temperature dependent changes of d1, d2 and θ in MIL-68(In)

Temp. (K)	d1 ($In_1'...In_1$) (Å)	d2 ($In_1...In_2$) (Å)	$\angle \theta$ (°) ($In_1'...In_1...In_2$)
125	10.946	10.946	60.00
150	10.942	10.942	60.00
175	10.936	10.939	60.01
200	10.930	10.936	60.02
225	10.926	10.933	60.02
250	10.921	10.929	60.03

Lama *et. al* [11] present a proficient work towards the synthesis of 3D polar MOF {[Zn(BTC)(HBPP)].H_2O}ₙ that crystallises in orthorhombic space group *Pna2₁* having small isolated voids (zero dimensional). The MOF exhibits anomalous thermal expansion behaviour owing to the guest water molecule from 100 K to 260 K. The MOF is composed of rigid 1,3,5-benzetricarboxylate (BTC) unit and flexible protonated 1,3-bis-(4-pyridyl)propane (HBPP) ligand. The asymmetric unit of $\{[Zn(BTC)(HBPP)].H_2O\}_n$ comprises of single Zn^{2+} ion, one BTC^{3-}, one HBPP$^+$ ligand. VTSCXRD data reveals that $\{[Zn(BTC)(HBPP)].H_2O\}_n$ shows contraction in the crystallographic b axis with fall in the temperature from 260 to 100 K, whereas a and c axes elongate steadily, with linear thermal expansion coefficients (α) -6.0(1), 53.0(7) and – 7.9(5) MK^{-1} along three crystallographic axes a, b, c respectively. This phenomenon of thermal expansion is reversible in nature in the operational temperature range between 260 K-100 K. The BTC ligand having two carboxylate groups bridged the Zn(II) ions to generate helices that are hexagonal in nature and runs along the b axis. The HBPP$^+$ units occupy the discrete helical columns in which the protonated pyridyl groups sits parallel to the BTC groups. These protonated pyridyl unit (HBPP$^+$) along with the carboxylate group forms hydrogen bonding interactions (N-H...O) with of the nearest helices and runs along c axis (**Figure 11**).

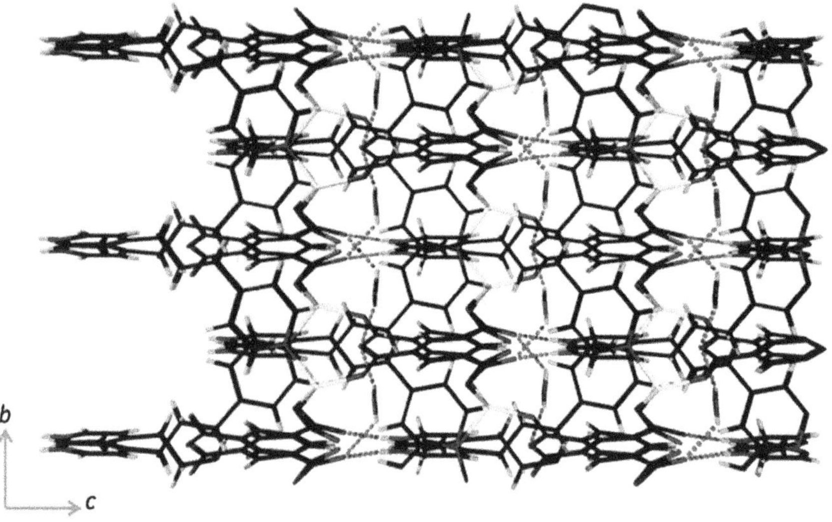

Figure 11. Packing diagram of $\{[Zn(BTC)(HBPP)].H_2O\}_n$ displaying interactions (N-H...O and O-H...O). Reproduced from Reference [11] with permission from Royal society of chemistry.

On lowering the temperature, the flexible HBPP$^+$ units undergoes conformational changes, and the HBPP$^+$ units bends causing the simultaneous movement of BTC units which leads to scissor/hinge-like motion in the helical columns. This causes PTE and NTE along b direction and *c* direction respectively. The scissor-like motion that is present in the helical columns also gets transferred to *a* axis due to the fact that the aqua molecules which linked the carboxylate groups between two adjacent helical chains forms weak C-H...O and strong O-H...O interactions leading to NTE along *a* axis. Now upon activation at 140 °C, the crystals of $\{[Zn(BTC)(HBPP)].H_2O\}_n$ gets dehydrated to form water free framework. The dehydrated framework $(\{[Zn(BTC)(HBPP)]\}_n)$ al so depicted similar thermal expansion phenomena along *b* and *c* axes as that of its hydrated form $\{[Zn(BTC)(HBPP)].H_2O\}_n$ with the only exception that *a* axis experiences zero thermal expansion (ZTE). From the SCXRD data, α along three crystallographic axes *a*, *b*, c were found to be -0.20(7), 47.8(7) and -7.4(7) MK^{-1} respectively. This results from the fact that there was no change in the N-H...O hydrogen bonding interactions with the temperature variation. However, due to the absence of guest water molecule the C-H...O and O-H...O interactions do not exist in the structure. Thus, the scissor/hinge-like motion did not get conveyed into *a* axis.

Kepert and co-workers [12] presents the detailed study on the mechanism involved for the NTE in $[Cu_3(BTC)_2]$ (BTC = 1,3,5-benzenetricarboxylate). $[Cu_3(BTC)_2]$, a metal organic framework consist of di-copper tetracarboxylate $\{Cu_2(carboxylate)_4\}$ "paddle wheel" unit and aromatic ring motifs. Spectroscopic tool (inelastic neutron scattering), crystallography (SCXRD) and molecular dynamics (MD) simulations present the evidence that NTE in $[Cu_3(BTC)_2]$ was favoured by 2 unique components: local molecular vibrations within the framework and transverse vibration of BTC. Neutron powder diffraction data was collected for the deuterated sample of $[Cu_3(BTC)_2]$ over the temperature span 20 to 300 K. The data clearly depicted linear NTE with α = - 4.9(1) * 10^{-6} K^{-1} which was analogous to that reported for protonated phase by XRD.[13] Temperature-dependent inelastic neutron scattering (INS) spectra showed comparable result to that obtained from simulated molecular dynamics study. The similarity in both the results arises from the local nature of vibrations exhibited by the aromatic ring of the BTC unit. The BTC ligand can perform two types of motions: 1) translation orthogonal to the BTC plane **(Figure 12 a)** and 2) dynamic tilting of aromatic ring **(Figure 12 b)**.[14]

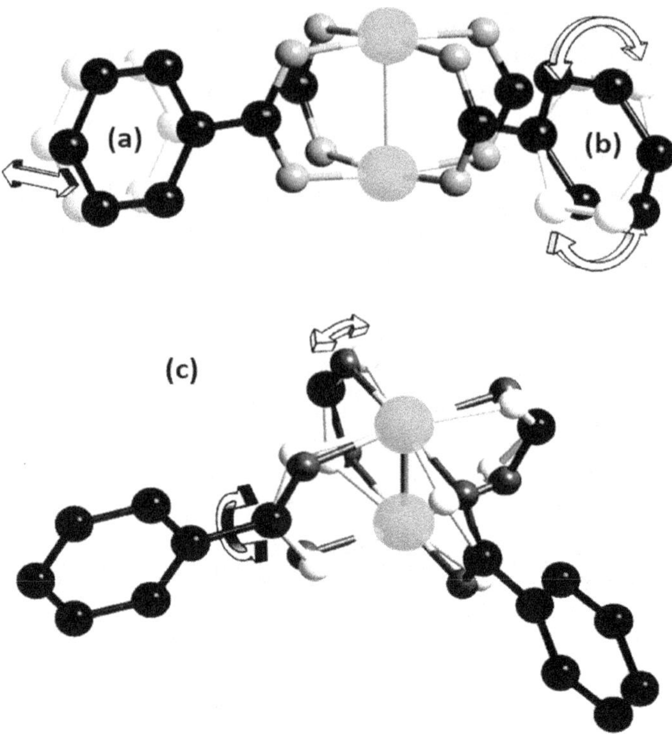

Figure 12. Vibrational motions accountable for NTE in [Cu$_3$(BTC)$_2$]: a) Translation of the aromatic ring; b) libration (dynamic tilting) of the aromatic ring; (c) local twisting vibrations of the paddlewheel unit.

Molecular dynamics simulation data reveals that the paddle wheel vibrations result in the deformation of the dicopper tetracarboxylate "paddle wheel" unit from square prismatic geometry to square antiprismatic geometry. The paddle wheel distortion leads to change in Cu...O distances (1.9832 Å) in the undistorted arrangement (square prismatic) to 1.9818-1.9866 Å (square antiprismatic) thereby indicating that paddlewheel deformation facilitates more interaction between Cu and O atoms. Temperature dependent ADP (atomic displacement parameters) value indicated that the extreme rate of thermal displacement took place for the carboxylate oxygen atom orthogonal to the BTC plane which reflects that the carboxylate displacement was dynamic instead of static. This large thermal displacement for the oxygen atom was consistent with both inelastic neutron scattering data and associated modelling which   depicted that carboxylate rotation causes paddlewheel deformation to occur at very low energies. The previous reported data [13] by X-ray diffraction in the higher temperature range suggested that low energy motion of the carboxylate units, associated both with the paddlewheel deformation and translation perpendicular to the plane of the benzene unit, was principally responsible for NTE in this material.

Baxter *et. al* [15] presents a unique study about the transition in the thermal expansion phenomenon from NTE in Zn-DMOF to PTE in a mixed Zn-DMOF-TM solid solution. Zn-DMOF correspond to $Zn_2(BDC)_2(dabco)$ that comprises of Zn_2 paddlewheels unit linked by 1,4-benzenedicarboxylate (BDC^{2-}) in their equatorial plane producing $Zn_2(BDC)_2$ layers that display NTE. These layers are further linked by 1,4-diazabicyclo[2.2.0]octane (dabco) that acts as a pillar to join $Zn_2(BDC)_2$ layers to one another (**Figure 13**). Zn-DMOF-TM i.e. ($Zn_2(TM- BDC)_2(dabco)$) is a functionalized MOF of Zn-DMOF that crystallizes in the tetragonal system having $P4/mmm$ space group, incorporating 2,3,5,6-tetramethyl-1,4-benzenedicarboxylate (TM-BDC^{2-}) ligand along with 1,4-benzenedicarboxylate (BDC^{2-}) and dabco. In the solid solution, the TM-BDC/BDC ratio has a noticeable influence on the thermal expansion behaviour of the framework. To describe the variation in the thermal expansion phenomenon of Zn-DMOF-TM_x i.e. ($Zn_2(BDC)_{2-2x}(TM- BDC)_{2x}(dabco)$), MOF was synthesized with in various ratio such as x = 0.0, 0.17, 0.45, 0.67, 1.0; {(x = [TM- BDC^{2-}]/(TM- BDC^{2-}]+[BDC^{2-}])}.

VTPXRD data in the temperature range from 150-375 K exhibited that Zn-DMOF shows NTE in the *ab* plane, whereas with the increase in the TM-BDC content this NTE decreases and becomes more positive for higher values of x (**Table 7**). The temperature at which ZTE take place advances in parent structure (Zn-DMOF) from 186 K (x = 0) to 325 K in Zn-DMOF-TM (x = 1.0).

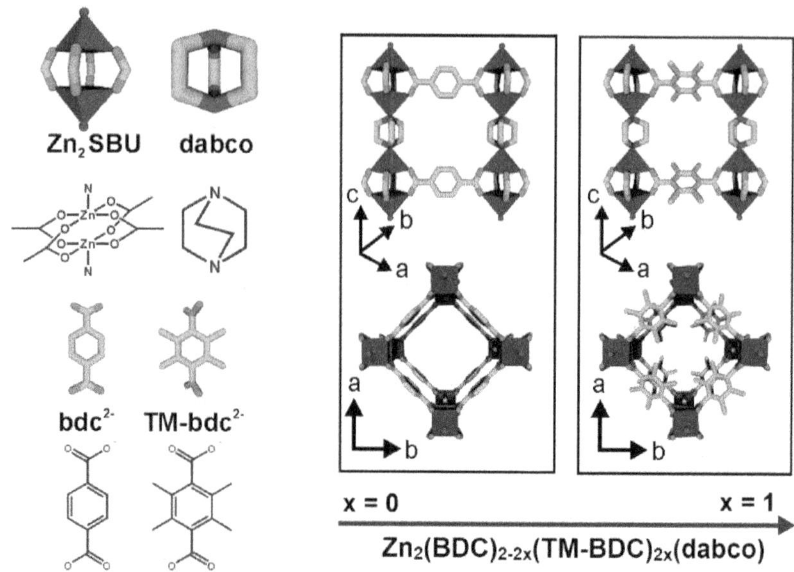

Figure 13. Representation of solid solution of Zn-DMOF-TM$_x$ is shown here (O, N, and C are represented in red, blue and gray colour respectively). The coordination mode around Zn^{2+} is represented by blue polyhedra. Reproduced from Reference [15] with permission from American chemical society.

Table 7. Variation in the lattice constants of Zn-DMOF and Zn-DMOF-TM with temperature

Temp. (K)	Zn-DMOF a/b (Å)	c (Å)	Zn-DMOF TM$_{0.17}$ a/b (Å)	c (Å)	Zn-DMOF TM$_{0.45}$ a/b (Å)	c (Å)	Zn-DMOF TM$_{0.67}$ a/b (Å)	c (Å)	Zn-DMOF-TM$_{1.0}$ a/b (Å)	c (Å)
376	10.939(1)	9.625(2)	10.9342(6)	9.630(1)	10.9334(5)	9.6204(6)	10.9224(5)	9.603(1)	10.9043(5)	9.580(1)
366	10.940(1)	9.623(2)	10.9354(6)	9.627(1)	10.9338(5)	9.6183(6)	10.9216(5)	9.602(1)	10.9034(5)	9.578(1)
352	10.941(1)	9.621(2)	10.9369(6)	9.625(1)	10.9335(5)	9.6153(6)	10.9214(5)	9.598(1)	10.9025(5)	9.575(1)
334	10.943(1)	9.619(2)	10.9378(6)	9.624(1)	10.9341(6)	9.6134(6)	10.9216(5)	9.596(1)	10.9018(5)	9.573(1)
318	10.943(1)	9.617(2)	10.9385(6)	9.622(1)	10.9337(6)	9.6110(6)	10.9215(5)	9.595(1)	10.9002(5)	9.570(1)
303	10.945(1)	9.614(2)	10.9392(6)	9.619(1)	10.9338(6)	9.6081(6)	10.9212(6)	9.593(1)	10.8991(5)	9.568(1)
288	10.946(1)	9.612(1)	10.9397(6)	9.619(1)	10.9338(6)	9.6044(6)	10.9213(6)	9.591(1)	10.8972(5)	9.567(1)
270	10.948(1)	9.611(2)	10.9403(6)	9.617(1)	10.9332(5)	9.6020(6)	10.9213(6)	9.588(1)	10.8966(5)	9.565(1)

256	10.950(1)	9.610(2)	10.9408(6)	9.615(1)	10.9329(5)	9.6001(6)	10.9205(5)	9.584(1)	10.8935(5)	9.562(1)
241	10.951(1)	9.608(2)	10.9412(6)	9.613(1)	10.9332(6)	9.5953(7)	10.9205(6)	9.579(6)	10.8922(5)	9.560(1)
225	10.952(1)	9.606(2)	10.9414(6)	9.610(1)	10.9322(5)	9.5919(5)	10.9197(5)	9.577(1)	10.8921(5)	9.557(1)
210	10.952(1)	9.603(2)	10.9412(6)	9.607(1)	10.9328(5)	9.5883(5)	10.9187(5)	9.577(1)	10.8901(5)	9.553(1)
195	10.953(1)	9.601(2)	10.9411(6)	9.603(1)	10.9318(5)	9.5850(5)	10.9180(5)	9.574(1)	10.8882(5)	9.549(1)
181	10.954(1)	9.599(2)	10.9403(6)	9.599(1)	10.9308(5)	9.5807(6)	10.9168(6)	9.571(1)	10.8854(6)	9.547(1)
167	10.952(1)	9.591(2)	10.9395(6)	9.595(1)	10.9301(6)	9.5760(6)	10.9158(6)	9.566(1)	10.8819(6)	9.548(1)
154	10.952(1)	9.594(2)	10.9384(6)	9.591(1)	10.9280(6)	9.5717(6)	10.9100(6)	9.564(1)	10.8799(6)	9.547(1)

With an increase in TM-BDC fraction in parent Zn-DMOF, the linear α increases from -10 ppm/K (x = 0) at 288 K and becomes positive for the *ab* plane. Although the expansion is parallel to the *c* axis, it was found to be positive for all compositions (x = 0 to 1.0). The thermal expansion along *c* axis with change in composition is unsurprising because the methyl groups present on the TM-BDC linkers have revealed earlier *via* molecular simulations and solid state NMR to take nominal interaction with dabco ligands.[16]

Das and co-workers [17] reported an important work towards the synthesis of mixed-metal mixed-linker two dimensional (2D) MOF $[Co_1Zn_1(\mu_3\text{-OH})_1(HBTC)_1(H_2BTC)_1(bpy)_{1.5}]$, (BTC = benzene tricarboxylic acid, bpy = 4,4′-bipyridine) that crystallizes in triclinic space group P-1 and exhibits all three kinds of thermal expansions (PTE,NTE,ZTE) ranging from 100 K to 450 K owing to the flexible coordination of organic linkers with Zn (II) and Co (II) ions in tetrahedral and octahedral geometry, respectively. Both Zn(II) and Co(II) ions including hydroxide ions and carboxylate groups formed a cluster in this 2D MOF (**Figure 14**). The 2D layers in MOF are interlocked by O-H...O interactions (**Figure 15**) along the *b* axis which leads to the generation of zip-like structure when viewed down [1 0 1].

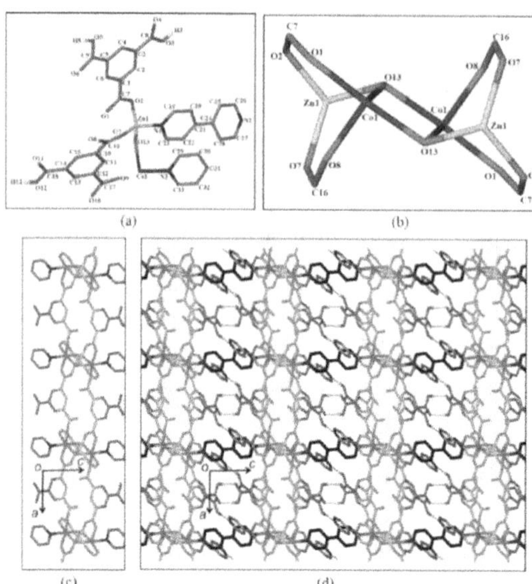

Figure 14. (a) Asymmetric unit of MOF [Co₁Zn₁(μ3-OH)₁(HBTC)₁(H2BTC)₁(bpy)₁.₅]; (b) Secondary building units (SBU) of MOF; (c) 1D linear range of SBU s generating along *a* axis; (d) 2D layers viewed on *ac* plane in the structure of MOF. Hydrogen bonds are indicated by red dotted line. Except the carboxylic hydrogen atoms, the remaining hydrogen atoms are removed for clarity. In the crystal structure, 1D linear design is displayed using yellow shade. Reproduced from Reference [17] with permission from American chemical society.

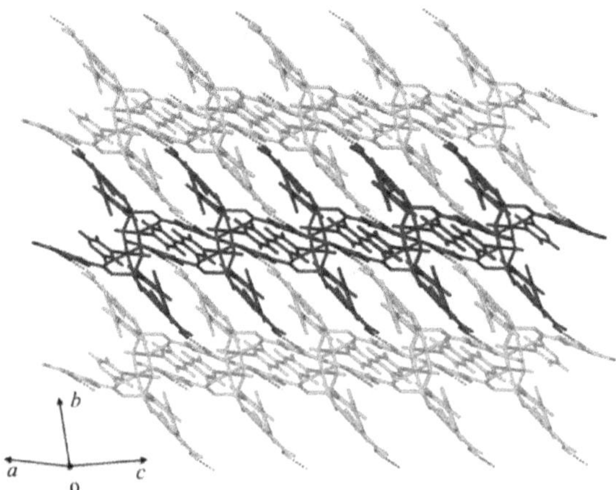

Figure 15. View of zip-like structure formed by interlocking of stacked two dimensional layers linked with O-H...O interactions. Reproduced from Reference [17] with permission from American chemical society.

The VTSCXRD reveals that the MOF undergoes PTE {α = 152(3) MK^{-1}} along the X3 axis in direction of [-1 -2 1] , NTE {α = - 36(2) MK^{-1}} along the principal axis X1 in direction of [-3 2 -1] and ZTE {α = 2(1) MK^{-1}} along the principal axis X2 in direction of [0 1 1]. This thermal expansion phenomena of the MOF also holds its reversible nature within the range of collected temperature.

Table 8. Change in bond angles with increasing temperature from 100 K to 450 K

Temp. (K)	Angle (°)					
	∠Co(1)-O(13)-Zn(1) (°)	∠Co(1)-O(13)-Co(1) (°)	∠Zn(1)-O(13)-Co(1) (°)	∠C(16)-O(7)-Zn(1) (°)	∠C(16)-O(8)-Co(1) (°)	∠C(17)-O(10)-Co(1) (°)
100	121.41(6)	100.63(5)	99.12(5)	111.7(1)	143.0(1)	121.7(1)
150	121.48(6)	100.70(5)	99.02(5)	111.8(1)	143.2(1)	122.0(1)
200	121.58(6)	100.78(5)	98.94(6)	112.1(1)	143.5(1)	122.4(1)
250	121.78(7)	100.89(6)	98.95(6)	112.3(1)	143.7(1)	122.8(1)
300	122.00(8)	100.85(6)	98.84(6)	112.4(1)	144.0(2)	123.2(1)
350	122.10(8)	100.91(7)	98.74(7)	112.6(2)	144.0(2)	123.4(1)
400	122.44(9)	100.97(7)	98.72(7)	112.8(2)	144.1(2)	123.8(1)
450	122.66(9)	101.0(8)	98.75(8)	113.1(2)	144.5(2)	124.0(2)

The variation in angle around the metal ions (**Table 8**) leads to the alteration of metal-metal distance in 2D layers which results in the increase of the crystallographic *b* axis by a large amount. This finally resulted in huge increase along the X3 axis. Here, the interlayer distances and the thickness of the 2D layers did not change with the variation in the temperature along the principal axis X2 that results in ZTE.

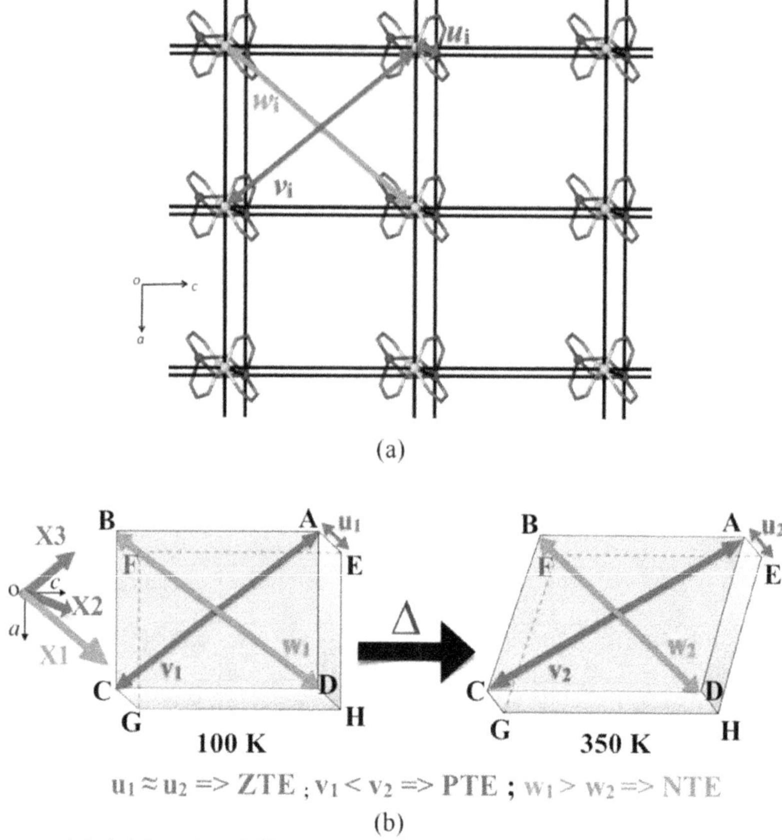

$$u_1 \approx u_2 \Rightarrow ZTE \; ; \; v_1 < v_2 \Rightarrow PTE \; ; \; w_1 > w_2 \Rightarrow NTE$$

(b)

Figure 16. (a) Graphical illustration of a 2D layer in MOF (solely SBUs in the layers are shown and linking ligands are removed for clarity. Part of the organic linkers that connects to SBUs are presented by bars. Purple and cyan color represents Co(II) and Zn(II) ions). (b) Diagram demonstration of a block in the 2D layer (**Figure 16a**) to designate the variation in the metal-metal distances are denoted by ui, vi, and wi along X2, X3, and X1 axes, respectively. Reproduced from Reference [17] with permission from American chemical society.

To describe the mechanism for this thermal expansion behaviour, authors presented the section of 2D layers by block having eight point edges (A-H). The points A, B, C, and D designated the SBUs having Zn(II) ions positions that are direct neighbours to each other and connected by translational symmetry. Similarly, the points E, F, G, and H represented the positions of translationally associated to Co ions of the same SBUs. The represented 2D layer's thickness (i.e., AE = BF = CG = DH), distances along the diagonal (i.e., BD = FH and AC = EG) are designated as u_i, v_i, and w_i (i = 1 or 2), respectively. Thermal expansion behaviour along X1, X2, and X3 axes can be rationalized by monitoring the change of u_i, v_i, and w_i with change in temperature The diagonal distance (v_1) increases from 16.653(1) Å to 16.872(1) Å (v_2) as the temperature rises from 100 to 350 K. Simultaneously, the distance ($w1$) deceases from 18.578(1) Å) to 18.527(1) Å ($w2$). The v_i expansion and w_i contraction of the layer was observed approximate along the X3 and X1 axes, respectively. As a result, PTE was found along X3 and NTE occurred along X1. Conversely, the layer's thickness, u_i (**Figure 16b**) remains almost constant i.e from 3.1139(5) Å to 3.1163(5) Å with temperature rise from 100-350 K. The u_i direction is approximate along the direction of the X2 axis and hence ZTE occurred along X2. Large thermal expansion of the crystallographic *b* axis in the compound was ascribed to the change in the coordination angle nearby the metal ions, which was significant in the large expansion along X3. From the structural analysis, it was reported that O–O distances of O–H⋯O bonding between 2D layers remain practically unaffected with the rise in temperature. Comparatively high variations that was observed in u_i, v_i, and w_i distances above 350 K were due to the large thermal vibrational motion of the atoms.

Lama *et. al* [18] reported another interesting thermal expansion property for two Zn(II)-coordination polymers, [Zn(tp)(BPP)]$_n$ (tp = terephthalate, desolvated form) and {[Zn(tp)(BPP)$_{0.5}$].0.5DMF}$_n$ (BPP = 1,3-bis(4-pyridyl)propane, solvated form). SCXRD analysis at 100 K shows that [Zn(tp)(BPP)]$_n$ possess orthorhombic crystal system with *Pbca* space group in which Zn^{2+} metal centres are connected by tp and BPP ligands in direction of *c* and *a* axes of crystal respectively to form 2D zig-zag layers (**Figure 17**).

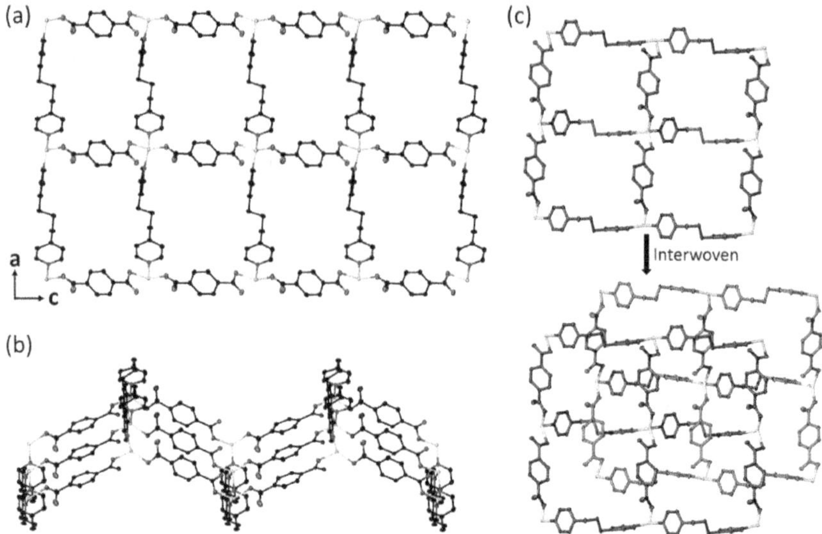

Figure 17. Perspective views of [Zn(tp)(BPP)]ₙ (tp=terephthalate). (a) Single two dimensional layer corresponding to (010). (b) Zig-zag layers made by bpp and carboxylate ligands. (c) Every two dimensional layer is interlaced with another layer to generate a two-fold interlaced sheet. Reproduced from Reference [18] with permission from Royal society of chemistry.

VTSCXRD analysis for [Zn(tp)(BPP)]ₙ in the temperature range 260K to 100K showed that when the temperature was cooled down, almost no change was observed along crystallographic axis *a*, whereas crystallographic *b* axis contracts but there was a little elongation along *c* axis in the above mentioned temperature range. The thermal expansion behaviour was found to be reversible in nature. From the crystal structure, it was observed that upon cooling the intermolecular interaction (C-H⋯π interaction) within the interwoven sheet get stronger with the additional strengthening of π⋯π interactions between the neighbouring twofold-interwoven sheets. These interactions are associated nearly in direction of *b* axis which results in the massive positive thermal expansion along the same axis. The Zn^{2+} ions distance joined by BPP ligand along *a* axis remains basically same that results in ZTE along *a* axis regardless of steady twisting and bending of the flexible alkyl unit of the BPP ligand with variation in temperature.

The non-bonding Zn1–Zn1 distance (d2) connected through tp linker (i.e. along *c*) experiences minor enlargement with decrease in temperature (**Figure 18**). As d1 remains practically same, extension of d2 was owing to minor complete rise in the angle θ1 upon cooling, and therefore NTE was observed along the crystallographic c axis. It was observed that within the two dimensional zig-zag layer, the framework remains mostly unaffected along *a* axis, however it experiences minor shrinkage along *c* with the rise in temperature which often designated as accordion motion.[19]

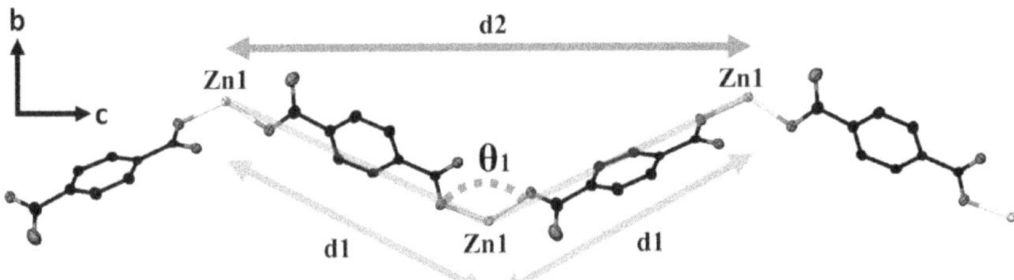

Figure 18. View presenting the angle (θ1) and non-bonding distances (d1 and d2) that are influenced by anomalous thermal expansion in [Zn(tp)(BPP)]$_n$. Reproduced from Reference [18] with permission from Royal society of chemistry.

In case of solvated form, SCXRD analysis at 100K shows that the polymer crystallises in monoclinic space group $P2_1/m$ where Zn^{2+} ion possess a distorted tetrahedral geometry in which three sites are occupied by three carboxylate oxygen atoms and fourth position is occupied by pyridyl group of BPP ligand. VTSCXRD experiments on solvated form from 260K to 100K similar to [Zn(tp)(BPP)]$_n$, along with PASCal program revealed that the α along principal axes X1, X2 , X3 were -14.7, 15.5 and 104.1 MK^{-1} respectively and the coefficient of volumetric thermal expansion was found to be 114.9 MK^{-1}. From the VTSCXRD, it was found that with decrease in temperature π⋯π separation among the tp and BPP ligand's aromatic ring between two adjacent layers decreases. This results in the stronger π⋯π interactions that leads to contraction of two dimensional layer along the *b* axis that finally contribute to PTE along X2 (**Figure 19**). This was obvious from the fact that the distance (L1) among the Zn^{2+} atoms linked by a BPP linker along the *b* axis shrinkages from 10.850Å to 10.776 Å upon cooling.

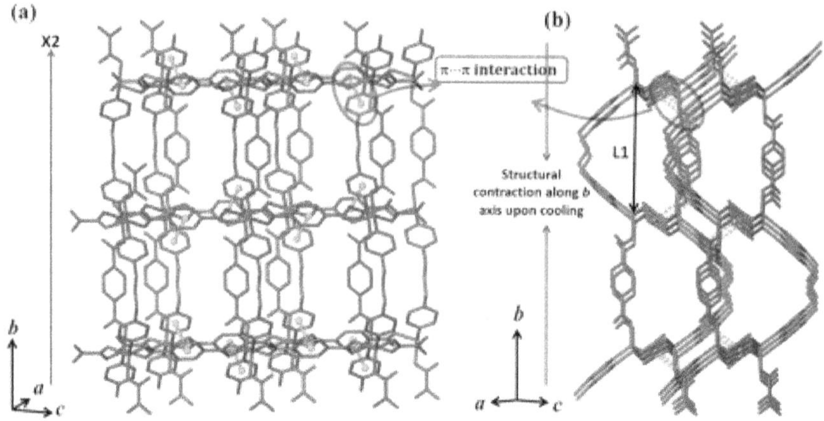

Figure 19. (a) Outlook of the π···π interactions among aromatic parts of the pyridyl and carboxylate of two neighbouring layers in [Zn(tp)(BPP)]$_n$. (b) Strength of π···π interactions increases upon lowering of temperature which leads to reduction along the *b* axis (X2) resulting in PTE. Reproduced from Reference [18] with permission from Royal society of chemistry.

With lowering in temperature, the distance among the Zn^{2+} atoms linked by the BPP linker along the *b* axis (L1) reduces. On lowering the temperature, the angle between a Zn and two centroids atom (∠i4–i5–Zn1; θ2, **Figures 20a and 20b**) reduces indicating that two adjacent two dimensional layers came nearer to each other. This results in the decrease of distance between the 2D layers along [−101] (**Figure 21**) which impact the thermal expansion along X1 and X3.

The C–H···O interactions between oxygen atom (O2) of the carboxylate group and hydrogen atoms of aromatic tp linkers (C7–H7···O2) and BPP (C9–H9···O2) become stronger with lowering in temperature (**Figures 20c** and **20d**). As an outcome of the collective results of stronger C–H···O and π···π interactions, two consecutive two dimensional layers comes nearer (along [−101]), with simultaneous sliding along [101]. The centroids distance between two SBUs (**Figure 21**) in two diverse layers (i.e. O and Q) rises with lowering in temperature that results in NTE along X1. Alternatively, the centroids separation between the SBUs (P and R) (**Figure 21**) lessens with lowering in temperature that adds to PTE along X3.

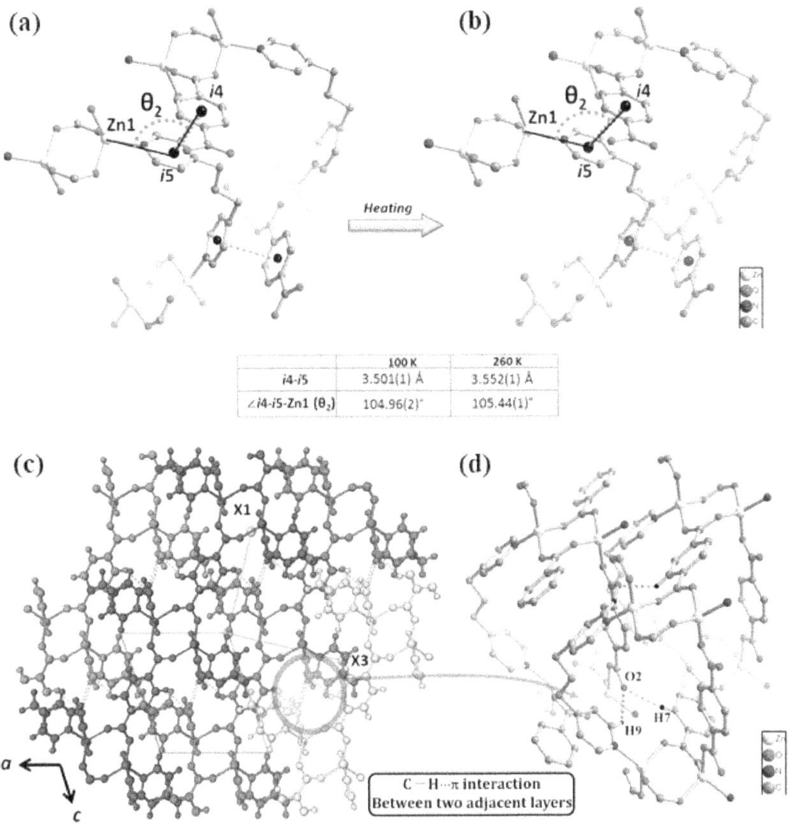

	100 K	260 K
i4-i5	3.501(1) Å	3.552(1) Å
∠*i4-i5-Zn1* (θ₂)	104.96(2)°	105.44(1)°

Figure 20. View showing the π···π interactions among the pyridyl units and aromatic groups of carboxylate at (a) 100 K and (b) 260 K in solvated form. View along [010] of solvated form showing C-H···O interactions (dotted lines) between the adjacent 2D layers (c) and (d) complete view of the C-H···O interactions (for clarity hydrogen atoms except those contributing in C-H···O bonding have been omitted). Reproduced from Reference [18] with permission from Royal society of chemistry.

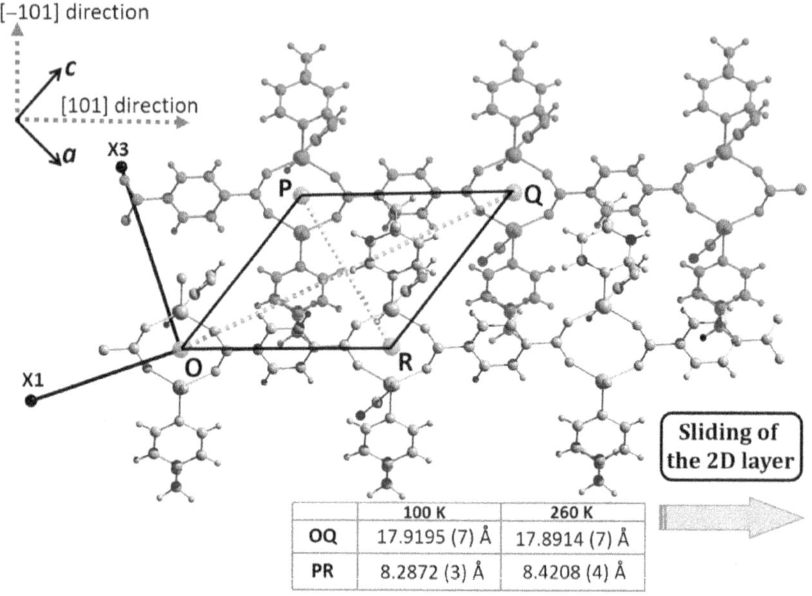

	100 K	260 K
OQ	17.9195 (7) Å	17.8914 (7) Å
PR	8.2872 (3) Å	8.4208 (4) Å

Figure 21. Outlook along [010] of the two consecutive 2D layers of solvated form. On lowering of temperature, the separation between the layers reduces, with simultaneous sliding along [101]. Both concurrent motions leads to PTE and NTE along X3 and X1, respectively. Reproduced from Reference [18] with permission from Royal society of chemistry.

Recently, Kaskel and co-workers [20] investigated the thermal expansion properties of flexible DUT-49(Cu) MOF using tetracarboxylate ligand [biphenyl bis(carbazole dicarboxylate), BBCDC] having Cu-based paddlewheel [21] SBU. In this framework, metal organic polyhedral (MOP) cores made by paddlewheel motif are interlinked by biphenyl linker (brown sphere in **Figure 22a**). In addition to MOP cavities, biphenyl spacer generates one octahedral and one tetrahedral cavity. The buckling of linker during the adsorption leads to decrease in the pore volume by approximately 61%, driving the gas molecules that was adsorbed to be excluded from the structure, which facilitates the distinctive negative gas adsorption activity of the DUT-49. Although several other factors such as crystallite size of the MOF, adsorbate composition, adsorption temperature etc intensely effects the flexibility of the MOF. However, in this work author reported the influence of temperature on the flexibility in DUT-49 by investigating the structural dynamics at atmospheric pressure in the existence of solvents and argon gas.

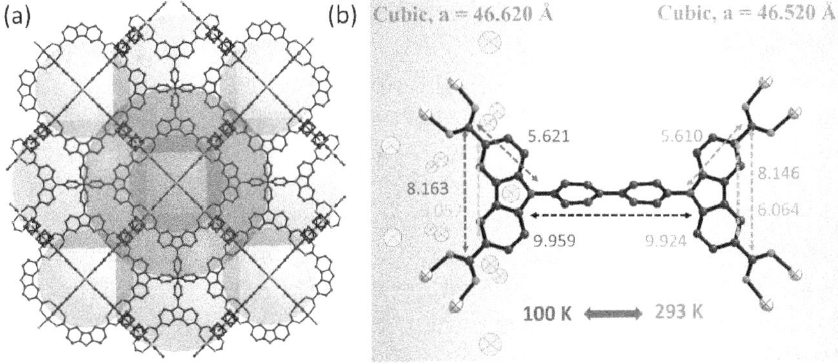

Figure 22. (a) View of three types of pores with three coloured spheres in the crystal framework of the MOF (DUT-49). (b) Structural variation of the solvent-free MOF is reversible (distances in Å) under the applied temperature conditions, introduced by the enlargement of the ligand at lower temperature. The ellipsoids (light brown, probability 50%) symbolise the Ar atoms location at 100K obtained by crystallography.

Amongst DUT-49(M) [M = Fe, Mn, Co, Cu, Ni, Zn, and Cd], only Cu and Ni comprising networks are proficient of offering adsorption-induced flexibility. Authors studied the thermally prompted flexibility in both desolvated and solvated states of DUT-49(Cu) with in situ synchrotron SCXRD. Desolvated DUT-49(Cu) MOF displayed a cubic unit cell (space group *Fm-3m*) with cell dimensions of 46.520 Å at 293 K and retain its single crystal nature.

After cooling to 100 K, desolvated crystal in an argon-filled capillary disclosed that the location of argon atoms adsorbed on the paddlewheel units having open metal sites. From crystallography, it was observed that absorption sites that are situated near to the copper atoms showed Cu–Ar distances of 3.040 Å and 2.565 Å for outer and inner sites of the framework (**Figure 22b**) where all additional remaining absorption sites are present in framework window. On cooling from 293K to 100K, the unit cell dimensions was increased to 46.620 Å with a rise in the unit cell volume by 651 Å. The volumetric negative thermal expansion coefficient was found to be -32.778 MK^{-1}. The space group remained unchanged during this contraction process but the volumetric NTE was related to factors such as increased M–O$_{carboxylate}$ lengths from 1.933Å to 1.941 Å and the extension of spacer distances among the carbazole moieties with decreases in temperature (**Figure 22b**). This thermal expansion/compression in unit cell of this crystal was found to be reversible upto 5 cycles. From the thorough analysis of the structure, it was elucidated that such NTE occurrence of DUT-49(Cu) initiates from the ligand where elongation took place with decrease in temperature. During thermal transition, the bond lengths in the carbazole core remain unaffected whereas the bonds with other parts of the ligand like the carboxylates and biphenyl spacer get lengthened due to the decrease in temperature. This results in a global increase in the N–N non-bonding distance from 9.924Å to 9.959 Å and the non- bonding C–C positioning concerning oppositely positioned carboxylates from 19.430 to 19.482 Å (**Figure 22b**). MOF solvated by DMF showed a different behaviour with variation in temperature. Unit cell decreases significantly from 46.620 Å to 45.660 Å on cooling the temperature from 298 K to 150 K. In case of solvated form, solvent DMF molecules present in the pores of MOF experience transition in phase as DMF get frozen below 212 K under ambient pressure leading to volumetric contraction.

Goodwin and co-workers [22] studied the thermal expansion phenomenon in Cd doped ZIF-8 MOF ($Zn_{1-x}Cd_x(mIm)_2$) (mIm= 2-methylimidazole), with Cd composition ranging from x = 0 to x = 1 in interval of x = 0.1. Structure of $Zn_{1-x}Cd_x(mIm)_2$ is analogues to sodalite except it was formed from tetrahedral Zn^{2+} centre coordinated with 2-methylimidazolate (mIm) linkers (**Figure 23**).

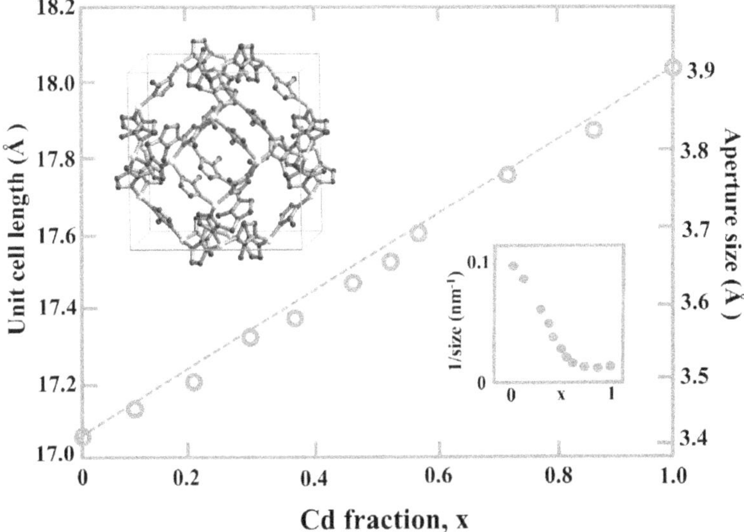

Figure 23. Dependence on the composition of the unit cell parameters and estimated size of the aperture in the ZIF-8/CdIF-1 solid solution ($Zn_{1-x}Cd_x(mIm)_2$). The Vegard limit is represented by the dashed line; we found that error bars are minor and smaller than the size of the symbols. The structure of ZIF-8 is shown in the inset (upper): the yellow colour spheres represents Zn atoms and grey colour stick denotes 2-methylimidazolate ligand. Lower inset presents the inverse crystallite size revealed by Scherrer analysis.

VTPXRD data over the range 100-310 K exhibited that Cd doped ZIF-8 MOF ($Zn_{1-x}Cd_x(mIm)_2$) (mIm= 2-methylimidazole) shows NTE, which decreases monotonically on increasing the Cd composition in $Zn_{1-x}Cd_x(mIm)_2$ samples as revealed by the decrease in the thermal expansion coefficient (α) values (**Table 9**).

Table 9. Variation in the thermal expansion coefficient (α) $Zn_{1-x}Cd_x(mIm)_2$ in with change in composition

x	α (MK^{-1})
0	11.9(17)
0.298(4)	11.2(19)
0.475(5)	6.3(3)
0.581(6)	4.4(3)
1	4.5(4)

With the introduction of Cd in ZIF-8 MOF, the flexibility of ZIF-8 framework increases due to the weaker interactions of Cd-mIm as compared to Zn-mIm which has been reported earlier by Cheetham and co-workers.[23] As a result of this weaker bonding interactions, Cd doped ZIF-8 MOF ($Zn_{1-x}Cd_x(mIm)_2$) (mIm= 2-methylimidazole) shows NTE, which decreases monotonically on increasing the Cd composition in ZIF-8 MOF. The thermal expansion behaviour is greatly influenced by the lattice modes of low-energy that commonly have characteristics [24,25] of NTE.

Another recent report was presented by Borguet and co-workers [26] where a detailed study for the thermal expansion behaviour of MOF i.e UiO-67 that crystallises in cubic crystal system and its amino/methyl functionalized MOF (UiO-67-NH_2 and UiO-67-CH_3) were analysed. All UiO-67 MOFs exhibited isotropic NTE as confirmed by both synchroton PXRD data and temperature-programmed IR spectroscopy. UiO MOFs consists of inorganic node having Zr_6 octahedron linked by oxo groups, μ_3-bridging hydroxyl and 12 carboxylates from bicarboxylate linkers to generate a three dimensional framework. [27,28] On heating upto activation temperature of 473K, each UiO-67 MOFs exhibit thermal behaviour (NTE) that is reversible in nature due to distortion of the carboxylate unit of the ligand. However, thermal treatment above 473K established that MOF showed irreversible NTE due to material's loss arising from the degradation of the MOF. The authors reported that there occurs a decrease in the unit cell edge length in UiO-67 and its analogues form (UiO-67-NH_2 and UiO-67-CH_3) with rise in the temperature (**Figure 24**).

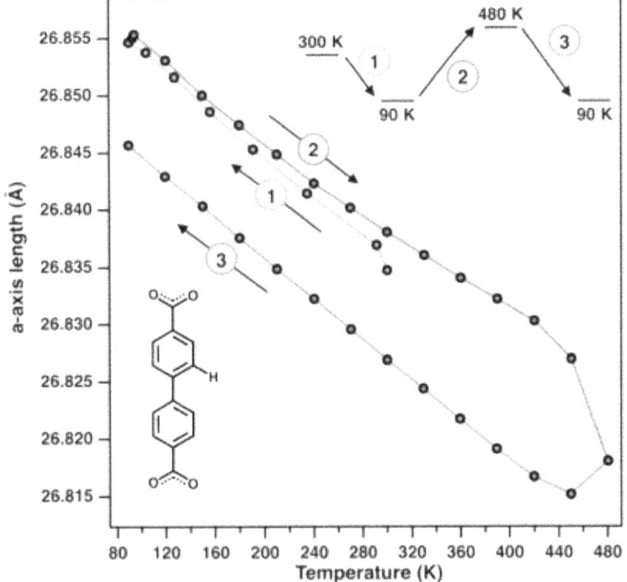

Figure 24. Change in unit cell length with temperature in UiO-67 between two temperature range, (1) 300 - 90 K (blue), (2) 90 - 480 K (magenta), and (3) 480 - 90 K (green). Reproduced from Reference [26] with permission from Royal society of chemistry.

The variation of edge length in the unit cell of UiO-67 with the temperature cycling from 90-480-90 K noticeably revealed a negative linear association between the temperature and length of the axes. UiO-67 MOF and its analogue forms exhibits isotropic NTE. This isotropic NTE is mainly attributed to the variation of bond length and bond angles that occurred in the carboxylate groups (**Figure 25**).

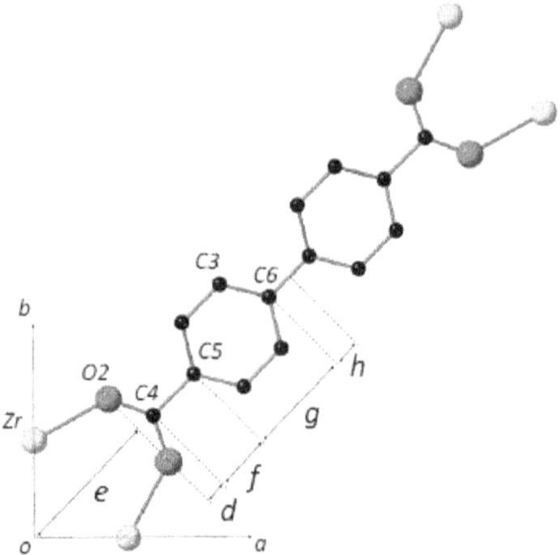

Figure 25. Part of UiO-67 structure showing the connection of Zr node and BPDC linker along the ab diagonal. Reproduced with permission from [26] American chemical society.

The bond length (C-O) and bond angle (O-C-O) changes from 1.259(5) Å and 129(1)° to 1.227(6) Å and 134(1)° on going from 90 K to 480 K, which leads to compression of the triangle (O-C-O) by 0.07 Å. The apparent decrease of C-O bond length on heating (**Figure 26**) likely resulted from the associated vibrational motion between carbon (C) and oxygen (O). With the variation in temperature, reversible change in the dihedral angle (ϕ) of the aromatic rings in the BPDC linker was observed (**Figure 27**).

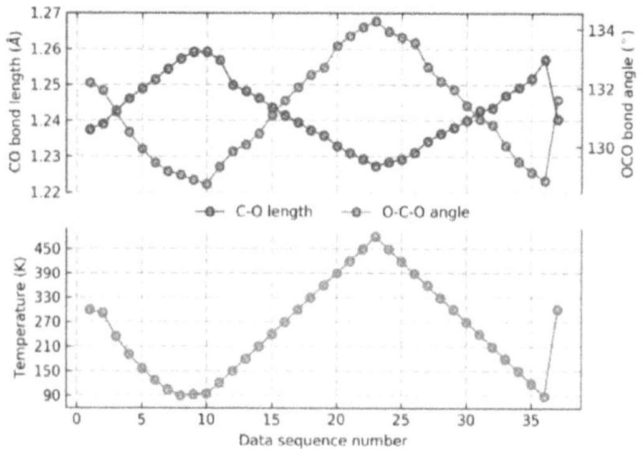

Figure 26: Evolution of the bond length (C-O) and bond angle (O-C-O) of UiO-67 during the temperature cycling experiment. Reproduced from Reference [26] with permission from Royal society of chemistry.

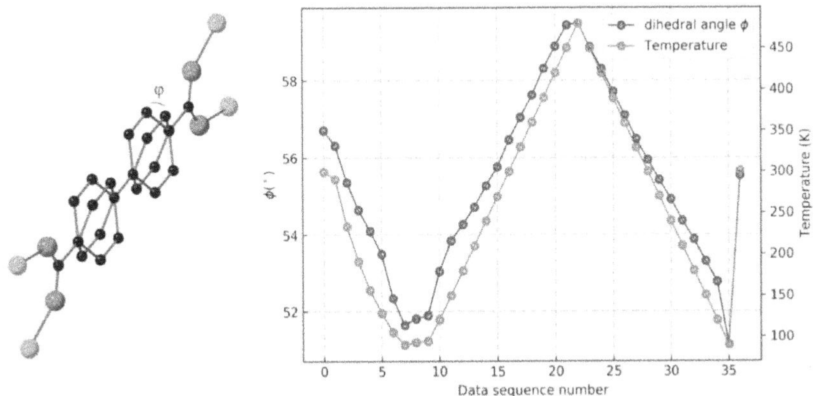

Figure 27. Evolution of the linker dihedral angle (φ) for UiO-67-NH$_2$ during the temperature cycling experiment. Reproduced from Reference [26] with permission from Royal society of chemistry.

The NTE behaviour for all MOFs is influenced by the different functional group attached on the organic ligand. Both amino and methyl substituted UiO-67 MOFs have lesser coefficients of NTE as compared to UiO-67 (**Table 10**). This is mainly attributed to the fact that the addition of different functional groups on the BPDC ligand did not allow the framework to compress more on heating to 480 K which is also reported by Burtch et al.[29]

Table 10. α of UiO-67 and its ligand functionalized MOFs

MOFs	Temperature cycling		Thermal degradation	
	Calculated range	α	Calculated range	α
UiO-67	90 K-420 K	-3.29(1)	570 K-790 K	-8.7(1)
UiO-67-NH$_2$	120 K-300 K	-2.48(7)	580 K-740 K	-9.5(3)
UiO-67-CH$_3$	120 K-300 K	-1.06(7)	580 K-770 K	-3.13(6)

The thermally induced structural changes in UiO-67 MOFs was revealed from the infra-red (IR) data. In case of UiO-67-NH$_2$, the difference in the IR spectra in the range 1750-1050 cm^{-1} delivered understandings into structural alteration linked with carboxylate, aromatic ring stretching, N-H bending and C-H bending etc. induced by thermal treatment. [30,31] Numerous IR peaks in this area were moved in the direction of lower wavenumber for UiO-67 (**Figure 28**) and its analogue MOFs respectively. These spectral alterations occurred due to thermal treatment are assumed to be associated with the variations in the dihedral angle of biphenyl ring and contraction of the lattice.

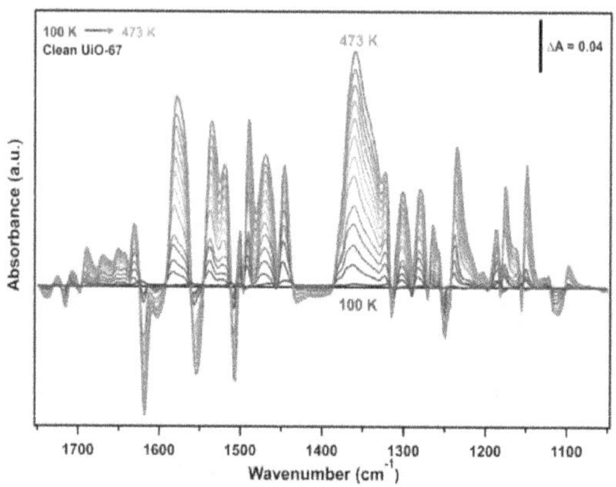

Figure 28. IR region (1750-1050 cm⁻¹) for UiO-67 taken in-situ while heating in the range of 100 K (blue) to 473 K (red). The colour legend displaying variations in carboxylate and aromatic rings and C-H bending types. Reproduced from Reference [26] with permission from Royal society of chemistry.

Overall, from the data it was observed that due to the disorderness of the carboxylate units of the organic ligand induced by the thermal treatment, it mainly causes NTE behaviour for all UiO-67 MOFs. On heating above 473 K, all MOFs resulted in the degradation as revealed by the absence of $\upsilon(OH)_{free}$ frequency thereby indicating the start of cluster dehydroxylation. The decrease in the typical bond length among Zr and μ_3–O(H) was observed (from 2.14(1) to 2.10(1) Å). In addition, site occupancy of μ_3–O(H) reduces from 1 to 0.75, implying the removal of hydroxyl groups thereby leading to irreversible NTE (**Figure 29**). These outcomes support the dehydroxylation process that was suggested based on X-ray absorption spectroscopy (XAS) result and computation analysis of UiO-66 and UiO-67 where $Zr_6O_4(OH)_4$ cluster get converted to distorted Zr_6O_6 cluster. [30,32]

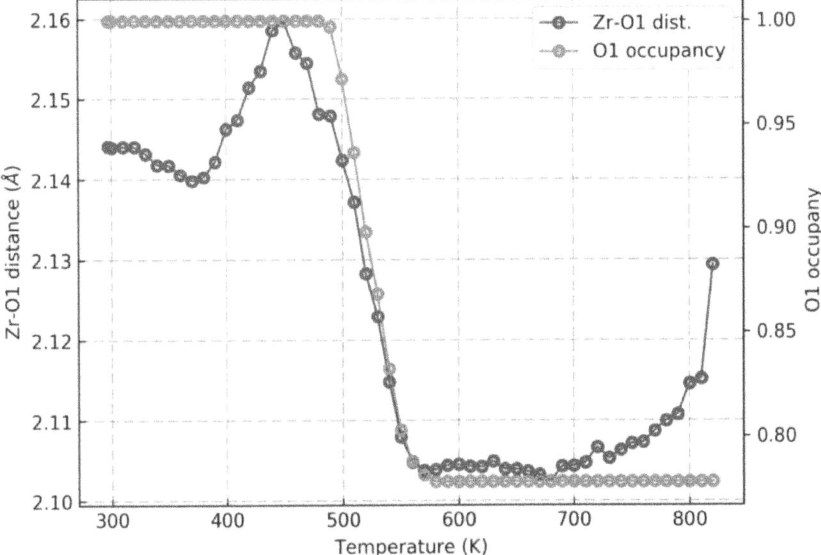

Figure 29. Change of μ_3–O(H) site occupancy (O1 occupancy) and the typical bond length among Zr and μ_3–O(H) as a function of temperature. Reproduced from Reference [26] with permission from Royal society of chemistry.

As observed from the above examples, thermal expansion behaviour in various MOFs is principally attributed to the flexible nature of the ligands, metal ions and the weaker interactions that are present between the host and the guest molecules. Degree of guest loading and tuning the functionalities in organic ligands can affect the thermal expansion properties of MOFs. In addition, low frequency transverse molecular vibrations such as bending of ligand unit/bond connected to the metal centre can contribute to the thermos-responsive property. Synthesizing new MOF and understanding the different properties of MOF along with the mechanism for their thermal expansion behaviour will definitely help to enrich the literature.

6. Conclusions and Perspectives

This chapter clearly puts forward the comprehensive assessment of the temperature dependent structural variations that occurred in MOFs. The occurrence of tuneable thermal expansion principally includes employing variations in interactions that are weak in nature by means of external incitements. The variation in the lattice parameters of MOFs with temperature presents a distinctive study that can result to numerous possible devices based on single crystal for example thermo-mechanical sensing device or actuators, optical fiber systems, cookware, extremely precise optical devices, and aerospace manufacturing. Elucidation of the mechanisms of anomalous thermal expansion offers information for the design principle of new thermo-responsive materials that can open up an application of such MOF based materials in new scientific/engineering fields. In recent years, copious consideration has been dedicated to completely understand such thermos-responsive performance of materials that can show PTE, NTE, ZTE and the efforts are being taken by researchers to navigate more about such properties specially in case of MOF materials. This thermal expansion behaviour of MOFs removes the barrier for the study towards the rise in temperature in relation with the adsorption for principal greenhouse and energy-related gases (CO_2 and CH_4) and it enables us to make use of this thermal property for the adsorption and desorption of gases. An example is the elimination of CO_2 from flue gas streams by means of temperature swing adsorption (TSA). Thus, the thermos-responsive behaviour of MOFs proves to be very essential tool for maintaining the ecological balance and sustainability through gas storage and separation. As our knowledge remains confined towards the materials exhibiting only expansion with the rise in the temperature, MOFs place themselves one step ahead and shows interesting contraction behaviour with rise in the temperature. Thus, the thermal engineering of MOFs for adsorption-based applications is also very interesting in vision of their industrial potential.

Acknowledgements

P.L. is thankful to Department of Science and Technology, New Delhi for DST-INSPIRE Faculty award [DST/INSPIRE/04/2017/000249]. S. D. and A. R. acknowledge the financial support from CSIR (JRF) and UGC (JRF) respectively. The authors declares no conflict of interest.

References

1. (a) Cliffe, M. J., and Goodwin, A. L. (2012) PASCal: a principal axis strain calculator for thermal expansion and compressibility determination. *J. Appl. Cryst*, **45**, 1321-1329; (b) The Win_strain programme is unpublished but available for download from http://www.rossangel.com/text_strain.htm.

2. (a) Ashcroft, N. W., and Mermin, N. D. (1976) *Solid State Physics*; Holt, Rinehart & Winston: 9th volume, 1st edition, New York ,XXll . (b) Krishnan, R. S., Srinivasan, R., and Devanarayanan, S. (1979) *Thermal Expansion of Crystals,* Pergamon, 12th volume, 1st edition, Elsevier.

3. (a) Evans, J. S. O. (1999) Negative thermal expansion materials. *J. Chem. Soc. Dalton Trans.*, **19**, 3317-3326; (b) Kepert, C. J. (2006) Advanced functional properties in nanoporous coordination framework materials. *Chem. Commun.*, **7**, 695-700; (c) Wu, Y., Kobayashi, A., Halder, G. J., Peterson, V. K., Chapman, K. W., Lock, N., Southon, P. D., and Kepert, C. J., (2008) Negative thermal expansion in the metal-organic framework material $Cu_3(1,3,5$-benzenetricarboxylate$)_2$. *Angew. Chem. Int. Ed.*, **47**(46), 8929-8932.

4. Grobler, I., Smith, V.J., Bhatt, P.M., Herbet, S.A., and Barbour, L.J. (2013) Tunable Anisotropic Thermal Expansion of a Porous Zinc(II) Metal–Organic Framework. *J. Am. Chem. Soc.*, **135**(17), 6411-6414.

5. Yang, C., Wang, X., and Omary, M.A. (2009) Crystallographic observation of dynamic gas adsorption sites and thermal expansion in a breathable fluorous metal-organic framework. *Angew. Chem., Int. Ed.*, **48**(14), 2500-2505.

6. Lama, P., Das. R.K., Smith.V.J.,and Barbour, L.J. (2014) A combined stretching–tilting mechanism produces negative, zero and positive linear thermal expansion in a semi-flexible Cd(II)-MOF. *Chem. Commun.*, **50**(49), 6464-6467.

7. DeVries, L. D., Barron, P.M., Hurley, E.P., Hu, C., and Choe, W. (2011) "Nanoscale Lattice Fence" in a Metal_Organic Framework: Interplay between Hinged Topology and Highly Anisotropic Thermal Response. *J. Am. Chem. Soc.*, **133**(38), 14848-14851.

8. Temperature-dependent PXRD scans were conducted at the 11-BM beamline at Argonne National Laboratory (April 9, 2011). http://11bm.xor.aps.anl.gov/description.html.

9. Liu, Z., Li, Q., Zhu, H., Lin, K., Deng, J., Chen, J., and Xing, X. (2018) 3D negative thermal expansion in orthorhombic MIL-68(In). *Chem. Commun.*, **54**(45), 5712-5715.

10. Volkringer, C., Meddouri, M., Loiseau, T., Guillou, N., Marrot, J., Ferey, G., Haouas, M., Taulelle, F., Audebrand, N., and Latroche, M. (2008) The kagome topology of the gallium and indium metal-organic framework types with a MIL-68 structure: Synthesis, XRD, solid-state NMR characterizations, and hydrogen adsorption. *Inorg. Chem.*, **47**(24), 11892-11901.

11. Lama, P., Alimi, L. O., Das, R. K., and Barbour, L. J. (2016) Hydration-dependent anomalous thermal expansion behaviour in a coordination polymer. *Chem. Commun.*, **52**(15), 3231-3234.

12. Peterson,V. K., Kearley, G. J.,Wu, Y., Cuesta, A.J.R., Kemner, E., and Kepert, C.J. (2010) Local Vibrational Mechanism for Negative Thermal Expansion: A Combined Neutron Scattering and First-Principles Study. *Angew. Chem. Int. Ed.*, **49**(3), 585-588.

13. Wu, Y., Kobayashi, A., Halder, G.J., Peterson, V.K., Chapman, K.W., Lock, N., Southon, P.D., and Kepert, C.J. (2008) Negative thermal expansionint the metal-organic framework material Cu_3(1,3,5-dibenzenetricarboxylate)$_2$. *Angew. Chem. Int Ed.*, **47**(46), 8929-8932.

14. Liu, L., Geng, B., Sayed, S. M., Lin, B.-P., Keller, P., Zhang, X.-Q., Sun, Y., and Yang, H. (2017) Single-layer dual-phase nematic elastomer films with bending, accordion-folding, curling and buckling motions. *Chem. Commun.*, **53**(11), 1844–1847.

15. Baxter, S.J., Schneemann, A., Ready, A.D., Wijeratne, P., Wilkinson, A.P., and Burtch, N.C. (2019) Tuning thermal expansion in metal-organic frameworks using a mixed linker solid solution approach. *J. Am. Chem. Soc.*, **141**(32), 12849-12854.

16. Burtch, N.C., Knoop, A.T., Foo, G.S., Leisen, J., Sievers, C., Ensing, B., Dubbeldam, D., and Walton, K.S. (2015) Understanding DABCO nanorotor dynamics in isostructural metal-organic frameworks. *J. Phys. Chem. Lett.*, **6**(5), 812-816.

17. Shrivastava, A., and Das, D. (2019) Axial Positive, Negative, and Zero Thermal Expansion in a Mixed- Metal Mixed-Linker Coordination Compound: Role of 2D Layer in the Thermal Expansion Property. *Cryst. Growth Des.*, **19**(9), 4908-4913.

18. Lama, P., Hazra, A.and Barbour, L.J. (2019) Accordion and layer-sliding motion to produce anomalous thermal expansion behaviour in 2D-coordination polymers. *Chem. Commun.*, **55**(80), 12048-12051.

19. (a) Blegger, D., Liebig, T., Thiermann, R., Maskos, M., Rabe, J.P., and Hecht, S. (2011) Light-orchestrated macromolecular "Accordions": Reversible photoinduced shrinking of rigid-rod polymers. *Angew. Chem., Int. Ed.*, **50**(52), 12559-12563; (b) Yoshida, Y., Mawatari, Y., Motoshige, A., Hiraoki, T., Wagner, M., Mullen, K., and Tabata, M. (2013) Accordion-like oscillation of contracted and stretched helices of polyacetylenes synchronized with the restricted rotation of side chains. *J. Am. Chem. Soc.*, **135**(10), 4110-4116.

20. Garai, B., Bon, V., Efimova, A., Gerlach, M., Senkovska, I., and Kaskel, S. (2020) Reversible switching between positive and negative thermal expansion in a metal–organic framework DUT-49. *J. Mater. Chem. A.*, **8**(39), 20420-20428.

21. Stoeck, U., Krause, S., Bon, V., Senkovska, I., and Kaskel, S. (2012) A highly porous metal-organic framework, constructed from a cuboctahedral super-molecular building block, with exceptionally high methane uptake. *Chem. Commun.*, **48**(88), 10841-10843.

22. Sapnik, A.F., Geddes, H.S., Reynolds, E.M., Yeung, H.H.-M., and Goodwin, A.L. (2018) Compositional inhomogeneity and tuneable thermal expansion in mixed-metal ZIF-8 analogues. *Chem. Commun.*, **54**(69), 9651-9654.

23. Baxter, E.F., Bennett, T.D., Cairns, A.B., Brownbill, N.J., Goodwin, A.I., Keen, D.A., Chater, P.A., Blanc, F., and Cheetham, A.K. (2016) A comparision of the amorphization of zeolitic imidazolate frameworks (ZIFs) and aluminosilicate zeolites by ball-milling. *Dalton Trans.*, **45**(10), 4258-4268.

24. Coudert, F.-X. (2015) Responsive metal-organic frameworks and framework materials: under pressure, taking the heat, in the spotlight, with the friends. *Chem. Mater.*, **27**(6), 1905-1916.

25. Dove, M.T., and Fang, H. (2016) Negative thermal expansion and associated anomalous physical properties: review of the lattice dynamics theoretical foundation. *Rep. Prog. Phys.*, **79**, 066503.

26. Goodenough, I., Devulapalli, V.S.D., Xu, W., Boyanich, M.C., Luo, T.Y., Souza, M.D., Richard, M., Rosi, N.L., and Borguet, E. (2021) Interplay between Intrinsic Thermal Stability and Expansion Properties of Functionalized UiO-67 Metal–Organic Frameworks. *Chem. Mater.*, **33**(3), 910-920.

27. Rogge, S. M. J., Wieme, J., Vanduyfhuys, L., Vandenbrande, S., Maurin, G., Verstraelen, T., Waroquier, M., and Speybroeck, V.V. (2016) Thermodynamic insight in the high-pressure behaviour of UiO-66: Effect of linker defects and linker expansion. *Chem. Mater.*, **28**(16), 5721-5732.

28. Gutov, O. V., Bury, W., Gualdron, D.A.G., Krungleviciute, V., Jimenez, D.F., Mondloch, J.E., Sarjeant, A.A., Juaid, S.S.A., Snurr, R.Q., Hupp, J.T., Yildirim, T., and Farha, O.K. (2014) Water-stable zirconium-based metal-organic framework material with high-surface area and gas-storage capacities. *Chem.- Eur. J.*, **20**(39), 12389-12393.

29. Burtch, N. C., Baxter, S. J., Heinen, J., Bird, A., Schneemann, A., Dubbeldam, D., and Wilkinson, A. P. (2019) Negative Thermal Expansion Design Strategies in a Diverse Series of Metal-Organic Frameworks. *Adv. Funct. Mater.*, **29**, 1904669.

30. Valenzano, L., Civalleri, B., Chavan, S., Bordiga, S., Nilsen, M. H., Jakobsen, S., Lillerud, K. P., and Lamberti, C. (2011) Disclosing the Complex Structure of UiO-66 Metal Organic Framework: A Synergic Combination of Experiment and Theory. *Chem. Mater.*, **23**(7), 1700-1718.

31. Ragon, F., Campo, B., Yang, Q., Martineau, C., Wiersum, A.D., Lago, A., Guillerm, V., Hemsley, C., Eubank, J. F., Vishnuvarthan, M., Taulelle, F., Horcajada, P., Vimont, A., Llewellyn, P. L., Daturi, M., Vinot, S. D., Maurin, G., Serre, C., Devic, T., and Clet, G. (2015) Acid-functionalized UiO-66(Zr) MOFs and their Evolution After Intra-framework

Cross-linking: Structural Features and Sorption Properties. *J. Mater. Chem. A*, **3**(7), 3294-3309.

32. Shearer, G. C., Forselv, S., Chavan, S., Bordiga, S., Mathisen, K., Bjørgen, M., Svelle, S., and Lillerud, K. P. (2013) In Situ Infrared Spectroscopic and Gravimetric Characterisation of the Solvent Removal and Dehydroxylation of the Metal Organic Frameworks UiO-66 and UiO-67. *Top. Catal.*, **56**, 770-782.

Fluoroswitchable Metal-Organic Frameworks: Self-Adjustable Detection of Multifarious Pollutants via Reversible Host-Guest Interplay

Ranadip Goswami[*,a,b], *Nilanjan Seal*[a,b], *Sonal Rajput*[b] and *Subhadip Neogi*[*,a,b]
[a]*Academy of Scientific and Innovative Research (AcSIR), Ghaziabad, India*
[b]*Inorganic Materials & Catalysis Division, CSIR-Central Salt & Marine Chemicals Research Institute, Gujarat, India*

1. Introduction

In recent decades, metal-organic frameworks (MOFs), constructed from self-assembly of metal ions and organic struts, have become most fascinating and rapidly growing field of research for chemists and material scientists owing to their intrinsic merits of ordered structures and excellent structural attributes such as enormous structural diversity, high surface area, remarkable stability, tunable pore-functionality etc. [1-4]. MOFs have been utilized in a plethora of potential applications in several areas including luminescent sensing [5-8], heterogeneous catalysis [9-11], gas adsorption/separation [12,13], bioimaging [14], anti-counterfeiting devices [15,16] and so on (Fig. 1). Compared to several other sensory materials, luminescent metal-organic frameworks (LMOFs) are highly responsive towards external stimuli such as electronically and structurally diverse analytes (cations, anions, vapors and small molecules), temperature, solvent, light etc. Particularly, stimuli-responsive and switchable emission are of great significance as they evident two different physically stable states [17], capable of recognizing multiple analytes through reversible sensing. Moreover, the switchable and naked-eye detection of the targeted species with high selectivity is favorable for real-time applications [18]. Also, it is very difficult to envisage the on-off luminescent phenomenon within a single matrix. So far, organic dyes [19], carbon nanomaterials [20] and hydrogels [21] have exhibited fluoro-switchable properties, but the complex design and multi-step synthesis are their major shortcomings. Hence, development of advanced luminescent probes possessing on-off emission switching features [22] to execute non-destructive detection is extremely necessary.

Conversely, according to UN's report, there is increasing overlap between residential and industrial domains because of urbanization, which in turn results in severe environmental pollution [23]. Therefore, it is of urgent necessity to develop advanced sensors for monitoring of industrial wastes, food quality and toxic chemicals for ecological safety. Additionally, quick response time, selectivity, sensitivity, reusability and enduring stability are the key aspects of any chemical sensor for instant applicability with high efficiency. The switchable sensing phenomena of MOF is controlled by several features and possess diverse advantages such as (i) high crystallinity helps to determine the exact structure of MOF and thus corroborate atomic-level interaction between targeted guest and host, (ii) easy trapping of specific analyte inside the functionalized pores, (iii) efficient interaction between the analyte and luminescent probes depends upon the concentration of analytes and (iv) different luminescent mechanism (*vide infra*) can be directly anticipated by X-ray crystallography. These distinct characteristics solely render these hybrids porous materials as efficient and technically relevant sensory scaffolds that are easily accessible and cost-effective as compared to complex analytical equipment (ICP, HPLC and GCMS and so forth). In this chapter, we aim discussing the advances on LMOFs with particular emphasis on stimuli-triggered luminescent switching for monitoring lethal chemicals over practical platforms.

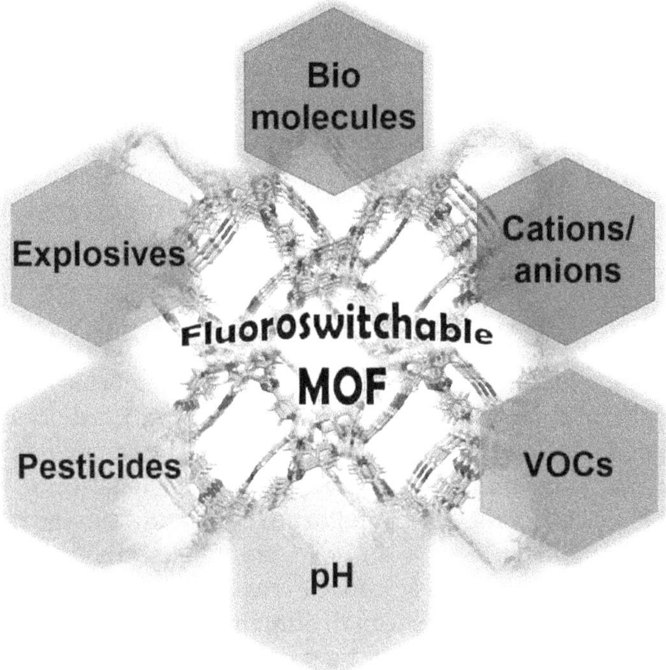

Figure 1. Classification of external stimuli those operate for luminescent de tection in MOFs.

2. MOFs as Switchable Chemosensors

The above-mentioned discussion brings into an essential interpretation regarding synthesis and important structural attributes of luminescent MOFs (LMOFs) which render them as potential candidates for sensing applications. It may be noted that receptor and transducer are the fundamental aspects of any sensors that are responsible for delivering the signal via several transduction mechanistic pathways. On the other hand, luminescent mechanism is less complicated, easily quantified by naked eye and has substantial practice in practical platforms [24]. Principally, MOFs are considered as giant "molecules" because of their extended structure, which possess definite energy gap between valence band (VB) and conduction band (CB). The electrons in MOFs are lifted from the ground state to the excited state by absorption of excitation energy, and then return to the ground state through fluorescence (short excited life time 1-100 ns) or phosphorescence (life time 1s or longer) emission. From the Jablonski diagram, the usual mechanisms involved in the absorption and emission of light within the energy levels, as well as the related lifetimes, are well known (Fig. 2a). The photoluminescence-based detection of luminous MOFs is based on changes in opto-electronic characteristics caused by contact with the targeted analyte under experimental conditions (temperature, solvent, etc) [25]. These emissions can be well justified by certain electron transfer phenomena such as Dexter energy transfer (DET) [26], photo-induced electron transfer (PET) [27] and Förster resonance energy transfer (FRET) [28] Charge transfer (CT) between metal-to-ligand (ML), ligand-to-metal (LM), ligand-to-ligand (LL), and metal-to-metal (MM), as well as pore occlusion of guest species inside the LMOFs are all possible mechanisms that affect the emission intensity of MOFs (guest sensitization and guest centred emission). (Fig. 2b) [29]. Porous cavity and sieving nature of MOF exhibit high sensitivity and exceptional selectivity [30]. Additionally, the sensing ability can be regulated at molecular level through variation of the pore environment i.e. polarizability, hydrophobicity etc. Most LMOFs detect a particular analyte by altering the PL intensity in terms of switch on/off responses, formation of a different peak, or changes in peak followed by emission. The type of interaction between the MOF and the analyte, whether through electron or energy transfer (see above), determines whether the molecule turns on or off. For example, although electron withdrawing aromatic chemicals normally reduce emission intensity, electron donatingaromatic compounds enable host MOF emission intensity to increase. Importantly, numerous characteristics including stability, acidity/basicity, hydrophilicity/hydrophobicity, ionic character, pore width, polarizability, and others must be evaluated with regard to the particular analyte when designing a luminous MOF for a targeted sensing application. However, in the vast majority of situations, analyte detection is mostly dependent on the "rational design approach," in which metal ions, linkers, and guests are carefully selected for a desired outcome. As the involved linkers are skilled to reduce the HOMO-LUMO (HOMO: highest occupied molecular orbital; LUMO: lowest unoccupied molecular orbital) energy gap and dragging the absorption and emission wavelengths to the visible region, highly conjugated organic struts are not required to form luminescent MOFs. Functionalization or post-synthetic alteration of the ligand has been shown to aid in the differentiation of target analytes by modifying the favourable interaction sites. [31]. Generally, paramagnetic metal ions (Cu^{2+}, Co^{2+}, Ni^{2+} etc.) are not used for constructing luminescent MOFs as they often quench the emission intensity by encouraging non-radiative emission through transfer of energy from the excited linker. Thus, d^{10} metal ions (Cd^{2+}, Zn^{2+} etc.) are more appropriate to form LMOFs that operate via ligand centered charge transfer route. Furthermore, lanthanide-ion based MOFs may produce sharp signals as well as an increase in emission intensity due to the antenna effect [32]. The emission properties of these MOFs may likewise be greatly influenced by the guest molecules. As guests, neutral molecules are less expected to interact with the MOF cavity and are more likely to escape or leach during the fluorosensing process. In contrast, considering charged frameworks, where ionic components already exist inside the cavity, host-guest interactions are much more beneficial for sensory experiments involving cations/anions as analytes. Due toobvious "space confinement effect" inside the MOFs, immobilization of nanoparticles within the framework cavity exhibits exceptional photophysical properties and acts as transducers. [33].

(a)

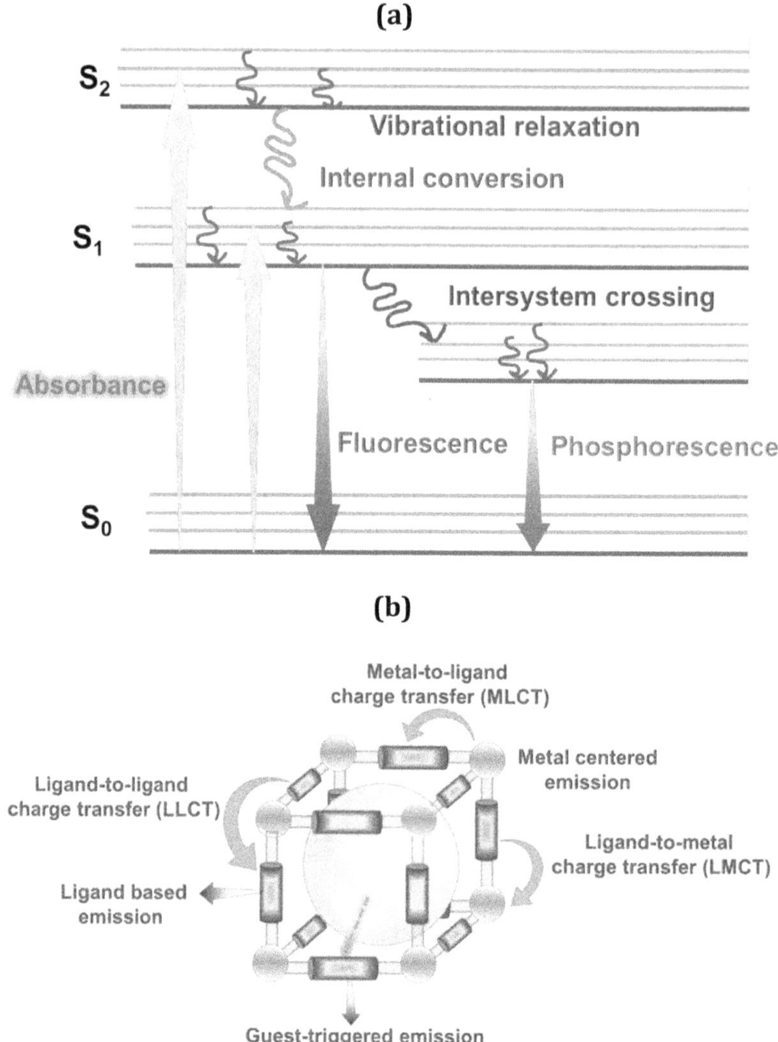

Figure 2. (a) the pictorial depiction of the Jablonski diagram. (b) The various emission pathways in MOFs.

2.1 Fluoroswitchable Detection of Volatile Organic and Nitro Explosive Compounds

The selective detection of toxic volatile organic compounds (VOCs) are of utmost importance because of their harmful effect on environment. Also, owing to their low vapour pressure, detection of these species under atmospheric condition is highly necessary [34]. MOFs have the potential to identify them effectively in both solid and solution phases [35], and several relevant literatures are discussed here. Gao *et. al.* demonstrated the selective identification of electron-rich volatile amines through space confinement in a series of charged MOFs. [36]. They synthesized Zr based, UiO-67-type of ionic MOFs, Zr-DQ-ANS and Zr-MQ-ANS, where MQ and DQ denotes N-alkylated 2,2′-bipyridinium components (MQ^{2+} or DQ^{2+}) as organic struts. The ANS (ANS^{-1}) is 8-anilino-1-naphthalenesulfonate that exists within the MOF pores (Fig. 3a). The later ionic guest acts as receptors (electron deficient in nature) while ANS^{-1} ion (electron rich) acts as an indicator. The basic advantages obtained due to the presence of cationic part, MQ^{2+}/DQ^{2+} are: (i) cationic sites seek to retain the indicator by absorbing its energy, rendering it non-fluorescent, and bestowing the MOF as a guest species. Due to the charge transfer (CT) between the cationic receptor and the anionic indicator, the naturally fluorescent ANS imparts emission to the whole framework (host-guest interaction).

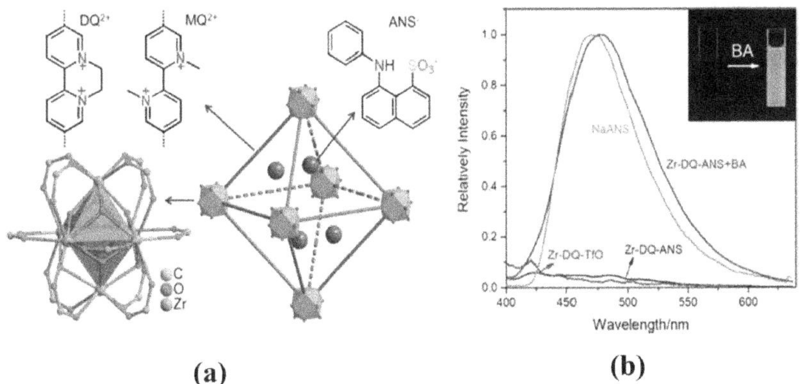

(a)   **(b)**

Figure 3. (a) $Zr_6O_4(OH)_4(COO)_{12}$ cluster with 2,2′-bipyridinium and indicator, ANS⁻.(b) The PL curve of Zr-DQ-ANS, Zr-DQ-ANS (solvent: dioxane) in presence of BA and NaANS. Inset shows enhanced emission of Zr-DQ-ANS by BA under 365 nm. Reproduced with permission from [36] American chemical society, Copyright 2019.

When Zr-DQ-ANS was treated with butyl amine (BA), the increment in PL response could be detected through naked eye with emission at 480 nm (Fig. 3b). In comparison to the primary amines, aromatic amine such as aniline and pyridine could not generate such strong turn on PL response owing to the superior electron-donating compe-tency of alkyl amines. The butyl amine disrupts the host-guest inter-action between MOF and ANS-1, causing in emission of ANS-1 with a peak shifting of 5 to 7 nm relative to free NaANS, and indicating that fluorescence originates from free ANS-1. Fluorescence experiment in unconfined homogeneous solution validated the confinement effect's contribution to the abrupt off-on switching characteristic of fluores-cence in MOF. Checking the PL intensity by adding $[Me_2DQdc]^{2+}$ (di-methyl ester of DQ) to ANS-1, a 0.5 ratio of $[Me_2DQdc]^{2+}$ to ANS quenches ANS-1 by only 50 percent. In contrast, in the same stoichi-ometric ratio between DQ^{2+} (Zr-DQ-ANS) and ANS, ANS-1 is quenched by 99 percent. (Fig. 4a).

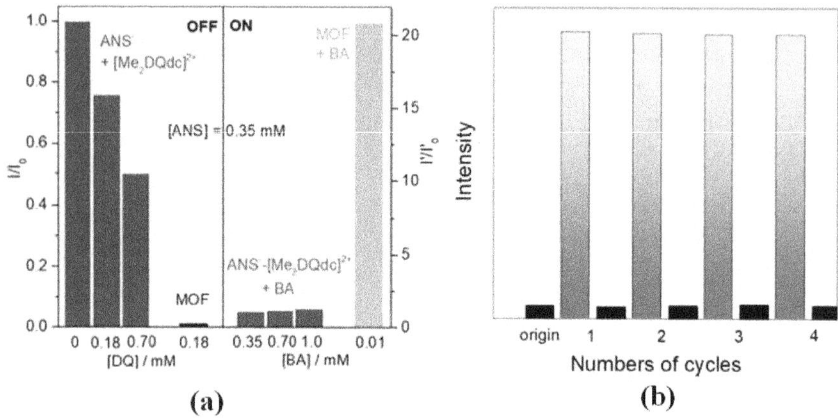

(a)   (b)

Figure 4. (a) Off state is the decrease of PL emission of ANS⁻ by $[Me_2DQdc]^{2+}$ and compares the intensity with Zr-DQ-ANS and the on state indicates the increase in PL intensity of ANS⁻ & $[Me_2DQdc]^{2+}$ solution in presence of butyl amine and compares with the emission response of Zr-DQ-ANS. (b) In pres-ence and absence of butyl amine, the recyclability of MOF with turn on and off state are shown in cyan and black color, respectively. Reproduced with permission from [36] American chemical society, Copyright 2019.

This phenomenon occurs because the solvation around both ions in unconfined homogeneous solution prevents the formation of a charge transfer complex between the ions in the former scenario, re-sulting in poor quenching efficiency. The confined environment pro-vided by MOF, on the other hand, prevents this type of solvation and results in the formation of a strong charge transfer complex with high quenching efficacy. The authors further demonstrated that this off-on switching behaviour may be four times reversed (Fig. 4b) without any alternation in emission properties. After every experiment, the MOF was extracted via centrifugation, which demonstrated perfect structural integrity as validated by PXRD analysis. The second MOF (Zr-MQ-ANS), like Zr-DQ-ANS, is sensitive to butyl amine, with detection limits of 0.43 and 0.47 nM, respectively. In addition, both MOFs demonstrated on-off switching behaviour in the solid state.

Very recently, the same group reported another Eu-based MOF, containing dual-interpenetrated networks built from $[Eu_4(\mu_3\text{-}OH)_4]$ cluster and a new zwitterionic linker tripyridinium-tricarboxylate [37]. The framework exhibits reversible fluoro-switching behaviour to electron rich volatile amine compounds with vivid colorimetric changes (Fig. 5a), owing to the presence of electron deficient pyridinium moiety that makes the MOF as an efficient host. In presence of smaller amine molecules, the color change takes place immediately while amines having larger molecular dimension, produce negative responses; which indicates that such chromic response is dependent on size and/or shape of guest molecules. Also, such colorimetric changes are completely reversible (Fig. 5b). The colored samples get immediately faded, returning to their original color, when exposed to air or washed with water and/or organic solvents. Quite in contrast, when the MOF sample is exposed to aniline, it turned into brilliant orange color but it does not readily revert back upon exposure to air. Decolourization occurs readily on washing with ethanol.

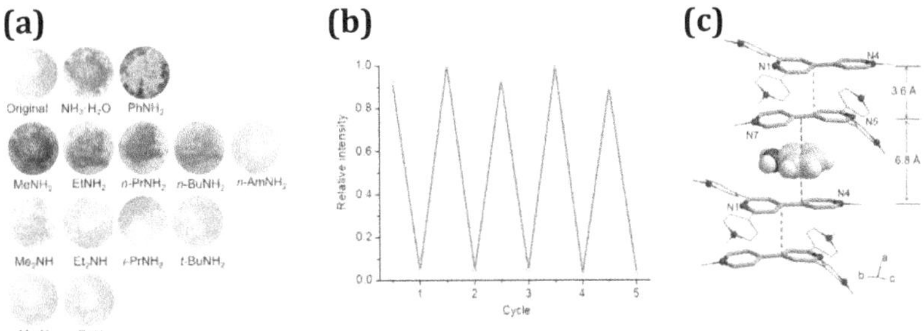

Figure 5. (a) Pictures of the pristine material before and after exposure to different amines. (b) Five cycles of fluorescence off–on switching. (c) Pictorial representation of the π–π stacking array after adsorption of aniline. Reproduced with permission from [37] American chemical society, Copyright 2020.

In validating the mechanism of such unique and reversible colorimetric change, the authors carried out UV-visible and EPR spectroscopic measurements. For smaller amine guests, electron transfer process occurs from electron rich amine to electron deficient pyridinium moiety, thus these amines-loaded samples showed strong signal in EPR spectrum. On the other hand, a charge-transfer π-π complexation takes place between aniline and MOF (Fig. 5c), which is a strong interaction, thus it is difficult for aniline to leave the framework readily. However, the complex gets destroyed upon solvation with ethanol, so the color gets removed on ethanol-washing. [38].

Zang *et. al.* reported a hydrogen-bonded, silver-chalcogenolate cluster-based MOF (SCC-MOF) by introducing phenylphosphonic acid as functional hydrogen-bond donor [39]. The MOF is able to distinguish among different chloromethanes such as chloroform ($CHCl_3$), dichloromethane (CH_2Cl_2) and carbon tetrachloride (CCl_4) by a simple luminescence spectral response and their vividly different luminescent color. Upon exposure to these different solvents, CH_2Cl_2 displayed no change in color i.e. the green color of pristine MOF is retained, CCl_4 showed a 15 nm red shift (yellow-green), whereas $CHCl_3$ displayed the highest red shift of 52 nm (yellow) (Fig. 6b and 6c). Surprisingly, such chemo-responsive luminescence of these three chloromethanes deviated from their trend of dielectric constant and solvent polarity order (CH_2Cl_2 > $CHCl_3$ > CCl_4), which are common criteria for evaluating the interaction of solvent–chromophores. Thus, to check the origin of this unusual solvatochromism, the authors further investigated CH_2Cl_2 and $CHCl_3$ induced structural dynamics of the MOF. Recording of PL spectra in presence of $CHCl_3$ vapor showed a slight increase in emission intensity up to $P/P_0 = 0.2$ and thereafter PL intensity sharply increases up to $P/P_0 = 0.7$ before attaining a stable value (Fig. 6d). Moreover, the observed PL variation closely follows the adsorption progress pattern (Fig. 6a), inferring that the chloroform interaction with host opened the pores and locked them in a specific site, enabling progressive improvement of host-guest interactions, which vividly modulated the emission till adsorption saturation. On the other hand, the framework emission peak wavelength was unaffected by CH_2Cl_2 vapour pressure.

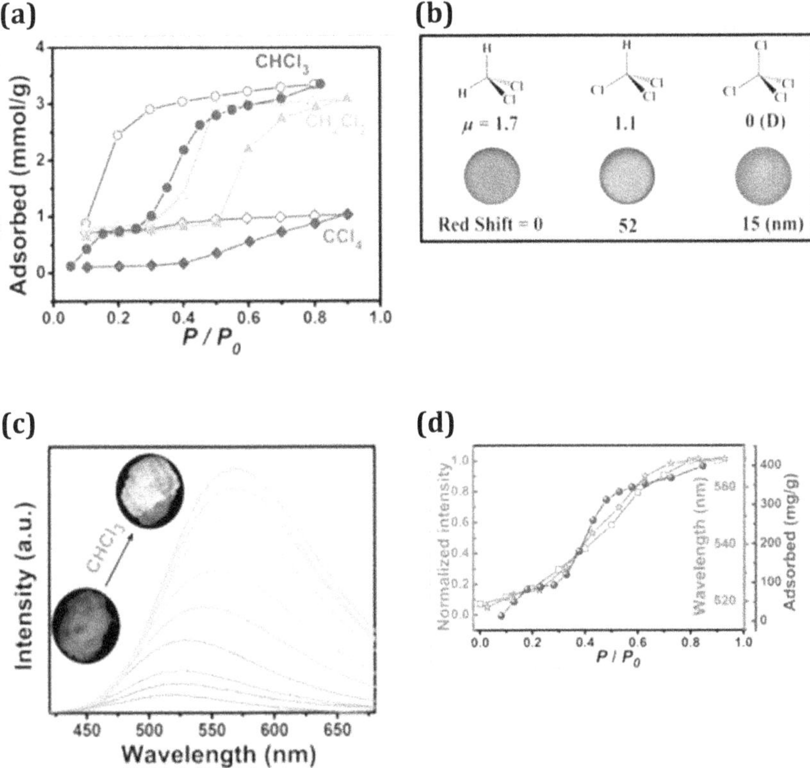

Figure 6. (a) The gate-opening adsorption of chloroform, dichloromethane and carbon tetrachloride for **Ag$_{10}$bpy** at room temperature. (b) The structures of CHCl$_3$, CH$_2$Cl$_2$ and CCl$_4$ with their respective luminescent responsive images (excited at 365 nm). (c) PL spectra of **Ag$_{10}$bpy** at various vapor pressures (P/P_0 = 0–1) of CHCl$_3$ at room temperature and luminescent images of single crystals. (d) The maximum emission peak (red) and normalized PL intensity (green) and adsorption plot (blue) as a function of CHCl$_3$ vapor pressure. Reproduced with permission from [39] American chemical society, Copyright 2018.

Fluorescent MOFs are widely used to detect explosive analytes, this was initially described by Li *et. al.* in 2009 using vapour phase sensing of nitro-explosive chemicals such as 2,3-dimethyl-2,3-dinitrobutane (DMDNB) and 2,4-dinitrotoluene (DNT)[40]. Subsequently, they reported the synthesis of a novel MOF, [Zn$_2$(oba)$_2$(bpy)]$_3$DMA (where, H$_2$oba = 4,4′-oxybis(benzoic acid); bpy = 4,4′-bipyridine; DMA = N,N′-dimethylacetamide) [41]. The framework is comprised of one-dimensional channel and shows luminescence signature at 420 nm upon exciting at 280 nm. Interestingly, upon exposure to electron deficient nitro-aromatic compounds, the MOF revealed a decrease in its emission intensity but on the other hand, when exposed to electronically rich aromatic compounds, an increase in luminescence intensity is observed. (Fig. 7a and 7c).

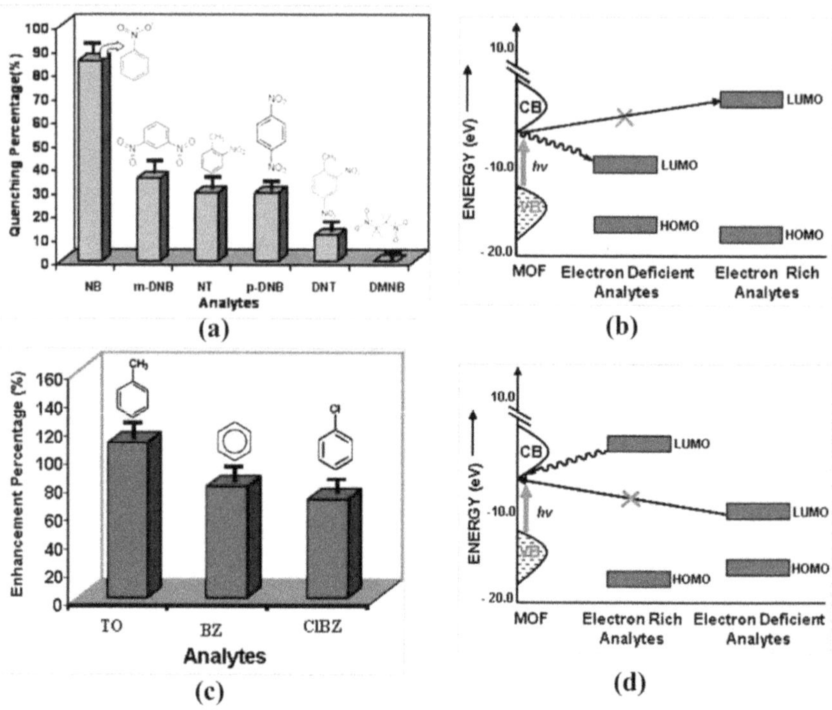

Figure 7. (a) The percentage of MOF quenching in the presence of electron withdrawing analytes is shown by the bar graph. (b) Quenching mechanism for electron removing analytes. (c) Fluorescence improvement of MOF (in %) in the existence of electron-rich analogues and the turn-on mechanism by them (d) the fluorescence enhancement process by aromatic analyte having an electron-donating functional group. Reproduced with permission from [41] American chemical society, Copyright 2011.

The mechanism of such analyte-induced fluorescence switching is attributed to the photo-induced electron transfer (PET) process between the LUMO and HOMO of donor and acceptor, respectively. When the electron is transferred from the LUMO of MOF to the LUMO of electron deficient analyte, decrease in emission intensity occurs, leading to fluorescence quenching (Fig. 7b). On the other hand, when the reverse phenomenon i.e. electron transfer occurs from electron rich analyte to the LUMO of MOF, increase in emission intensity takes place which in turn leads to luminescence enhancement (Fig. 7d).

In 2013, Ghosh *et. al.* reported a novel 3D porous luminescent MOF [Cd(NDC)$_{0.5}$(PCA)]·G$_x$ (G = guest molecules) which could selectively detect highly explosive 2,4,6-trinitrophenol (TNP) in a reversible fashion through quenching of emission intensity [42]. In 2019, Neogi *et. al.* reported the solvothermal synthesis of a cationic Ni(II)-MOF **CSMCRI-3** (Fig. 8b) which also was able to selectively recognize TNP among other nitro-aromatics in aqueous medium [43]. The emission intensity at 395 nm ($\lambda_{\text{excitation}}$=314 nm) gradually de- creases upon incremental addition of aqueous TNP solution and 99.1
% quenching takes place upon addition of 0.1 (M) TNP. The Stern-Volmer constant for quenching was calculated to be 1.3×10^5 M^{-1} (Fig. 8d), with very low LOD (0.25 ppm). Further, DFT calculation

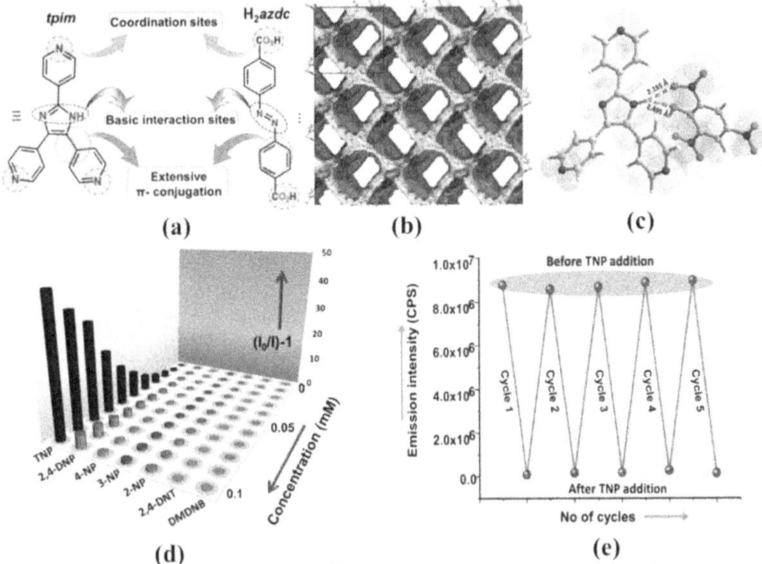

Figure 8. (a) Schematic representation of different struts and functionality in MOF. (b) Porous channel of the framework (crystallographic *b* axis), (c) H-bonding interaction of ligand and TNP. (d) The superior turn-off efficacy of TNP with regard to rest of the nitro analytes, and (e) switchable off-on luminescence of MOF upto five cycles. Reproduced with permission from [43] American chemical society, Copyright 2019.

showed that photo-induced electrons can easily transfer from the LUMO of MOF to the LUMO of TNP, where the nitrogen atom of the imidazole moiety (Fig. 8a) inside the pore surface forms effective H-bonding interaction with TNP (Fig. 8c), resulting in high quenching. The mechanistic scenario is additionally corroborated through resonance energy transfer (RET) between activated MOF and TNP. The turn off-on detection of TNP was found to be recyclable up to five times with trivial change in MOF PL intensity (Fig. 8e).

The same group published a dual functionalized Zn(II)–framework $[Zn_3(bpg)_{1.5}(azdc)_3]\cdot(DMF)_{5.9}\cdot(H_2O)_{1.05}$ (**CSMCRI-1**) [44] with an OH group-integrated linker and –N=N– moiety possessing ligand (Fig. 9a), which demonstrates ultrasensitive and recyclable fluorescence quenching by TNP and nitrofurazone (NZF) (Fig. 9b and 9c). Exquisitely, the MOF demonstrated the turn-on sensing of poisonous and electron-rich aromatic 4-aminophenol in the simultaneous presence of isomeric counterparts. DFT simulations give further evidence of supramolecular host-guest interactions, validating the significance of pillar functionalization in such "turn-off" or "turn-on" sensing of electrically divergent hazardous organics.

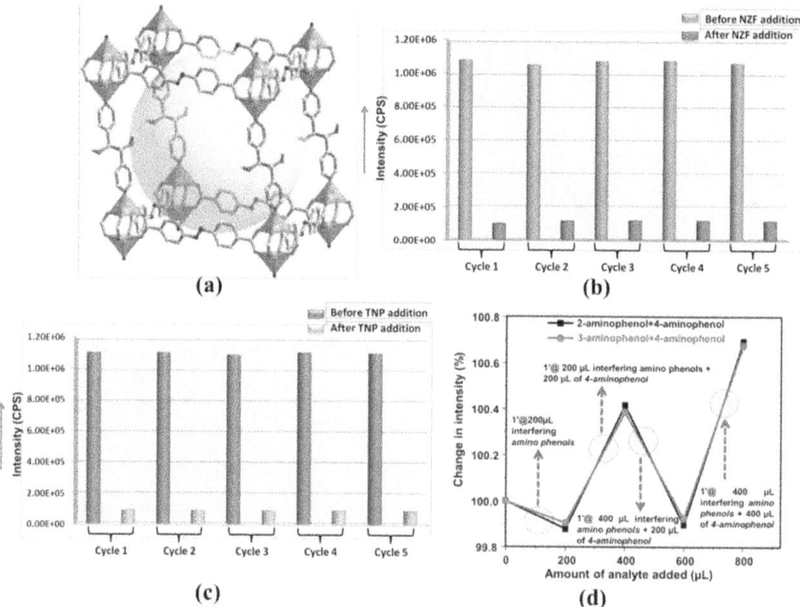

Figure 9. (a) Box shaped structure with large pores in the pillar-layer framework of **CSMCRI-1** (b) Recyclability of **1** towards 1 mM NZF and (c) 1 mM TNP solution for a maximum of 5 cycles. (d) "turn-on" behaviour of 4-aminophenol in the presence of other congeners. Reproduced with permission from
[44] American chemical society, Copyright 2019.

Recently Neogi *et. al.* further reported the synthesis of a Li(I)–ion containing anionic framework (**CSMCRI-4**) [45] for luminescent sensing of dichloran and NZF (Fig. 10a and 10b), where astute involvement of highly π-conjugated tri-carboxylate ligand instigates high emissive nature to the non-interpenetrated structure, and assists creating unbound oxygen atom decorated one-dimensional (1D) porous channels (Fig. 10e). The remarkable quenching constant and low limit of detection renders the MOF as one of the best sensory probes for two aforementioned nitro-organics. To the best of sensory system, reversibility of host-guest interplay between the framework and respective studied analyte initiates rarest fluorescence "off–on–off" switching that is certified by colorimetric variation as well as titration experiment in solution phase, and visibly demonstrated by paper-strip method, benefitting real-time monitoring of these detrimental nitro-organics (Fig. 10c and 10d).

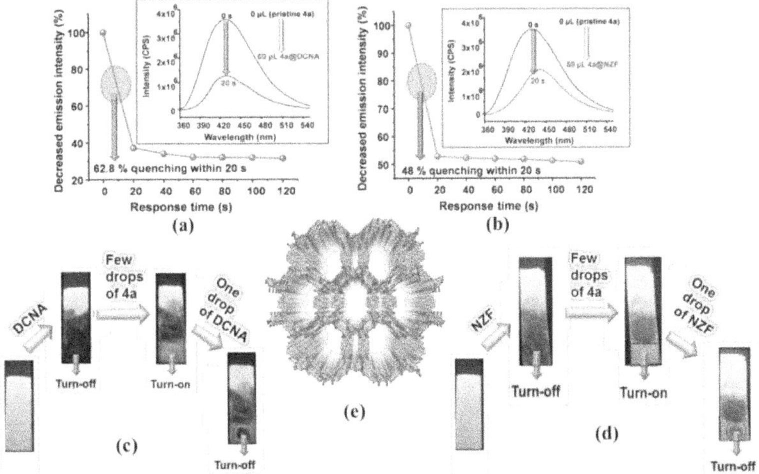

Figure 10. (a) Plot illustrating the Percent reduction in emission intensity at various time intervals for DCNA. The inset shows the fluorescence spectra of 4a before to (0 s) and after (20 s) the addition of 60 µL DCNA. (b) A plot of the % reduction in emission intensity over time for NZF. Fluorescence spectra of 4a before (0 s) and after (20 s) the addition of 60 µL NZF are shown in the inset. Pictures of **4a** coated paper strips, showing recurrent "off–on–off" fluoroswitching in the presence of (c) DCNA and (d) NZF. e) Perspective view of one-dimensional channel along *c* axis. Reproduced with permission from [45] Royal Society of Chemistry, Copyright 2021.

Very recently, the same group reported two isostructural diamondoid MOFs **CSMCRI-7** and **CSMCRI-8**, built from azolate struts and

Cd(II) as nodes [46]. The presence of extra N-atom in the triazole ring of the latter, makes it a pore-functionalized framework. Both the MOFs show excellent luminescent signature in DMF dispersion. Interestingly, the MOF displayed unprecedented fluorescence turn-on detection of 3-(diethylamino)phenol (DEP) (Fig. 11a) with 13.4 fold enhancement of emission intensity for activated **CSMCRI-7 (7a)** while 17.7-fold for **8a** (Fig. 11d). The pore-functionalization in the latter showed the added advantage in sensing ability through additional host-guest interactions. In addition to this DEP sensing, the MOFs also show pesticide triggered fluoroswitching i.e. turn on detection for isoproturon (IPT) (Fig. 11b) while turn-off detection for dichloran (DCNA) (Fig. 11c). IPT sensing takes place with 1.5-fold intensity enhancement while 99.6 % quenching transpires in case of DCNA using **8a** as fluorescent probe (Fig. 11e). The luminescence enhancement constant (K_{EC}) value for IPT detection was calculated as 1.82×10^3 M^{-1} with LOD of 110 ppb. The sensing mechanism is attributed to the PET and RET mechanism.

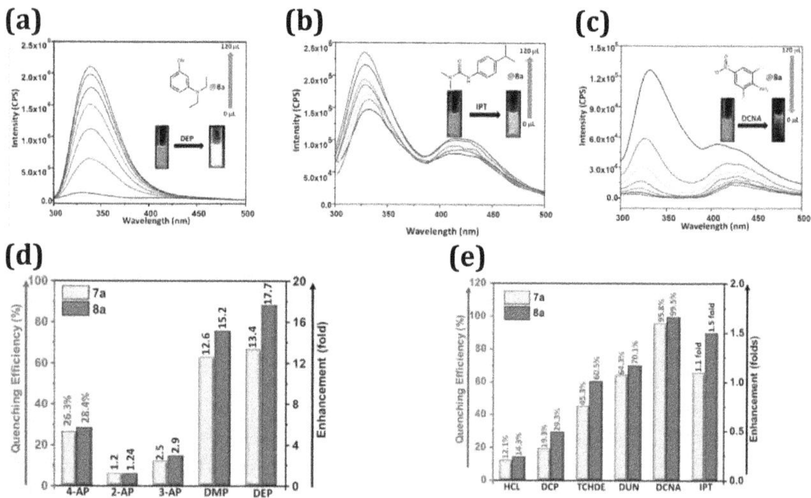

Figure 11. (a) The luminescence spectrum of **8a** with the addition of 1 mM DEP solution (the inset shows the colorimetric changes of the MOF dispersion before and after the addition of DEP under 365 nm UV light). The luminescence spectrum of **8a** after the addition of 5 mM solutions of (b) IPT and (c) dichloran (DCNA). (d) Quenching/enhancement efficiencies of **7a** and **8a** (1 mg/2 mL DMF) for various aminophenols. (e) Fluorescence quenching/enhancement efficiencies of **7a** and **8a** (1 mg/2 mL DMF) for several pesticides. Reproduced with permission from [46] Royal Society of Chemistry, Copyright 2021.

Another highly chemo-robust, contrasting pore-functionalized pillar-layer framework [Zn$_2$(BTTB)(dpta)]·2DEF·1.5EtOH·1.5H$_2$O (**CSMCRI-6**) (Fig. 12a), has been reported from the same group [47]. The two-fold interpenetrated MOF represents one-of-a-kind tetra-sensory platform for highly specific, recyclable and ultra-fast detection of four classes of organo-toxins with record-high quenching in water. The nanomolar limits of detection for 2,4,6-trinitrophenol (67 nM), dichloran (91 nM), aniline (6 nM) and nicotine (51 nM) stand among the lowest values reported so far.

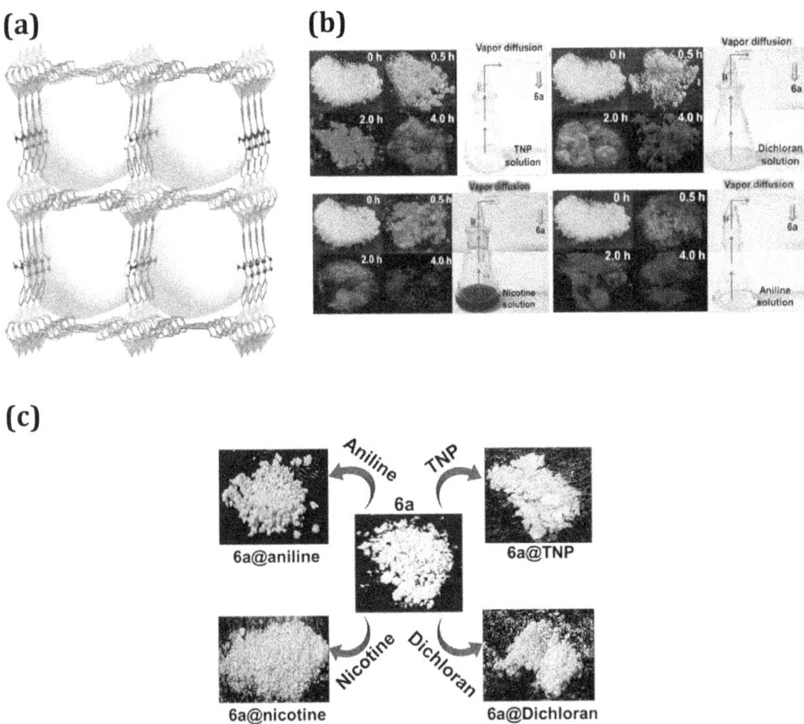

Figure 12. (a) The large voids of **CSMCRI-6** along crystallographic *c* axis. (b) Digital photographs (0 to 4 h) of pristine and analyte vapor incorporated framework under the UV light with fixed time interval (left side); prototypical experimental setup for vapor phase detection of four diverse classes of contaminants (right side). (c) Solid state colorimetric variation of analyte incorporated framework after dipping in 1 mM aqueous solution of individual analyte solutions. Reproduced with permission from [47] American chemical society, Copyright 2021.

Apart from liquid-phase fluorosensing, vapor-phase plus solid-state luminescent recognition of these harmful organo-aromatics, involving prominent colorimetric changes (Fig. 12b and 12c) justifies this MOF as hitherto unreported triphasic sensory material for a wide variety of organo-toxins. The MOF was further utilized towards real-time applicability for acute monitoring of four classes of toxic organics by simple and handy paper-strip based repetitive detection (Fig. 13b). In addition, visible colorimetric detection of dichloran pesticide over food sample authenticates potential of the framework towards in-field sensory application (Fig. 13a). Photo-induced electron transfer for sensing was supported by comprehensive DFT validation involving the 3D framework, which clearly portrays changes in the energy levels of MOF in presence of toxic aromatics. In addition, geometry optimizations of the polymeric network were performed with structurally diverse organo-toxins; which validated the supramolecular interactions with the framework.

(a)

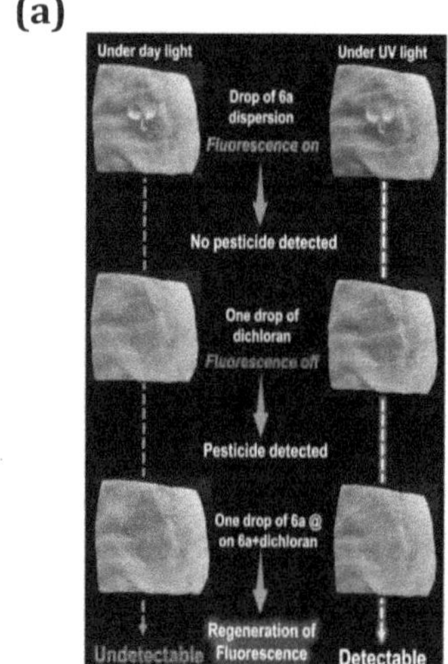

(b)

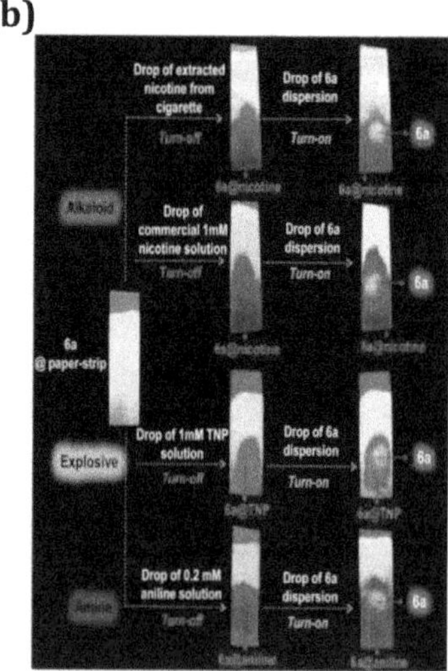

Figure 13. (a) visible recognition of ultra-low concentration of dichloran on cabbage leaf containing few drops of pristine MOF solution along with regeneration of original emission intensity of the framework under UV light. **(b)** Demonstration of real-time recyclable detection of alkaloid (commercial nicotine and extracted from cigarette), explosive (TNP), amine (aniline) on **6a** aqueous solution coated handy paper strip via "off-on" fluoroswitching manner. Reproduced with permission from [47] American chemical society, Copyright 2021.

2.2 Reversible Fluoro-sensing of Various Ions

As of volatile organic compounds and nitro-aromatic explosives, recognition of charged analytes has gained tremendous importance because these species play critical role in natural systems. However, some of the ions are extremely poisonous to human health [48]. In this milieu, MOFs have been progressively regarded as a suitable host matrix for detection of these ionic congeners owing to formation of efficient host-guest interaction and fine tuning of electronic properties. Recently, Yoon *et. al.* demonstrated the synthesis of a highly stable, water-dispersible, nano-sized Ti-based MOF (NH$_2$-MIL-125), which selectively recognizes toxic, heavy metal ion Pb^{2+} in an ultra-sensitive manner [49]. Due to the presence of highly hydrophilic functional group (-NH$_2$), the MOF showed unusual reversible fluorescence attributes. The framework exhibited an emission peak at 435 nm upon 335 nm excitation and proved to be an appropriate fluorosensor by virtue of the ligand-to-cluster charge transfer (LCCT) mechanism. This was due to its good water-dispersibility, outstanding stability, and high quantum yield. The luminescent screening was performed with a number of other metal ions, which resulted in the selective suppression of the emission peak with the addition of just 11 nM of Pb^{2+} ion to the aqueous dispersion of MOF. Interestingly, when EDTA is added to the mixed solution, the MOF luminescence intensity is regenerated from turn-off to turn-on. The significant chelating capacity of EDTA, which removes Pb^{2+} from the MOF cavity, is credited with the recovery of fluorescence intensity. Furthermore, the fluorescence on-off-on cycle may be reversed up to five times without affecting the MOF sensing capabilities. From a molecular standpoint, the author claimed that high quenching aptitude arises because of energy transfer from the MOF to the Pb^{2+}ions, wherein the presence of free –NH$_2$ moiety benefits LCCT with the Pb^{2+} ion as well. The detection limit is discovered to be as low as 7.7 ppm since the permissible concentrations of Pb^{2+} in tap water and river water are 152 ppm and 186, respectively. Because of its exceptional sensing capabilities, this MOF may be utilized for data encryption and decryption.

Lin *et. al.* reported the synthesis of a dual-emissive luminescent MOF as a ratiometric sensor for toxic hypochlorite ion via fluorescent on-off-on switching process [50]. The probe was synthesized by a two-step process; of which the first step is the solvothermal synthesis of BPyDC@MOF-253-NH$_2$ and the latter step involves grafting of Eu^{3+}

ion within the framework through post-synthetic modification. The 2,2'-bipyridyl units around the pore wall act as the anchoring site for Eu^{3+}. (Fig. 14a).

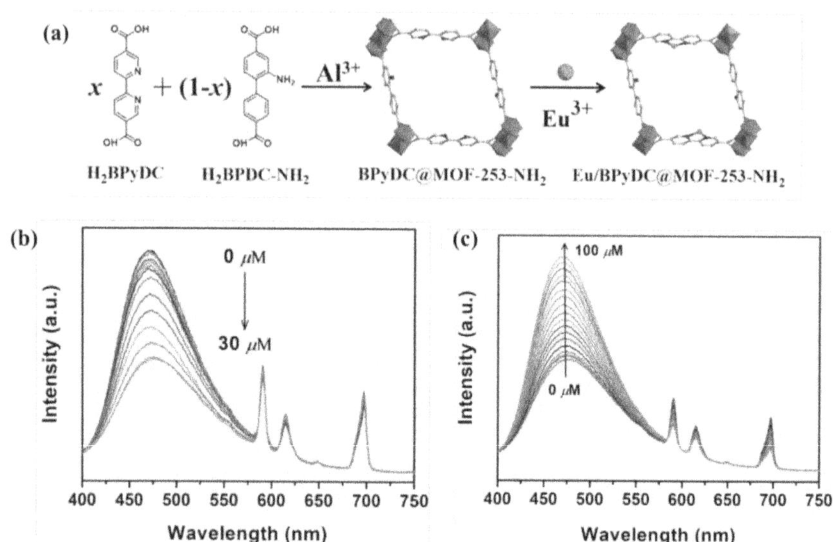

Figure 14. (a) Schematic illustration for MOF synthesis. (b) The reduction of PL response upon regular addition of ClO⁻ (c) the recovery of PL intensity upon ascorbic acid addition. Reproduced with permission from [50] American chemical society, Copyright 2019.

Upon excitation at 305 nm, the post-modified MOF displayed four emission peaks. The peak at 471 nm originates from blue emission of H_2N-carboxylate ligand. Rest of the peaks are characteristics of Eu^{3+} emission i.e. 698 nm ($^5D_0{\rightarrow}^7F_4$), 614 nm ($^5D_0{\rightarrow}^7F_2$) and 591 nm ($^5D_0{\rightarrow}^7F_1$). During fluorescence titration, the blue emission serves as the detection signal whereas the red emission acts as reference signal. With incremental addition of hypochlorite solution, the intensity of the blue emission at 471 nm steadily decreases without any change to red emission peak at 614 nm (Fig. 14b). Contrastingly, the intensity ratio of both the emission peak remains constant for a number of other ions including Br^-, SO_4^{2-}, Cl^-, NO_3^-, IO_3^-, NH_4^+, Na^+, K^+, Mg^{2+} and Ca^{2+}, and signifies specific sensing of hypochlorite anion in ratiometric fashion. The LOD was found to be 0.094 μM. Quite amusingly, addition of 100 μM of ascorbic acid to the solution immediately regenerated pristine blue emission of original MOF (Fig. 14c). This phenomenon is attributed to the redox neutralization reaction between hypochlorite and ascorbic acid which revived the original blue emission of the MOF. The limit of detection for ascorbic acid was found to be 0.73 µM. The authors further investigated that hydrogen bonding interaction between the $-NH_2$ group of the ligand and hypochlorite ion (N–H··· OCl⁻) (Fig. 15) facilitates the charge transfer between ClO⁻ and the MOF, resulting in such excellent fluoswitchable behaviour of the framework.

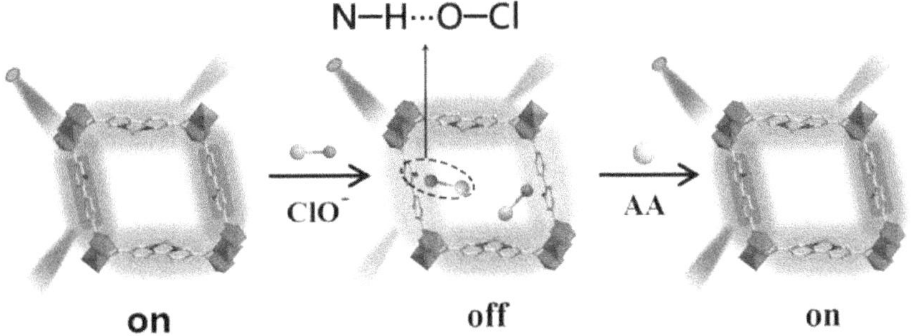

Figure 15. Cartoon diagram for probable mechanistic route to switchable on-off-on sensing. Reproduced with permission from [50] American chemical society, Copyright 2019.

Very recently, Li *et. al.* reported the solvothermal synthesis of a new water stable luminescent MOF In(tcpp) i.e. LMOF-651, which revealed a strong luminescence signature at 400 nm upon exciting at 280 nm [51]. The MOF was able to detect both fluoride (F⁻) ions and perfluorooctanoic acid (PFOA) in aqueous phase. Interestingly, F⁻ detection occurs through turn-on fashion whereas PFOA could be recognized by the probe in a turn-off manner. The Stern-Volmer constant (K_{SV}) for F⁻ sensing was found to be 9.06×10^3 M⁻¹ with limit of detection as 1.3 µg L⁻¹. On the other hand, K_{SV} value for PFOA detection was came out as 1.446×10^4 M⁻¹ with detection limit of 19 µg L⁻¹. Given, the World Health Organization (WHO) recommended limit for optimal level of fluoride in drinking water is 1.5 mg L⁻¹, the LOD value for F⁻ sensing achieved by LMOF-651, has exceeded the WHO limit by three orders of magnitude. The detection process was recyclable up to five times with almost no change in emission intensity. Along with ultrasensitive recognition of F⁻ and PFOA, the MOF also acts as an efficient adsorbent for these two analytes. The adsorption capacity for F⁻ was 36.7 mg g⁻¹ whereas the same is 980 mg g⁻¹ for

PFOA, demonstrating its strong potential for drinking water purification. IR spectroscopic analysis and DFT calculations were carried out to explore the mechanism of F⁻ and PFOA detection and adsorption.

2.3 Reversible Sensing of Bio-markers and Bio-molecules

Biomarkers are the biological indicators that are used for the identification of specific pathological or physiological processes, or pharmacological responses to therapeutic intervention, disease and so on [52]. In recent years, biomarkers (such as antibiotic, bio molecules, amino acids, proteins etc.) have become an important part of medical science as they are used for monitoring the patients with an unusual condition and understanding the cause, help in early diagnosis, disease progress, and in curing disease [53]. Therefore, there is a clinical exigency to design and develop exceptionally sensitive biological sensors to comprehend or track the biological responses in human body and could pave the way for the new advanced medical facilities.

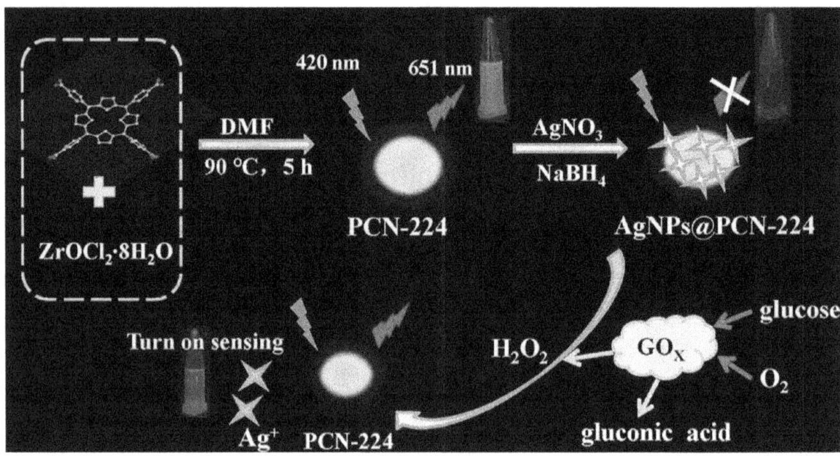

Figure 16. Synthetic scheme of AgNPs@PCN-224 and an overview of switchable fluorescence process. Reproduced with permission from [54] American chemical society, Copyright 2020.

In 2019, Lu *et. al.* designed porphyrin-based MOF (PCN-224) with the support of silver nanoparticles for the sensitive and selective detection of glucose (Fig. 16) [54]. Glucose serves as a primary fuel for energy production in the living cell and plays a vital role to stimulate both aerobic and anaerobic cellular respiration. Though PCN-224 is a fluorescent MOF combining it with Ag NPs, enhanced its detection sensitivity and specificity for glucose as switching of fluorescent signal from the "off" to "on" state takes place. Moreover, the fluorescence intensity of the composite material increases (switching from "off" to "on"), when treating with the buffer solution of glucose oxide in presence of oxygen (Fig. 17a). The mechanistic path for such enhancement is the oxidation of glucose via O_2 that results in H_2O_2 production which gives rise to the leaching of Ag^+ with switching fluorescence turn off to turn on state. Fluorosensing of glucose occurs through highly specific and sensitive manner with an admirable 0.078 µM LOD (Fig. 17b).

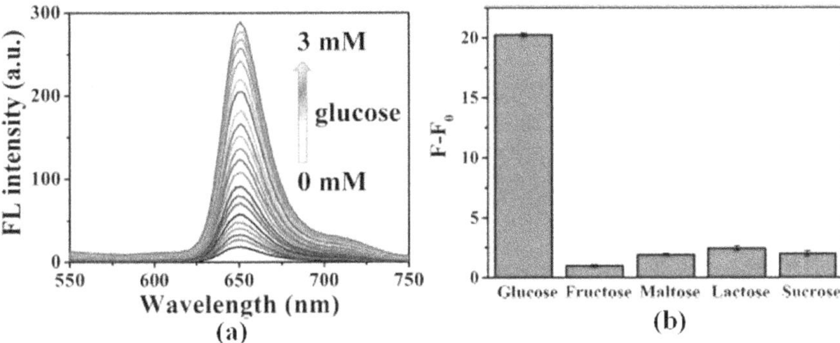

Figure 17. (a) Fluorescence intensity increase upon adding glucose to the MOF dispersion. (b) Selectivity glucose detection amid numerous other saccharides. Reproduced with permission from [54] American chemical society, Copyright 2020.

Cheng *et. al.* developed a water-stable Europium based-MOF which exhibits high sensitivity and selectivity towards 5-hydroxytryptamine (HT) and 5-hydroxy indole-3-acetic acid (HIAA), which are potential biomarkers for carcinoid tumor [55]. Eu-MOF on excitation at 280 nm displayed five finely resolved emission peaks (697, 652, 616, 592 and 579 nm) due to the effective transition among the different electronic states of Eu^{3+} ions which corresponds to $^5D_0{\rightarrow}^7F_J$, where J = 0, 1, 2, 3, and 4 respectively. Nevertheless, the electronic transition from J=0 to 2 plays a key role in contributing red emission at 616 nm of Eu-MOF which is visible by naked eye. The Eu-MOF also shows outstanding sensitivity with a quick response within 1 min, excellent reusability, and estimated detection limits are 0.66×10^{-6} M and 0.54×10^{-6} M for HT and HIAA, respectively. Furthermore, the sensing functionality shows higher selectivity in the co-existence of other parallel species in blood plasma and urine. Moreover, the switching of fluorescence intensity takes place via ultrasonic centrifuged recovered MOF materials illustrating its recyclability.

Neogi *et. al.* reported a zinc based luminescent MOF (**CSMCRI-2**) (Fig. 18a), for highly sensitive detection of adenosine monophosphate (AMP) via fluorescent enhancement [56]. Fluorescence studies of **CSMCRI-2** exhibits ultrafast detection of AMP with 9 ppb as detection limit. In addition, this LMOF further shows fluoro-switchable detection for electron rich antibiotic sulfadiazine (SDZ) through fluorescence enhancement, while electron deficient antibiotic nitrofurazone (NZF) through luminescence quenching (Fig. 18b). Inspired by this on-off-on fluorescent switching, authors designed an electronic logic circuit as a technical security guard (Fig. 18d). Moreover, DFT calculations were performed and studied to elucidate the mechanism for the turn on phenomenon. The results reveal that the LUMO of the electron rich analyte resides above the LUMO of the linker, resulting in transfer of electron from analyte to the MOF. Such an operation leads to increase emission intensity of **CSMCRI-2**. In addition, there also exists strong H-bonding interaction between pendent $-NH_2$ group of pillaring linker and AMP molecules (Fig. 18a) that instigate facile PET effect.

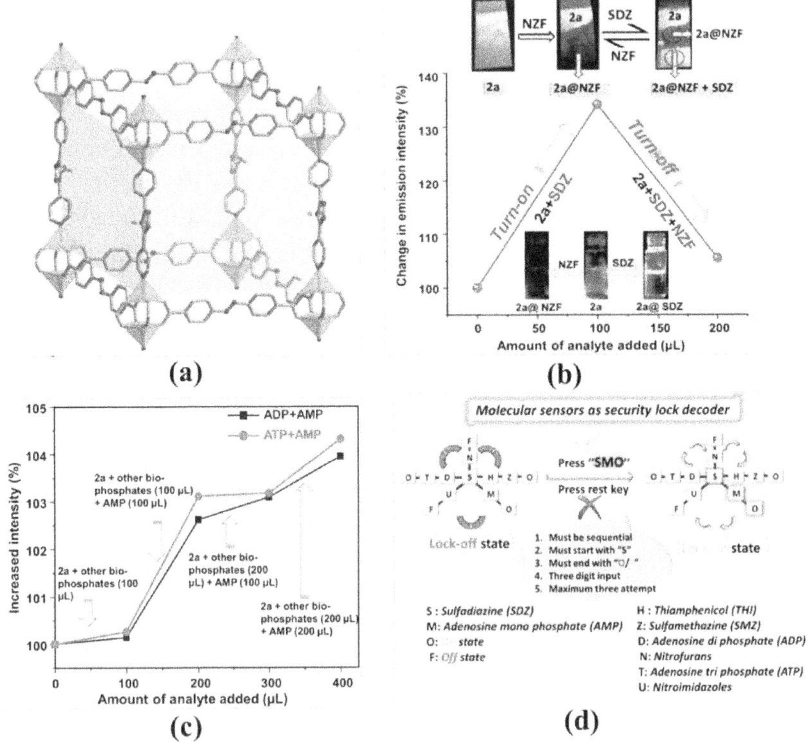

Figure 18. (a) The pillared layer 3D framework of **CSMCRI-2**. (b) The analyte-triggered activation and inactivation responses of **2a** in the presence of 100 µL SDZ and NZF, respectively. Photographs of a dispersed solution of **2a** taken under UV light (365 nm) with SDZ and NZF present (inset below). Pictures of MOF coated paper strip demonstrating the recurrent "off-on-off" switching of fluorescence in the presence of selective antibiotics (inset above). (c) CAT plot for various biophosphates in the presence of AMP, illustrating the anti-interference action of AMP. (d) Illustration of molecular sensors functioning as a security-lock decoder. Reproduced with permission from [56] Royal Society of Chemistry, Copyright 2019.

Wang *et. al.* designed a water-stable zirconium-MOF (UiO-68) containing anthracene and maleimide moiety, which exhibited switching of the fluorescent intensity from turn on/off for the potential detection of amino acids and bio-thiols [57]. Unlike the usual "fluorophore–spacer–receptor" format, here pseudo-intramolecular PET employed between donor anthracene and acceptor maleimide group due to which UiO-68-An/Ma shows very feeble fluorescence. Interestingly, proficient exclusion of the pseudo intramolecular fluorescent PET process through D-A reaction between maleimide and 3-furanmethanol within in UiO-68-An/Ma can turn on the fluorescence. Moreover, UiO-68-An/Ma exhibits very high sensitivity towards biothiols at very low concentration 50 µmol/L within 5 minutes in solid state experiments. Inspired by these results with biothiols, the flouroswitchable experiments were also performed with amino acids, and it was observed that MOF become highly selective in existence of glutathione (GSH), homocysteine (Hcy), and cysteine (Cys) even at a low concentration of 50 µmol/L as compared to other amino acids.

In 2017, Li *et. al.* designed a cadmium-based MOF that can serve as an "off-on" fluorescent switch for the highly selective detection of dopamine (neuromodulatory molecule) with a very low detection limit of 57 nM [58]. Similarly, Liu *et. al.* synthesized cadmium-based MOF for sensitive detection of ascorbic acid through the "on–off–on" method [59]. Initially, fluorescent MOF converted into non-fluorescent MOF in presence of CrO_4^{2-} i.e. CrO_4^{2-}@Cd-MOFs by post-synthetic modification, whose fluorescent intensity can be regenerated with the detection of the ascorbic acid and shows turn-on luminescence.

2.4 Acid-base Reversible Fluoroswitching

pH value is a major parameter in reflecting the health status, water quality and owing to its immense importance to environmental, biological, and reaction processes, the sensitive ratiometric detection of pH or acidic/basic group is an ongoing and essential need. Farha *et. al.* synthesized a post-synthetically modified MOF, NU-1000 that displays adjustable color changes within the pH range of 2 to 9. Moreover, CNF functionalized MOF catalytically detoxifies the nerve-agent simulant DMNP and also identifies acidic by-products [60].

Zhao *et. al.* designed a pH stable lanthanide-based MOF **V104**, which is the first self-calibrated ratiometric pH sensor that exhibits dual emission peaks at 390 and 480 nm and these peaks reversibly detect acid (H^+) and base (OH^-) among various ions respectively [61].

Yong *et. al.* synthesized the porous three dimensional Zn(II) MOF (USTC-1) which exhibited encapsulation of Methyl Orange (MO) and Rhodamine B (RB) dye [62]. In addition, the dye@USTC-1 showed switchable fluorescent aptitude upon contact to acid (HCl) and base

(NH₃) vapor. Thorough study divulged that lower uptake of N_2 and CO_2 by pristine MOF is due to vibration of amide chains at linker that blocks the pore of MOF which then opened by dye molecules. The RB dye encapsulated MOF showed two different emission peaks of USTC-1 (at 433 nm) and Rhodamine B (at 602 nm) when excited at 365 nm. Moreover, USTC-1 and dye@USTC-1 MOF phosphorescent naturewas confirmed from their microseconds lifetimes i.e. 27.2 µs and 29.0 µs respectively. When MO@USTC-1 and RB@USTC-1 were exposed to HCl vapor in a closed container for 30 min, the colorless MO@USTC-1 turns into weak pink color whereas pink color of RB@USTC-1 change to pale pink. Besides, the blue phosphorescenceof MO@USTC-1 and bright pink phosphorescence of RB@USTC-1changed into dim-blue and dim-pink, respectively. In addition, the in- tense phosphorescence is regenerated from their respective pale col-our when the probe are held within NH₃ vapour in a closed containerfor 30 minutes. Protonation of RB@USTC-1 in HCl vapour results in an increase in the emission peak at 433 nm, a decrease in intensity, and a shift of the emission peak from 602 nm to 588 nm. In the case of MO@USTC-1, the similar blue peak shift (emission at578 nm from602 nm) is seen. The deprotonation of the corresponding frameworkin the presence of NH₃ vapour allows the original colour to regener- ate. Importantly, such acid-base vapour phase activity had no effect on individual dyes or the framework USTC-1 in the solid form. As a result, switchable pH sensing behaviour is solely due to the protona-tion and deprotonation activity of well-dispersed encapsulated dye inside the framework cavity.

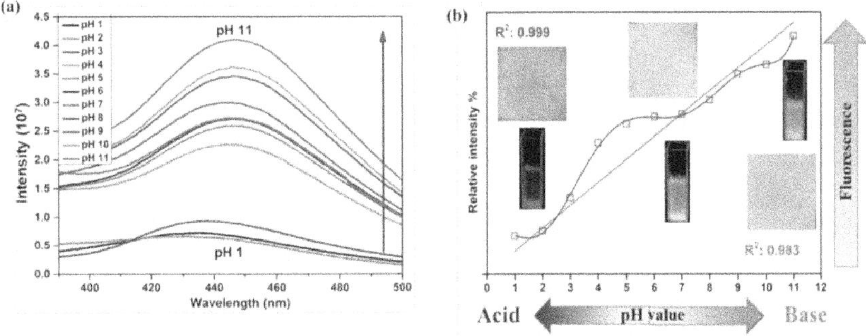

Figure 19. (a) Variation in Zn-MOF fluorescence intensity with varying pH 1 to 11. (b) Linear Fitting and polynomial between fluorescence intensity and pH range; (inset: color of Zn-MOF under UV light). Reproduced with per-mission from [63] American chemical society, Copyright 2020.

In 2020, Mandal *et. al.* reported the zinc based autofluorescent MOF for pH sensing ranging from (1 to 11) and which shows emission at 443 nm (Fig. 19a) [63]. Fluorescent studies revealed that highest and lowest fluorescent intensity was observed at pH=11 and pH=1, respectively (Fig. 19b). In addition, the change in fluorescent intensity was further studied under UV light for naked eye detection. In acidic and basic media, the variation of luminescence intensity caused by protonation and deprotonation of the secondary amine that enables the electron transfer within MOF. In fact, in highly acidic medium the protonation of the secondary amine stops the electron transfer between ligand and metal ions which in turn leads to quenching of the emission intensity. It was observed that the MOF was recyclable as pH sensor up to three times with insignificant change in its activity and negligible deviation in its structural integrity.

(a)   **(b)**

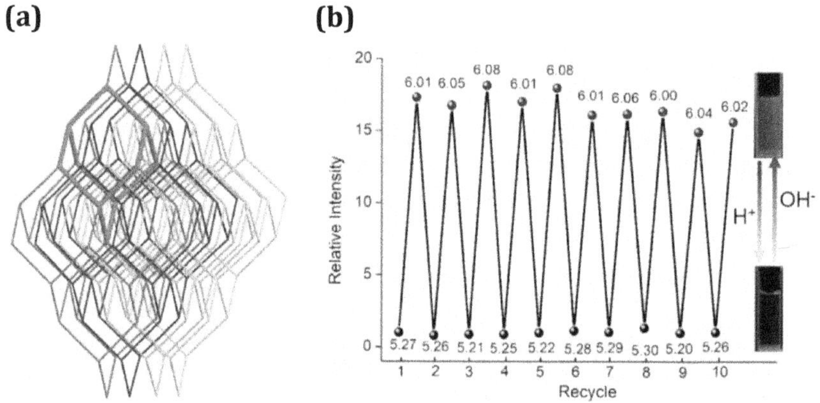

Figure 20. (a) Topological representation of the 4-fold interpenetrated structure of the MOF. (b) Ten cycles of fluorescence on–off switching. The figures refer to the pH. Reproduced with permission from [64] American chemical society, Copyright 2020.

In 2020, Gao *et. al.* reported a three-dimensional Zn(II) based MOF derived from a bipyridyl-tetracarboxylic ligand and composed of four-fold interpenetrated diamond type framework (Fig. 20a) [64]. Because of the steric hindrance around the pyridyl group, the uncoordinated basic N-sites are present for better interaction with H+ ions and show pH-responsivity. Due to the protonation of the uncoordinated N-sites, the aqueous dispersion of MOF shows reversible on-off fluorescence transition and can be used as pH-triggered fluorescent switch (Fig. 20b).

3. Conclusion

As a burgeoning class of porous crystalline materials, MOFs put its unique signature in terms of highly porous architecture, superior stability, remarkable emission attributes, easy functionalization, which in turn assist in luminescent detection of several analytes through efficient host-guest interaction. The present chapter is aimed at potential application of various fluoro-switchable MOFs for selective and sensitive detection of different analytes (volatile organic compounds, ions, nitro-aromatic explosives, biomarkers and acid-base) at the interface of astute design principles. The chapter further provides light on the fundamental sensing mechanism that can help for constructing a devise-based sensor for commercialization in future. Also, since turn-on sensing is more reliable than the corresponding turn-off process, the former will be an important upcoming research motif. Although effective host-guest interaction, rapid response time, selectivity, high sensitivity, robustness, reusability, and viable manufacturing make them into innovative fluorescent nanoscale devices, several aspects must be critically identified and analysed for developing advance MOF sensors. MOF should be able to adapt to a wide range of working circumstances (pressure, temperature etc.), and the development of aqua-robust MOF sensor is critical considering biological sensing (bio imaging), which is sporadically investigated compared to other typical analyte sensing. However, the toxicity of severalMOFs is a major barrier to clinical use that has yet to be overcome. Due to the difficulty of acquiring a single crystal data of the host-guestmatrix, the sensing mechanism, particularly the host-guest interac- tions, is frequently hypothesised based on computational studies. Asa result, there is a strong need for theoretical and experimental studyto be in agreement. Despite several advancements, a variety of combinations of metal ions and organic linkers, as well as changing the programming of experimental conditions, may lead to substantial progress in MOF sensing.

Acknowledgement

S. N. and R. G. acknowledge DST-SERB (Grant No. CRG/2021/002529) and the CSIR, New Delhi, India for research fellowship. S.R acknowledges CSIR (grant no. MLP-0054) for Research fellowship grant. N.S. thanks the CSIR, New Delhi, India, for a research fellowship (CSIR Grant No: 31/028(0252)/2019-EMR-I).

References

1. O'Keeffe, M., and Yaghi, O. M. (2012) Deconstructing the crystal structures of metal-organic frameworks and related materials into their underlying nets. *Chem. Rev.*, **112**, 675– 702.
2. Furukawa, H., Cordova, K. E., O'Keeffe, M., and Yaghi, O. M. (2013) The chemistry and applications of metal-organic frameworks. *Science*, **341**, 1230444– 1230456.
3. Kreno, L. E., Leong, K., Farha, O. K., Allendorf, M., Van Duyne, R. P., and Hupp, J. T. (2012) Metal-organic framework materials as chemical sensors. *Chem. Rev.*, **112**, 1105– 1125.
4. Yuan, S., Feng, L., Wang, K., Pang, J., Bosch, M., Lollar, C., Sun, Y., Qin, J., Yang, X., Zhang, P., Wang, Q., Zou, L., Zhang, Y., Zhang, L., Fang,Y., Li, J., and Zhou, H.-C. (2018) Stable metal-organic frameworks: design, synthesis, and applications. *Adv. Mater.*, **30**, 1704303– 1704337.
5. Seal, N., Palakkal, A. S., Singh, M., Goswami, R., Pillai, R. S., and Neogi, S. (2021) Chemically robust and bifunctional Co(II)-framework for trace detection of assorted organo-toxins and highly cooperative deacetalization–Knoevenagel condensation with pore-fitting-induced size-selectivity. *ACS Appl. Mater.* Interfaces, **13**, 28378– 28389.
6. Seal, N., Singh, M., Das, S., Goswami, R., Pathak, B., and Neogi, S. (2021) Dual-functionalization actuated trimodal attribute in ultra-robust MOF: exceptionally selective capture and effectual fixation of CO_2 with fast-responsive, nanomolar detection of assorted organo-contaminants in water. *Mater. Chem. Front.*, **5**, 979– 994.
7. Chand, S., Mondal, M., Pal, S. C., Pal, A., Maji, S., Mandal, D., and Das, M. C. (2018) Two azo-functionalized luminescent 3D Cd(II) MOFs for highly selective detection of Fe^{3+} and Al^{3+}. *New J. Chem.*,**42**, 12865– 12871.
8. Goswami, R., Pal, T. K., and Neogi, S. (2021) Stimuli-triggered fluoro-switching in metal–organic frameworks: applications and outlook. *Dalton Trans.*, **50**, 4067–4090.

9. Seal, N., Karthick, K., Singh, M., Kundu, S., and Neogi, S. (2022) Mixed-ligand-devised anionic MOF with divergent open Co(II)-nodes as chemo-resistant, bi-functional material for electrochemical water oxidation and mild-condition tandem CO_2 fixation. *Chem. Eng. J.*, **429**, 132301– 132311.

10. Seal, N., and Neogi, S. (2021) Intrinsic-unsaturation-enriched biporous and chemorobust Cu(II) framework for efficient catalytic CO_2 fixation and pore-fitting actuated size-exclusive hantzsch condensation with mechanistic validation. *ACS Appl. Mater. Interfaces*, **13**, 55123–55135.

11. Singh, M., Solanki, P., Patel, P., Mondal, A., and Neogi, S. (2019) Highly active ultrasmall Ni nanoparticle embedded inside a robust metal–organic framework: remarkably improved adsorption, selectivity, and solvent-free efficient fixation of CO_2. *Inorg. Chem.*, **58**, 8100–8110.

12. Zheng, B., Luo, X., Wang, Z., Zhang, S., Yun, R., Huang, L., Zeng,W., and Liu, W. (2018) An unprecedented water stable acylamide-functionalized metal–organic framework for highly efficient CH_4/CO_2 gas storage/separation and acid–base cooperative catalytic activity. *Inorg. Chem. Front.*, **5**, 2355– 2363.

13. Guo, X., Zhou, Z., Chen, C., Bai, J., He, C., and Duan, C. (2016) New rht type metal-organic frameworks decorated with acylamide groups for efficient carbon dioxide capture and chemical fixation from raw power plant flue gas. *ACS Appl. Mater. Interfaces*, **8**, 31746–31756.

14. Wei, S. C., Pan, M., Li, K., Wang, S., Zhang, J., and Su, C. Y. (2014) A multistimuli- responsive photochromic metal-organic gel. *Adv. Mater.*, **26**, 2072– 2077.

15. Wang, Y.-M., Tian, X.-T., Zhang, H., Yang, Z.-R., and Yin, X.-B. (2018) Anticounterfeiting quick response code with emission color of invisible metal–organic frameworks as encoding information. *ACS Appl. Mater. Interfaces*, **10**, 22445–22452.

16. Pan, A. Z., Wang, J. L., Jurow, M. J., Jia, M. J., Liu, Y., Wu, Y. S., Zhang, Y. F., He, L., and Liu, Y. (2018) General strategy for the preparation of stable luminous nanocomposite inks using chemically addressable $CsPbX_3$ perovskite nanocrystals. *Chem. Mater.*, **30**, 2771–2780.

17. Daly, B., Ling, J., and de Silva, A. P. (2015) Current developments in fluorescent PET (photoinduced electron transfer) sensors and switches. *Chem. Soc. Rev.*, **44**, 4203– 4211.

18. Xu, R., Wang, Y., Duan, X., Lu, K., Micheroni, D., Hu, A., and Lin, W. Nanoscale metal-organic frameworks for ratiometric oxygen sensing in live cells. (2016) *J. Am. Chem. Soc.*, **138**, 2158– 2161.

19. Fukaminato, T. J. (2011) Single-molecule fluorescence photoswitching: Design and synthesis of photoswitchable fluorescent molecules. *Photochem. Photobiol., C*, **12**, 177-208.

20. Zhu, L., Wu, W., Zhu, M. Q., Han, J. J., Hurst, J. K., and Li, A. D. Q. (2007) Reversibly photoswitchable dual-color fluorescent nanoparticles as new tools for live-cell imaging. *J. Am. Chem. Soc.*, **129**, 3524– 3526.

21. Wei, H., Ratchford, D., Li, X., Xu, H., and Shih, C.-K. (2009) Propagating Surface Plasmon Induced Photon Emission from Quantum Dots. *Nano Lett.*, **9**, 4168– 4171.

22. Lustig, W. P., Mukherjee, S., Rudd, N. D., Desai, A. V., Li, J., and Ghosh, S. K. (2017) Metal-organic frameworks: functional luminescent and photonic materials for sensing applications. *Chem. Soc. Rev.*, **46**, 3242– 3285.

23. World Urbanization Prospects, UN Department of Economic and Social Affairs, 2014.

24. You, L., Zha, D., and Anslyn, E. V. (2015) Recent advances in supramolecular analytical chemistry using optical sensing. *Chem. Rev.*, **115**, 7840– 7892.

25. Anichini, C., Czepa, W., Pakulski, D., Aliprandi, A., Ciesielski, A.,and Samorì, P. (2018) Chemical sensing with 2D materials. *Chem. Soc. Rev.*, **47**, 4860– 4908.

26. Rutledge, L. R., and Wetmore, S. D. (2011) Modeling the chemical step utilized by human alkyladenine DNA glycosylase: a concerted mechanism aids in selectively excising damaged purines. *J. Am. Chem. Soc.*, **133**, 16258– 16269.

27. Andronov, A., and Fréchet, J. M. J. (2000) Light-harvesting dendrimers. *Chem. Commun.*, 1701-1710.

28. Zhang, J., Fu, Y., and Lakowicz, J. R. (2007) *J. Phys. Chem. C*, **111**, 50– 56.

29. Rocha, J., Carlos, L. D., Almeida Paz, F. A., and Ananias, D. (2011) Luminescent multifunctional lanthanoids-based metal–organic frameworks. *Chem. Soc. Rev.*, **40**, 926– 940.

30. Lin, X., Hong, Y., Zhang, C., Huang, R., Wang, C., and Lin, W. (2015) Pre-concentration and energy transfer enable the efficient luminescence sensing of transition metal ions by metal-organic frameworks. *Chem. Commun.*, **51**, 16996– 16999.

31. Eddaoudi, M., Kim, J., Rosi, N., Vodak, D., Wachter, J., O'Keeffe, M., and Yaghi, O. M. (2002) Systematic design of pore size and function-ality in isoreticular MOFs and their application in methane stor- age. *Science*, **295**, 469– 472.

32. Zhao, Y., and Li, D. (2020) Lanthanide-functionalized metal-organic frameworks as ratiometric luminescent sensors. *J. Mater. Chem. C*, **8**, 12739-12754.

33. Moon, H. R., Lim, D.-W., and Suh, M. P. (2013) Fabrication of metal nanoparticles in metal–organic frameworks. *Chem. Soc. Rev.*, **42**, 1807– 1824.

34. Hu, M. L., Razavi, S. A., Piroozzadeh, M., and Morsali, A. (2020) Sensing organic analytes by metal–organic frameworks: a new way of considering the topic. *Inorg. Chem. Front.*, **7**, 1598–1632.

35. Cui, Y. J., Yue, Y. F., Qian, G. D., and Chen, B. L. (2012) Luminescent functional metal-organic frameworks. *Chem. Rev.*, **112**, 1126–1162.

36. Yin, X.-M., Gao, L.-L., Li, P., Bu, R., Sun, W.-J., and Gao, E.-Q. (2019) Fluorescence turn-on response amplified by space confinement in metal-organic frameworks. *ACS Appl. Mater. Inter-faces*, **11**, 47112–47120.

37. Guo, M. Y., Li, P., Yang, S. L., Bu, R., Piao, X. Q., and Gao, E. Q. (2020) Distinct and selective amine- and anion-responsive behaviors of an electron-deficient and anion-exchangeable metal-organic framework. *ACS Appl. Mater. Interfaces*, **12**, 43958–43966.

38. Zhao, S. S., Yang, J., Liu, Y. Y., and Ma, J. F. (2016) Fluorescent aromatic tag-functionalized MOFs for highly selective sensing of metal ions and small organic molecules. *Inorg. Chem.*, **55**, 2261–2273.

39. Dong, X.-Y., Huang, H.-L., Wang, J.-Y., Li, H.-Y., and Zang, S.-Q. (2018) A flexible fluorescent SCC-MOF for switchable molecule identification and temperature display. *Chem. Mater.*, **30**, 2160–2167.

40. Lan, A. J., Li, K. H., Wu, H. H., Olson, D. H., Emge, T. J., Ki, W., Hong, M. C., and Li, J. (2009) A luminescent microporous metal-organic framework for the fast and reversible detection of high explo- sives. *Angew. Chem. Int. Ed.*, **48**, 2334–2338.

41. Pramanik, S., Zheng, C., Zhang, X., Emge, T. J., and Li, J. (2011) New microporous metal-organic framework demonstrating unique selectivity for detection of high explosives and aromatic compounds. *J. Am. Chem. Soc.*, **133**, 4153–1455.

42. Nagarkar, S. S., Joarder, B., Chaudhari, A. K., Mukherjee, S., and Ghosh, S. K. (2013) Highly selective detection of nitro explosives by a luminescent metal-organic framework. *Angew. Chem., Int.Ed.*, **52**, 2881–2885.

43. Goswami, R., Seal, N., Dash, S. R., Tyagi, A., and Neogi, S. (2019) Devising chemically robust and cationic Ni(II)-MOF with nitrogen-rich micropores for moisture-tolerant CO_2 capture: highly regenerative and ultrafast colorometric sensor for TNP and multiple oxo-anions in water with theoretical revelation. *ACS Appl. Mater. Inter-faces*, **11**, 40134–40150.

44. Goswami, R., Mandal, S. C., Pathak, B., and Neogi, S. (2019) Guest-induced ultrasensitive detection of multiple toxic organics and Fe^{3+} ions in a strategically designed and regenerative smart fluores-cent metal-organic framework. *ACS Appl. Mater. Inter-faces*, **11**, 9042–9053.

45. Seal, N., Goswami, R., Singh, M., Pillai, R. S., and Neogi, S. (2021) An ultralight charged MOF as fluoro-switchable monitor for assorted

organo-toxins: size-exclusive dye scrubbing and anticounterfeiting applications *via* Tb^{3+} sensitization. *Inorg. Chem. Front.*, **8**, 296– 310.

46. Singh, M., Palakkal, A. S., Pillai, R. S. and Neogi, S. (2021) N-functionality actuated improved CO_2 adsorption and turn-on detection of organo-toxins with guest-induced fluorescence modulation in isostructural diamondoid MOFs, *J. Mater. Chem. C*, **9**, 7142-7153.

47. Goswami, R., Das, S., Seal, N., Pathak, B., and Neogi, S. (2021) High-performance water harvester framework for triphasic and synchronous detection of assorted organotoxins with site-memory-reliant security encryption via pH-triggered fluoroswitching. *ACS Appl. Mater. Interfaces*, **13**, 34012– 34026.

48. Cui, Y., Chen, B., and Qian, G. (2014) Lanthanide metal-organic frameworks for luminescent sensing and light-emitting applications. *Coord. Chem. Rev.*, **273–274**, 76– 86.

49. Venkateswarlu, S., Reddy, A. S., Panda, A., Sarkar, D., Son, Y., and Yoon, M. (2020) Reversible fluorescence switching of metal-organic framework nanoparticles for use as security ink and detection of Pb^{2+} ions in aqueous media. *ACS Appl. Nano Mater.*, **3**, 3684– 3692.

50. Zeng, Y.-N., Zheng, H.-Q., Gu, J.-F., Cao, G.-J., and Lin, Z.-J. (2019) Dual-emissive metal-organic framework as a fluorescent "switch" for ratiometric sensing of hypochlorite and ascorbic acid. *Inorg. Chem.*, **58**, 13360– 13369.

51. Yin, H.-Q., Tan, K., Jensen, S., Teat, S. J., Ullah, S., Hei, X., Velasco, E., Oyekan, K., Meyer, N., Wang, X.-Y., Thonhauser, Yin, T., X.-B., and Li, J. (2021) A switchable sensor and scavenger: detection and removal of fluorinated chemical species by a luminescent metal–organic framework. *Chem. Sci.*, **12**, 14189-14197.

52. Naylor, S. (2003) Biomarkers: current perspectives and future prospects, Taylor & Francis, **3**, 525–529.

53. Mayeux, R., (2004) Biomarkers: potential uses and limitations. NeuroRx, **1**, 182–188.

54. Du, P., Niu, Q., Chen, J., Chen, Y., Zhao, J., and Lu, X. (2020) "Switch-on" fluorescence detection of glucose with high specificity and sensitivity based on silver nanoparticles supported on porphyrin metal-organic frameworks. *Anal. Chem.*, **92**, 7980– 7986.

55. Wu, S., Lin, Y., Liu, J., Shi, W., Yang, G., and Cheng, P. (2018) Rapid detection of the biomarkers for carcinoid tumors by a water stable luminescent lanthanide metal-organic framework sensor. *Adv. Funct. Mater.*, **28**, 1707169.

56. Goswami, R., Mandal, S. C., Seal, N., Pathak, B., and Neogi, S. (2019) Antibiotic-triggered reversible luminescence switching in amine-grafted mixed-linker MOF: exceptional turn-on and ultrafast nanomolar detection of sulfadiazine and adenosine monophosphate with molecular keypad lock functionality. *J. Mater. Chem. A*, **7**, 19471– 19484.

57. Gui, B., Meng, Y., Xie, Y., Tian, J., Yu, G., Zeng, W., Zhang, G., Gong,S., Yang, C., Zhang, D., and Wang, C. (2018) Tuning the photoinduced electron transfer in a Zr-MOF: toward solid-state fluorescent molecular switch and turn-on sensor. *Adv. Mater.*, **30**, 1802329.

58. Cheng, Y., Wu, J., Guo, C., Li, X.-G., Ding, B., and Li, Y. (2017) A facile water-stable MOF-based "off-on" fluorescent switch for label-free detection of dopamine in biological fluid. *J. Mater. Chem.B*, **5**, 2524– 2535.

59. Xiao, J., Liu, J., Liu, M., Ji, G., and Liu, Z. (2019) Fabrication of a luminescence-silent system based on a post-synthetic modification Cd-MOFs: a highly selective and sensitive turn-on luminescent probe for ascorbic acid detection. *Inorg. Chem.*, **58**, 6167– 6174.

60. Moon, S.-Y., Howarth, A. J., Wang, T., Vermeulen, N. A., Hupp, J. T., and Farha, O. K. (2016) A visually detectable pH responsive zirconium metal-organic framework. *Chem. Commun.*, **52**, 3438– 3441.

61. Hou, S.-L., Dong, J., Tang, M.-H., Jiang, X.-L., Jiao, Z.-H., and Zhao, B. (2019) Triple-interpenetrated lanthanide-organic framework as dual wave bands self-calibrated pH luminescent probe. *Anal.Chem.*, **91**, 5455– 5460.

62. Li, B., Jiang, W., Xu, Y., Xu, Z., Yan, Q., and Yong, G. (2020) Dyes encapsulated in a novel flexible metal–organic framework show tunable and stimuli-responsive phosphorescence. *Dyes Pigm.*, **174**, 108017.

63. Das, P., and Mandal, S. K. (2020) Nanoporous Zn-based metal-organic framework nanoparticles for fluorescent pH sensing and thermochromism. *ACS Appl. Nano Mater.*, **3**, 9480– 9486.

64. Yang, S.-L., Liu, W.-S., Li, G., Bu, R., Li, P., and Gao, E.-Q. (2020) A pH-sensing fluorescent metal-organic framework: pH-triggered fluorescence transition and detection of mycotoxin. *Inorg.Chem.*, **59**, 15421– 15429.

6

Metal-Organic Frameworks for Catalysis

Neha Antil, Rajashree Newar, Wahida Begum, Kuntal Manna
Department of Chemistry, Indian
Institute of Technology Delhi, New Delhi, India

1. Introduction

Homogeneous catalysts offer high selectivity and have well-defined structures and mechanisms. The poor thermal stability, molecular decompositions, low TON and high cost make them less commercialized in industrial catalysis. Easy separation from the reaction system, recyclability and high TON provided by heterogeneous catalysts makes it more effective in industrial catalytical reactions. Research has been focusing on overcoming the limitations and combining the advantages associated with both conventional homogeneous and heterogeneous catalysis. MOFs provide an opportunity in bridging the gap between these two types of catalysts. MOF catalysts offer the advantages of both heterogeneous catalysts, such as high stability, facile catalyst separation and recovery, and less leaching, and homogeneous catalysts, such as homogeneity of the active sites, reproducibility, and selectivity.

Since the early example of cadmium MOF having a 2-dimensional structure constructed from coordination of Cd^{2+} with the nitrogen of 4,4'-bipyridine, catalysed cyanosilylation of aldehyde in 1994, the development of heterogeneous catalysts using MOFs has become a continuously growing research interest due to a number of key features offered by MOFs including chemical mutability, readily accessible internal surface areas and tunability of all its components (pores, linkers and nodes) [1]. Owing to the modular and crystalline nature, and well-defined pores with very high surface area and pore volume [7839 m^2g^{-1} and 5.02 cm^3g^{-1} reported for DUT-60 MOF ($Zn_4O(bcpbd)(bbc)_{4/3}$); bcpd:1,4-bis-p-carboxyphenylbuta-1,3-diene; bbc: 4,4',4''-(benzene-1,3,5-triyl-tris(benzene-4,1-diyl))tribenzoate)], [2] MOFs offer tunable support to develop highly robust, active and chemoselective catalysts via site-isolation. Larger pores facilitate the diffusion of substrate and product, and tunable pores offer the advantages of shape selective catalysis. As MOFs are constructed from metal nodes and organic linkers, one strategy for tuning the pore sizes of MOF is varying the organic linkers with its geometrical analogue of larger sizes. This is known as isoreticular synthesis, as the MOF of same topology as of original compound is synthesized but with varying pore sizes. In 2002, Yaghi reported a series of MOF-5 (IRMOF series) showing the easy tunabiltity of MOF by its isoreticular synthesis [3]. IRMOF series were built from Zn_4O as the inorganic cluster node and organic linkers based on alkyne, phenyl, biphenyl and terphenyl moieties with varying length. Depending on the requirements, the catalyst can be developed by functionalizing the nodes or linkers or both sites of a MOF. In some instance, the as-synthesized MOF itself acts as a catalyst. Examples for each type has been discussed in next sections. Herein, an overview of designing of MOF-based catalysts is provided.

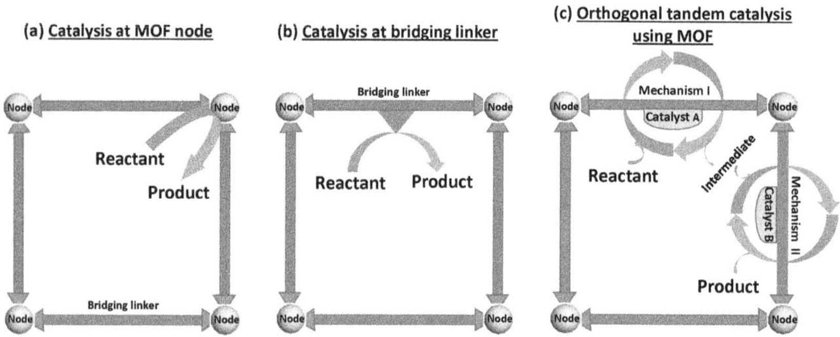

Figure 1. Development of catalysts based on MOFs using various strategies; (a) Node or SBU of MOFs as the catalytically active site; (b) bridging linker of MOFs bearing catalytic species. (c) Tandem catalytic systems based on MOFs.

1.1 SBUs or Nodes of MOFs as the Catalyst

SBUs or nodes of MOFs composed of metal ions or metal-oxo clusters may act as an active catalytic sites without any modification. The metal ion at SBUs coordinates with the substrate and catalyzes a reaction keeping framework intact. For example, vanadium-based MIL-47 MOF is reported for the oxidation of methane to acetic acid [4]. The MIL-47 is built up from linear chain of vanadium oxide linked to the terephthalate organic linker resulting in one-dimensional rhombic pores. In MIL-47, V^{+4} is a redox active species which undergoes reversible oxidation to V^{+5} during the catalytic oxidation reactions, via a reversible cleavage of terephthalate-vanadium bond.

1.2 SBUs or Nodes of MOFs as the Coordinatively Unsaturated Metal Sites for Catalysis

Metal ions at SBUs or nodes of MOFs are often coordinated with labile ligands such as solvent molecules, which can be removed thermally to create vacant coordination sites at metal ions for substrate binding and its activation. For example, HKUST-1 MOF with the formula of $Cu_3(BTC)_2 \cdot 3H_2O$, constructed from dimeric copper unit linked by four carboxylate moieties from four benzenetricarboxylate linkers [5, 6]. Removal of the three water molecules coordinated to the Cu^{+2} ion provides Lewis-acid copper sites, which is an active catalyst for cyanosilylation of carbonyl compounds catalysed by Lewis-acid copper sites [5]. A wide range of MOFs are well known that consist of bridging hydroxyl groups in their SBUs. These bridging hydroxyl groups, coordinated with the metal ions of SBUs can behave as bronsted acid sites which is inspired by the acidified zeolites. Farrusseng and coworkers reported MOF-69C which is constructed from $Zn_3(OH)_2$ as metal cluster and benzenedicarbocylic acid (bdc) as a dicarboxylate linker and has formula of $Zn_3(OH)_2(bdc)_2$ [7]. It consists of μ_3-OH as bridging species which has a key role in generating catalytic activity. MOF-69C catalyzed shape selectively the alkylation of large polycyclic aromatics, such as biphenylene.

1.3 SBUs or Nodes of MOFs as the Support for Developing Catalysts

The metal oxide nodes or SBUs of MOFs can serve as the support to graft or anchor catalytic species. The inorganic nodes of many MOFs feature functional OH groups that mimic metal oxides/hydroxides. Due to the presence of highly dispersed and uniform OH groups at the nodes, tunable and regular pores, the nodes of MOFs offer a unique platform to develop robust single-site catalysts for various applications. For example, Lin and co-worker have reported various cobalt and iron catalysts supported by nodes of Zr-MOFs for various catalytic reactions such as hydrogenation of nitro, nitrile, isocyanides, and C–H functionalization [8,9]. In this approach, OH groups of

SBUs were deprotonated first followed by metalation to afford SBU-supported single-site metal catalysts. Another technique used to functionalize SBUs is atomic layer deposition (ALD), where metal ions are uniformly and precisely installed at the node of MOFs. Farha and co-worker developed Ni-AIM catalysts using ALD technique for hydrogenation of ethene to ethane [10]. In Ni-AIM, Ni ions were uniformly deposited onto the nodes of NU-1000 MOF constructed from $Zr_6(\mu_3\text{-}O)_4(\mu_3\text{-}OH)_4(H_2O)_4(OH)_4$ nodes and tetratopic 1,3,6,8-(p-benzoate)pyrene (TBAPy^{4-}) linkers. A molecular catalyst can also be anchored at SBUs, as exemplified by the Ferey group. The authors reported the coordination of ethylenediamine and diethylenetriamine to the SBUs of MIL-101(Cr), resulting the formation of amine-functionalized MIL-101(Cr), which exhibited high catalytic activity in Knoevenagel condensation reactions [11].

1.4 Catalysts at Organic Linkers of MOFs

The organic linkers of MOFs can serve as the support to develop single-site catalysts by attaching catalytic subunits with the linkers. MOFs built with linkers bear various functional groups that can serve as catalytically active sites, such as amino, pyridyl, amide, carboxylic or sulfonic acid groups, bipyridyl, sulfoxide had been reported. In some cases, these functional groups were further post-synthetically modified to afford single-site catalysts at the linkers. In addition, molecular catalysts based on organic molecules and organometallic compounds have been developed by functionalizing the organic linkers. For example, Kitagawa and co-workers synthesized -SO$_3$Na group functionalized MIL-101(Cr) via solvothermal reaction of chromium(VI) oxide, monosodium 2-sulfoterephthalic acid, and hydrochloric acid in water [12]. MIL-101(Cr) with SO$_3$Na is highly effective for cellulose hydrolysis catalytic reaction. Another example of the linker-based MOF-catalysts are UiO-MOFs based on bypyridine (bpy), BINAP (BINAP = 2,2'-bis(diphenylphosphino)-1,10-binaphthyl) derived dicarboxylate linkers [13,14].

1.5 Encapsulation of Catalytic Species in MOFs

The porous MOF system could be used as a host matrix for encapsulating various catalytic centres such as metal, metal oxide nanoparticles, and molecular catalysts. The resulting guest@MOF combines the properties of both host and guest, and due to their mutual inter-actions, it also serves new synergic effects in catalytic reaction. For example, Voort group in 2013 employed ship in a bottle approach to encapsulate a chiral Mn(III)salen complex within NH_2-MIL101(Al) MOF to obtain Mn-salen@MIL-101(Al) [15]. Mn-salen@MIL-101(Al) was an enantioselective and recyclable catalyst for the asymmetric epoxidation of olefins to afford epoxides with ee up to 70% [15]. Xu group reported CO oxidation using Au NPs deposited to ZIF-8 MOF via a simple solid grinding method [16]. The ZIF-8 and organogold complex $[(CH_3)_2Au(acac)]$ were grounded uniformly in an agate mor-tar at room temperature. Then, the sample was treated with a stream of 10 vol% H_2 in He, at 230 °C for 2.5 h to afford Au@ZIF-8. Incorpo-ration of Au NPs into MOFs is a highly effective catalysts for oxidation of CO.

1.6 MOF for Orthogonal Tandem Catalysis

Orthogonal tandem catalysis involves two or more independent cat-alysts for sequential chemical transformations. Tandem catalysis eliminates intermediate workup and isolation of the intermediates. However, detrimental catalyst-catalyst interactions are often ob-served in homogenous tandem catalyst systems. Due to their modu-lar nature, rapid mass transport through open channels, and isola-tion of active catalytic sites, MOFs provide a suitable platform for combining two compatible reactions together for efficient tandem catalysis. Lin and coworkers have reported a bifunctional FeX@Zr_6-Cu MOFs (X = Br^-, Cl^-, AcO^-, and BF_4^-) having cuprous photosensitiz-ing linkers (Cu-PSs) and catalytically active Fe^{II} centers linked to the μ_3-OH sites of $Zr_6(\mu_3$-O$)_4(\mu_3$-OH$)_4$ SBUs for highly effective photocata-lytic H_2 evolution with turnover numbers (TONs) of up to 33700 and turnover frequencies (TOFs) of up to 880 h^{-1} [17]. Same group has reported photocatalytic MOFs comprising of Cu-photosensitizer and Co for photocatalytic hydrogen evolution reaction (mPT-Cu/Co). mPT is a mixed MOF of UiO topology developed from Zr cluster and phenanthrolinedibenzoate (PT) and 2''-nitro-[1,1':4',1'':4'',1'''-qua-terphenyl]-4,4'''-dicarboxylate (TPHN) ligands. Similarly, mPT-Cu/Re catalyst was also synthesized for reduction of CO_2 to ester (CO$_2$RR) with TON of 1328 [18]. This chapter will give an overview of the development of various MOF-based heterogeneous catalysts for various applications. We will discuss the catalytic reaction along with the detailed synthesis and structure of MOF catalysts used. This chapter will be structured according to the type of reactions such as reduction, oxidation, C–H activation and asymmetric reactions.

2. Reduction Reactions

Reduction reactions play an important role in the various organic transformations such as the synthesis of amines from nitro compounds, hydrogenation of vegetable oils into solid or semi-solid fats and hydrogenolysis of ethers. Today most industrial reduction reactions rely on heterogeneous catalysts and hydrogen as the eco-friendly and economical reductant. In the petroleum industry, destructive hydrogenation of hydrocarbon is used in various processes to manufacture petrochemical products and gasoline. In the late 20th century, the hydrogenation of coal for liquid fuel production have become an attractive alternative to petroleum extraction. The development of MOF-supported heterogeneous catalysts for reduction reactions is discussed below.

2.1 Reduction of the Aromatic Nitro Compound to Amines

In 1854, Pierre J. A. Bechamp reported the reduction of nitrobenzene using metallic iron in acidic media [19]. Although this process is extensively applied for manufacturing various aniline-based compounds in industry, catalytic reduction of nitro groups is drawing much attention in recent years to avoid the usage of the stoichiometric amount of reducing agent [20]. In 2017, Lin and co-workers have reported single-site cobalt(II) catalyst supported by SBUs of a Zr_{12}-TPDC MOF (TPDC: triphenyldicarboxylate) for hydrogenation of nitro compounds [8]. Zr_{12}-TPDC MOF was synthesized via a solvothermal reaction of $ZrCl_4$ with H_2TPDC in DMF and water at 90 °C in the presence of acetic acid as a modulator. In Zr_{12}-TPDC MOF, $Zr_{12}O_8(\mu_3\text{-OH})_8(\mu_2\text{-OH})_6$ node connected with nine triphenyldicarboxylate organic linkers to give a 3D framework, which adopts trigonal P31c space group and has a surface area of 1967 m^2/g. The Zr_{12}-MOF was first treated with $LiCH_2SiMe_3$ to deprotonate both μ_2-OH and μ_3-OH sites present in the SBU followed by its post-synthetic metalation with cobalt dichloride in THF to give the corresponding cobalt-functionalized MOF (Zr_{12}-TPDC-CoCl) with the formula of $Zr_{12}(\mu_3\text{-O})_5[(\mu_3\text{-O-})CoCl]_8[(\mu_2\text{-O-})_2(\mu_3\text{-O})CoCl]_3Li_3(TPDC)_9$. Zr_{12}-TPDC-CoCl was treated with the $NaEt_3BH$ to afford Zr_{12}-TPDC-CoH, which was an active catalyst for hydrogenation reactions of nitroarenes, nitriles, and isocyanides under 30 bar H_2 at 110 °C with 0.5-2 mol % of Co (**Figure 2**). A range of aromatic nitro substrates bearing various electron-withdrawing and electron-donating functional groups, including heterocyclic nitro compounds, were chemoselectively reduced to the corresponding amines in high yields.

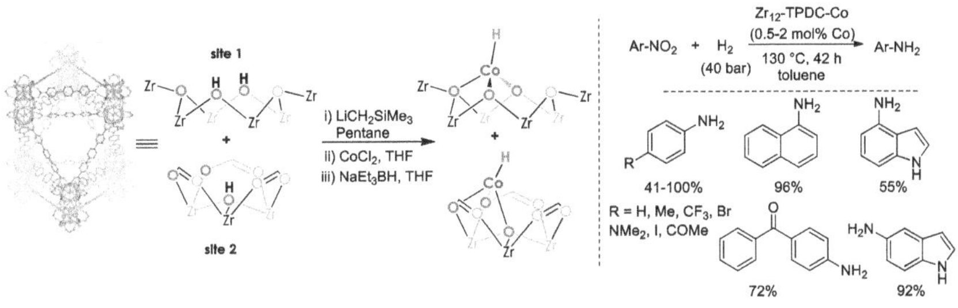

Figure 2. Synthesis of Zr$_{12}$-TPDC-CoH MOF (left) and its application in catalytic hydrogenation of aromatic nitro compounds to amines (right).

Manna and co-worker have developed a DUT-5(Al) supported nickel(II) hydride catalyst for the hydrogenation of a range of aromatic and aliphatic nitro compounds under 1 bar of H_2 with 1 mol % Ni-loading at 130 °C [21]. The solvothermal reaction between aluminum trichloride hexahydrate and 4,4′-biphenyldicarboxylic acid in DMF at 120 °C for 48 h afforded DUT-5 MOF with the formula of Al(OH)(bpdc) (bpdc:4,4′- biphenyldicarboxylate). DUT-5 is built from Al^{3+} cations that are octahedrally coordinated with μ_2-hydroxide and (bpydc^{2-}) bridging linkers to afford a three-dimensional porous framework with rhombic pores. The deprotonation of μ_2-OH with *n*-BuLi followed by its reaction with NiBr$_2$ afforded DUT-5-NiBr. The subsequent halide-hydride exchange by treating NaEt$_3$BH resulted in the formation of nickel hydride species supported by SBUs of DUT-5 MOF (DUT-5-NiH). DUT-5-NiH displayed excellent functional group tolerance in the hydrogenation of a range of nitro substrates bearing halogens, esters, amino, hydroxy, methoxy and amide functional groups, including heterocyclic and aliphatic nitro compounds to the corresponding amines with high chemoselectivity and excellent yields (**Figure 3**).

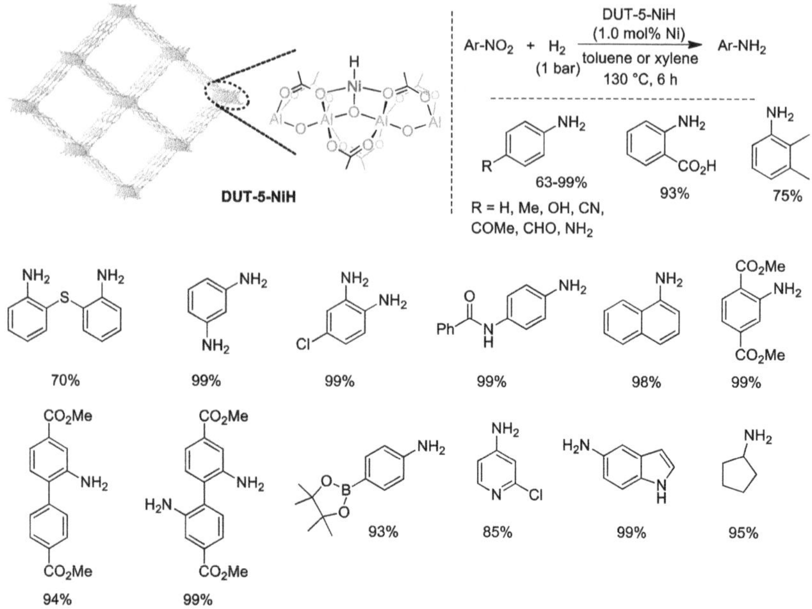

Figure 3. Schematic diagram showing the DUT-5 node-supported nickel(II) hydride (DUT-5-NiH) for catalytic hydrogenation of various aromatic compounds along with the examples of synthesized amine products.

2.2 Hydrogenolysis of Ethers

Chemoselective cleavage of C–O bond without affecting the C–C/C–H bond is an economically potential reaction for biomass conversion to value-added products [22]. Alcohol and ether bonds are ubiquitous in lignocellulosic biomass. Hydrogenolysis of ether bonds and hydrodeoxygenation of alcohols are efficient methods for depolymerization of lignin and its conversion to hydrocarbon fuels [23, 24]. β-O-4, α-O-4, and 4-O-5 linkages are the common type of ether bonds present in the structure of hardwood lignin contributing to 45–62%, 3–12% and 4–9% of ether bonds, respectively [23]. Lin group has reported a tandem catalytic system using DUT-5 MOF constituted of Lewis acid and palladium nano-particles catalysts for the hydrogenolysis of etheric, alcoholic, and esteric C–O bonds to generate saturated alkanes via a tandem dehydroalkoxylation–hydrogenation process. The MOF was synthesized via a solvothermal reaction of

Al(NO$_3$)$_3$.9H$_2$O and organic ligands, 2,2′-bipyridine-5,5′- dicarboxylate (dcbpy) and 1,4-benzenediacrylate (pdac), in DMF at 120 °C. Then, pdac ligands were removed selectively by cleavage of the olefinic groups via ozonolysis and created defects sites as Al$_2$(OH)(OH$_2$) in MOF. Lewis acidity of the resultant MOF nodes was further enhanced by treatment of Me$_3$SiOTf (trifluoromethanesulfonic acid trimethylsilyl ester, OTf is electron-withdrawing), resulting in the transformation of Al$_2$(OH)(OH$_2$) moieties to Al$_2$(μ_2-OTf) due to the oxophilic nature of Me$_3$Si groups. The resultant MOF was metallated with Pd(MeCN)$_2$Cl$_2$ in THF to afford the precatalyst (**1**-OTfPdCl$_2$) in which palladium coordinated with the two nitrogen atoms of 2,2′-bipyridine-5,5′-dicarboxylate. Treatment of (1-OTfPdCl$_2$) with hydrogen reduced Pd^{+2} to Pd-nanoparticles. Metal triflate salts and supported Pd catalysts in MOF effectively catalyzed the ether and alcohol C–O bond hydrogenolysis with the catalyst loading of 0.1-0.5 mol % and 20 bar H$_2$ at 100 °C in 24 h [25].

MOF-supported base-metal catalysts had also been developed for chemoselective hydrogenolysis of aryl ethers using Ti$_8$-BDC (MIL-125) MOF [26]. Lin and his co-workers reported the Ti^{IV}$_8$-BDC MOF supported nickel hydride catalyst with a formula of Ti^{III}$_2$Ti^{IV}$_6$(μ_2-O)$_8$(μ_2-OLi)$_{1.1}$(μ_3-ONiH)$_{2.9}$(BDC)$_6$. First, Ti$_8$-BDC MOF was synthesized by solvothermal reaction of Ti^{IV}(O^{i}Pr)$_4$ and 1,4-benzenedicarboxylic acid in DMF and methanol at 150 °C followed by deprotonation of its SBU's μ_2-OH using LiCH$_2$SiMe$_3$ to afford Ti^{IV}$_8$(μ_2-O)$_8$(μ_2-OLi)$_4$(BDC)$_6$. The lithiated MOF was post-synthetically modified with NiBr$_2$ to afford Ti^{IV}$_8$-BDC-NiBr, and its treatment with NaEt$_3$BH resulted in the formation of Ti^{III}$_2$Ti^{IV}$_6$-BDC-NiH via reduction of some Ti^{IV} to Ti^{III} and halide-hydride exchange. 1 mol % loading of Ti^{III}$_2$Ti^{IV}$_6$-BDC-NiH (based on Ni) completely cleaved benzyl phenyl ether, a model substrate for α-O-4 linkage in lignin, in 6 h at 140 °C under 1 bar of H$_2$ producing toluene and cyclohexanol as the sole products (**Figure 4**). Phenylethyl phenyl ether as a model compound for β-O-4 linkage was also reduced to cyclohexanol, ethylbenzene and ethylcyclohexane at 160 °C and 1 bar H$_2$ with Ti-NiH MOF (1 mol % of Ni). However, diphenyl ether containing 4-O-5 linkage gave only 19% conversion to form cyclohexanol and benzene under similar reaction conditions. The rate of hydrogenolysis of aryl ethers was as follows: β-O-4 > α-O-4 > 4-O-5, which is in accordance with their bond energies followed as 218 kJ/mol, 289 kJ/mol and 314 kJ/mol, respectively.

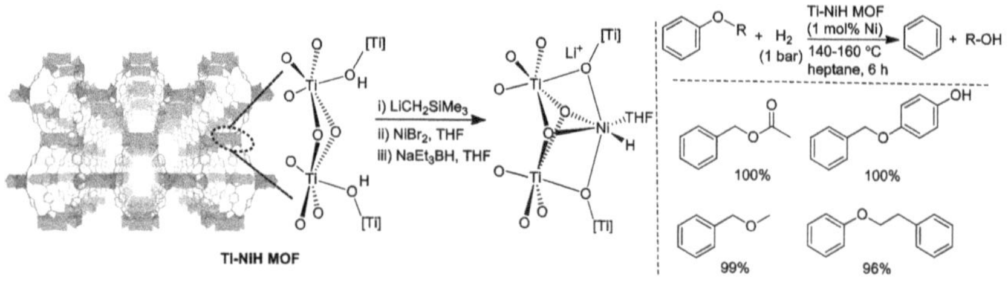

Figure 4. Scheme for the synthesis of Ti_8-BDC (MIL-125) node-supported nickel(II) hydride and its application in catalytic hydrogenolysis of ethers with examples of ether models.

2.3 Hydrodeoxygenation of Carbonyls and Alcohols

Chemoselective deoxygenation of carbonyls and alcohols to the corresponding saturated compounds using H_2 by heterogeneous base-metal catalysts is important for the sustainable production of fine chemicals and biofuels [27, 28]. Classical methods for the reduction of carbonyls and alcohols to the alkanes such as those based on the Barton–McCombie (R_3SnH) [29], borohydrides [30], boranes [31], Clemmensen (Zn/Hg, HCl) [32], HI/red-P [33] and Wolff–Kishner–Huang (N_2H_4, KOH) reductions [34] are generally associated with rigorous reaction conditions, poor functional group tolerance and the utilization of stoichiometric amounts of reagents, and the generation of undesirable byproducts. Manna and his coworkers in 2021 reported DUT-5 SBU-supported cobalt(II) hydride, which is a highly chemoselective and recyclable heterogeneous catalyst for deoxygenation of a range of aromatic and aliphatic carbonyls, and primary and secondary alcohols, including biomass-derived substrates under 1 bar H_2 (**Figure 5**). At a 0.4 mol % Co-loading, DUT-5-CoH is highly efficient in hydrodeoxygenation of ketones, aldehydes and alcohols to alkanes under 1 bar H_2 at 80-110 °C within 24 h [35]. DUT-5-Co catalyzed deoxygenation of carbonyl-groups had broad substrate scopes, including aryl alkyl ketones, diaryl ketones, heterocyclic ketones, naphthone and aliphatic ketones. Several heterocyclic ketones and aldehydes were also hydrogenated to corresponding alkanes without any hydrogenation of arenes. The kinetic and DFT calculation suggests reversible carbonyl coordination to cobalt followed by the turnover-limiting step involving 1,2-insertion of the coordinated carbonyl into the cobalt–hydride bond.

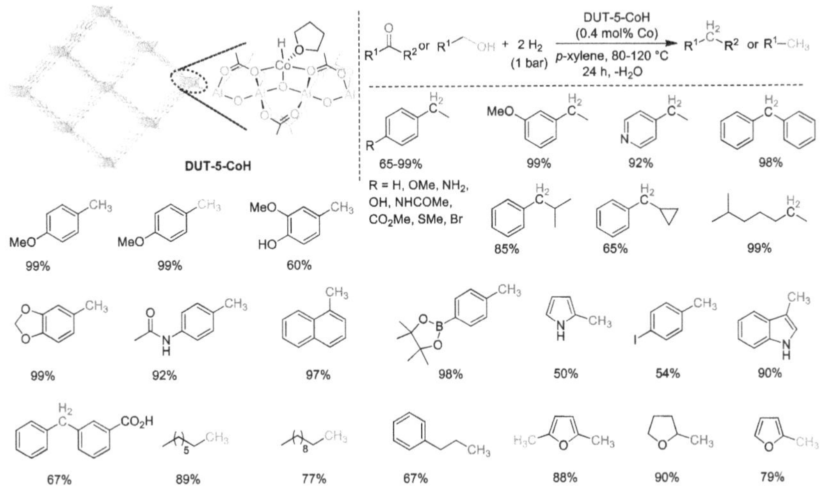

Figure 5. DUT-5-CoH-Catalyzed deoxygenation of aldehydes, ketones and primary alcohols by H_2.

2.4 Hydrogenation of Arenes and Heteroarenes

The chemoselective reduction of aromatic rings is the most facile pathway to prepare cycloalkanes and saturated heterocycles as aromatic compounds are naturally occurring and essential feedstocks for chemical synthesis. Cycloalkanes are common structural motifs for synthesizing resins, polymers and fine chemicals. They have a wide range of applications such as fuels and solvents and can be used as the starting material for the production of agrochemicals and pharmaceuticals. However, the reduction of heterocyclic compounds is challenging due to the ring stabilization and inhibition by heteroatoms [36-38]. Several MOF based catalysts have been reported for the arene hydrogenations, for example, $Ti^{III}_2Ti^{IV}_6$-BDC-$Co^{II}H$ [39]. The Ti_8-BDC MOF with a formula of $Ti^{IV}_8O_8(OH)_4(BDC)_6$ was constructed from $Ti^{IV}_8(\mu_2\text{-}O)_8(\mu_2\text{-}OH)_4$ metallic node and benzene dicarboxylic acid as the bridging linker. μ_2-OH of SBU was deprotonated using $LiCH_2SiMe_3$ followed by its metalation with cobalt chloride to form cobalt chloride species supported on SBU of Ti_8-BDC MOF (Ti_8-BDC-CoCl). $Ti^{III}_2Ti^{IV}_6$-BDC-$Co^{II}H$ was synthesized by treating Ti_8-BDC-CoCl with $NaEt_3BH$, which replaced chlorine to hydride and reduced some Ti^{IV} to Ti^{III}. Arenes and heteroarenes were hydrogenated to saturated cycloalkanes and saturated heterocyclic compounds at 120 °C under

50 bar H_2, at a 0.05 mol % loading of Co (**Figure 6**). Catalysts showed good productivity for reducing a wide range of aromatic compounds to their corresponding cycloalkanes, including disubstituted benzenes such as *p*-xylene, *o*-xylene and, trisubstituted benzenes such as mesitylene in quantitative yields. Importantly, fused aromatic rings such as naphthalene and 9H-fluorene were also hydrogenated to decahydronaphthalene and dodecahydro-1H-fluorene with 100% and 85% yields, respectively. Moreover, under similar reaction conditions, substituted benzenes with the polar functional groups such as amine, methoxy and ester were also reduced to the substituted cycloalkanes. Heterocyclic compounds such as pyridine and their derivatives were successfully hydrogenated to piperidine, with yields up to 99%. Method for selective dearomatization of only heterocyclic ring, keeping the other ring intact, has also been developed using $Ti^{III}_2Ti^{IV}_6$-BDC-Co^{II}H MOF catalyst. In the case of quinolines and indoles, only pyridine and pyrrole rings were hydrogenated to piperidine and pyrrolidine, respectively.

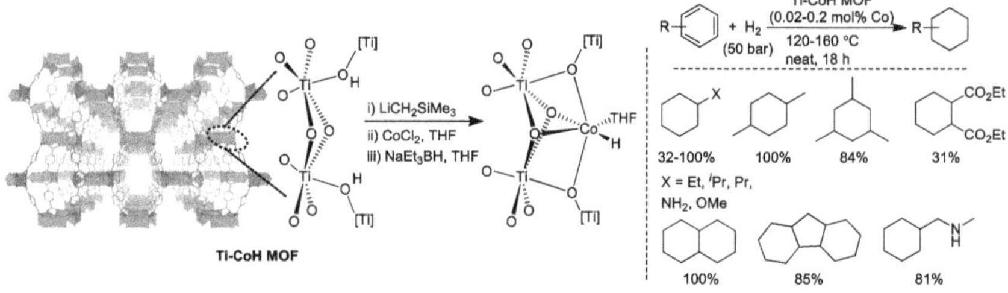

Figure 6. Scheme for the synthesis of Ti_8-BDC (MIL-125) node-supported cobalt(II) hydride and its application in catalytic hydrogenation of arenes to saturated cycloalkanes.

2.5 Hydrogenation of Olefins

Hydrogenation of olefins is one of the most important reactions in organic synthesis, though industrially important hydrogenation reactions mostly rely on precious late-transition metal catalysts. Lin group has developed a sal-MOF of UiO-topology constructed from Zr-oxo-hydroxo cluster-based SBUs and salicylaldimine-functionalized linear dicarboxylate bridging linkers (**Figure 7**) [40]. The sal-MOF having a formula of $Zr_6O_4(OH)_4(sal\text{-}TPD)_6$ has both tetrahedral and octahedral cages with pore sizes of 8.2 and 11.1 Å. Postsynthetic metalation of sal-MOF with $FeCl_2 \cdot 4H_2O$ followed by its treatment with $NaEt_3BH$ in THF resulted in Fe-functionalized MOFs (sal-Fe-MOF). At a 0.01 mol % of Fe-loading, sal-Fe-MOF completely hydrogenated 1-octene to *n*-octane at 23 °C and under 40 atm of H_2 in 18 h. sal-Fe-MOF gave a very high TON of 145000 for hydrogenation of 1-octene. Under identical reaction conditions, sal-Fe-MOF effectively hydrogenated aliphatic and aromatic terminal alkenes with 100% yields. Styrene and its alkoxy and halide derivatives were also hydrogenated to corresponding saturated products with 93-100% yields using 0.05-0.01 mol % catalyst loading. Gemdisubstituted alkenes, such as α-methylstyrene, and dialkenes such as 1,7-octadiene and allyl ether were also reduced to afford cumene, octane and saturated ether with 100% yields. The same group also developed a Co-NacNac MOF, which was synthesized via functionalization of UiO-68-NH_2 MOF having a formula of $Zr_6O_4(OH)_4(TPDC-NH_2)_6$ with β-diketiminate ligand, commonly called NacNac. Hydrogenation of 1-octene with 0.0005 mol % Co-NacNac MOF and 40 bar H_2 at room temperature afforded *n*-octane in 100% yield [41]. Several other aliphatic and aromatic alkenes such as styrene, cyclohexene and 4-methoxyallylbenzene were also hydrogenated to afford their corresponding alkanes in 72-100% yields using 0.0005-0.025 mol % Co loading and 40 bar hydrogen.

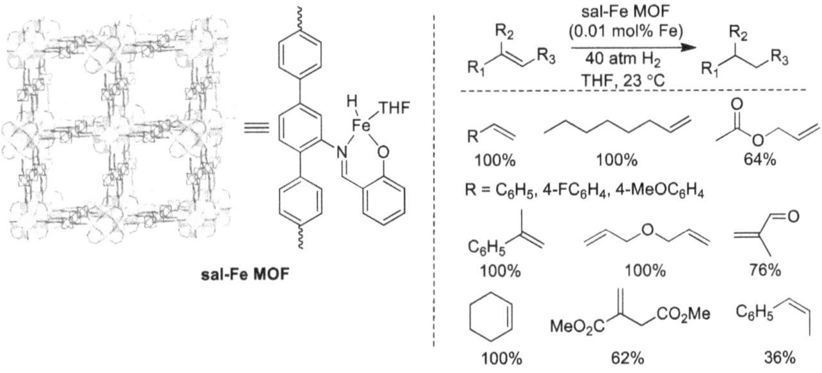

Figure 7. sal-Fe MOF built with linkers bearing salicylaldimine-coordinated iron subunit as heterogeneous Fe-catalyst for alkene hydrogenation.

3. C–C and C–X (X = N, B) Bond Formation

3.1 Suzuki-Miyaura Coupling and Heck Coupling Reaction

Li group employed UiO-67 MOF to incorporate organic palladium complex $Pd(H_2bpydc)Cl_2$ (H_2bpydc = 2,2'-bipyridine-5,5'-dicarboxylic acid), and this heterogeneous palladium system was more active than its homogenous analogue for Heck and Suzuki-Miyaura coupling reactions [42]. Pd(II) doped UiO-67 is a mixed MOF, constructed from Zr metal cluster linked with the para-biphenyldicarboxylate and (5,5'-dicarboxy-2,2'-bipyridine-)palladium(II)]- dichloride to form a 3D structure of UiO-67 topology with the molecular formula of $Zr_6O_4(OH)_4(L_1)_x(L_2)_{6-x}$ (**Figure 8**). It has a high surface area and pore volume of 2000 m^2g^{-1} and 0.79 cm^3g^{-1}, respectively. This catalyst was used for coupling reaction of various substituted aryl halides to give several biaryl products with yields up to 97%. In Heck coupling reaction, palladium (II) doped UiO-67 (0.46 mol % of palladium) was used for coupling of aryl halide (0.5 mmol) with olefin (0.75 mmol) in presence of K_2CO_3 (1.5 mmol) and tetrabutylammonium bromide (TBAB) (0.3 mmol). The biaryl was synthesized in an inert atmosphere at 100 °C with 99% yield in 24 hours. Similarly, Suzuki coupling was carried using palladium (II) doped UiO-67 (0.46 mol % of palladium), aryl chloride (0.5 mmol) and phenylboronic acid (0.75 mmol) in the presence of KOH (1.5 mmol) under inert atmosphere. Both electron-withdrawing and electron-donating group substituted aryl halides were explored for the coupling reactions, which suggested that electron-withdrawing substituted aryl halides are more active than electron-rich aryl halides.

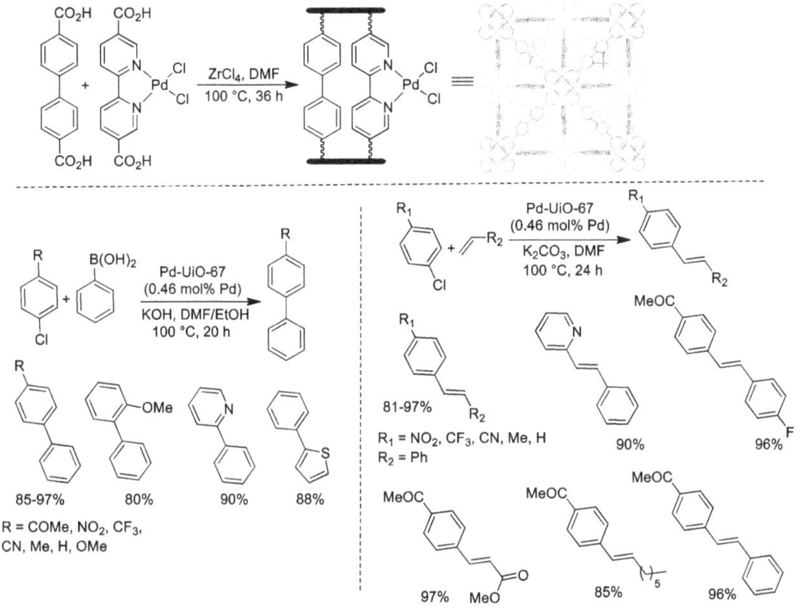

Figure 8. Scheme for the synthesis of Pd(II) doped UiO-67 mixed MOF for Heck and Suzuki-Miyaura coupling reactions.

3.2 Alkenylation of 2-Substituted Azaarenes

2-Alkenylazaarenes are important heterocyclic structural motifs present in several pharmaceutical compounds, agrochemicals, and natural products. Previously, the general synthesis of such compounds has been reported via condensation of 2-alkylazarenes with aldehyde in the presence of a stoichiometric amount of strong base or acid [43]. $Fe_3O(BPDC)_3$ (BPDC = biphenyl-4,4′-dicarboxylic acid) (MIL-126) was employed for C–H activation of azaarenes [44]. Solvothermal reaction between biphenyl-4,4′-dicarboxylic acid and iron(III) chloride hexahydrate in DMF in the presence of acetic acid at 120 °C for 48 h afforded 3D rigid porous framework with a formula of $Fe^{III}_3O(H_2O)_2(OH)[(O_2C)C_{12}H_8(CO_2)]_3 \cdot n(solv)$. MIL-126 exhibits a tetragonal symmetry and has BET surface area of 1700 m^2g^{-1}. Direct coupling reaction via C–H activation in the presence of transition metal catalysts offers an environmentally benign method for synthesizing these compounds. In general, experimental conditions iron-based MOF catalyst (0.012 g, 10 mol %) was used in the reaction of azaarene (0.057 g, 0.4 mmol) with benzaldehyde (0.064 g, 0.6 mmol) in the presence of acetic acid (0.002 g, 10 mol %) and 1 mL toluene as a solvent. The catalysis was investigated in inert atmosphere at 120 °C for 24 h. Alkenylation of 2-methylquinoline with benzaldehyde (E) under mentioned conditions produced 2-styrylquinoline with 60% yield.

3.3 Arylation of Heterocycles

The C-H bond of α position to a heteroatom in heterocyclic compounds can be easily activated for its arylation using transition metal-based homogenous or heterogeneous catalysts. Importantly prior functionalization of heterocycles is not required for this catalytic coupling reaction. Nickel MOF as $Ni_2(BDC)_2(DABCO)$ (H_2BDC = 1,4-benzenedicarboxylic acid, DABCO = 1,4-diazabicyclo[2.2.2]octane) was prepared for the direct arylation of heterocyclic compounds [45]. $Ni_2(BDC)_2(DABCO)$ is a paddle-wheel based porous MOFs with tetragonal framework structures synthesized solvothermally with H_2BDC, DABCO and $Ni(NO_3)_2 \cdot 6H_2O$ in DMF at 100 °C in 40 h and has BET surface area of 1473 m^2g^{-1}. This $Ni_2(BDC)_2(DABCO)$ MOF was found to be more active than other Ni MOFs such as $Ni_3(BTC)_2$ (BTC:1,3,5-benzenetricarboxylate) and $Ni(BTC)(BPY)$ (BPY:4,4′-bipyridine),and homogeneous systems like Ni^{2+} salts, such as $NiCl_2$, $Ni(NO_3)_2$, Ni_2SO_4 and $Ni(OAc)_2$ under identical reaction conditions. 2-phenylbenzoxazole was synthesized in 75% conversion by the coupling reaction between benzoxazole and phenylboronic acid with the molar ratio of 1:2, in the presence of a base K_3PO_4 and 10 mol % of $Ni_2(BDC)_2(DABCO)$ catalyst along with 20 mol % of 2,2′-bipyridine as a ligand at 100 °C for 3 h (**Figure 9**). $Cu_2(BPDC)_2(BPY)$ MOF was used as a catalyst for a similar type of coupling reactions, however, instead of boronic acid, aryl halide was used for C–C coupling with the heterocyclic compounds at α position C–H [46]. $Cu_2(BPDC)_2(BPY)$ MOF was synthesized by dissolving H_2BPDC(4,4′-biphenyldicarboxylic acid), BPY (4,4′-bipyridine), and $Cu(NO_3)_2.3H_2O$ in DMF and methanol in the presence of pyridine, and then heating mixture at 120 °C for 24 h. 2-Phenylbenzoxazole was synthesized via coupling reaction of phenylboronic acid (0.112 mL, 1 mmol) with benzoxazole (0.238 g, 2 mmol) in the presence of K_3PO_4(0.532 g, 2 mmol) in DMF (4 mL) at 120 °C under an argon atmosphere for 180 min using 7.5 mol % of $Cu_2(BPDC)_2(BPY)$ as a catalyst.

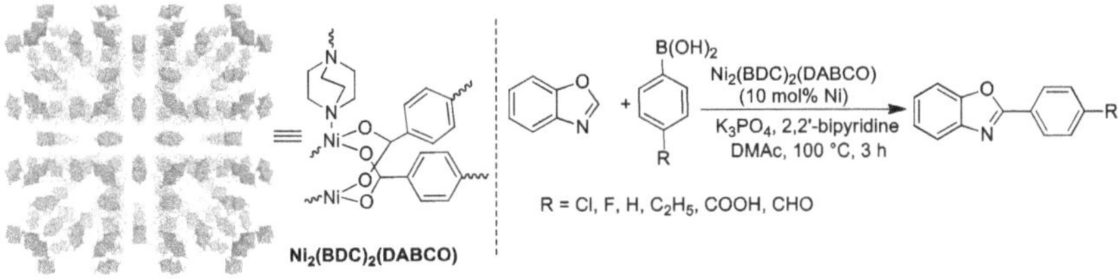

Figure 9. $Ni_2(BDC)_2(DABCO)$ for arylation of heterocyclic compounds using phenylboronic acid.

3.4 C–N Bond Formation

C–N bond formation is an important organic reaction for synthesis of amines, and it has a wide range of applications in the synthesis of various essential intermediates and products. Recently, MOF-based heterogeneous catalysts for C-N bond formation have been developed. For example, 2-morpholino-N-(quinolin-8-yl)benzamide was synthesized in quantitative yield at 90 °C using MOF-199 $[Cu_3(BTC)_2]$ catalysts via a coupling reaction between N-benzoyl-8-amino-quinoline and morpholine in presence of N-methylmorpholine oxide as an oxidant [47]. Several other organic compounds such as 2-phenylbenzimidzole using 537-MOF{$[Cu_2(TPPB)_2](DMF)_6$ [48] and 3-morpholinoquinoxalin-2(1H) using Cu-CPO-27 MOF [49] have been synthesized through a C–N coupling reaction. In 537-MOF Cu_1 and Cu_2 atoms are bridged by four bridging carboxylate groups from $TPPB^{2-}$ {3-(3-(4-(4H-1,2,4-triazol-4-yl)phenoxy)-5-(4-carboxylatophenoxy)phenoxy)benzoate} linkers, forming SBU as $Cu_2(O_2C)_4$. 537-MOF crystallizes in triclinic space group P1. 2-phenylbenzimidzole was successfully synthesized in 67% yield from N-phenylbenzimidamide (0.25 mmol) using 537-MOF (10 mg), in DMSO/DMF (2 mL, 1/1) in the presence of 5 equiv. of acetic acid at 100 °C under air atmosphere for 24 h. Cu-CPO-27 has one of the highest densities of Cu^{2+} sites (4.7 nm^{-3}) of any known copper MOF. It has a honeycomb-like structure constructed from tetradeprotonated 2,5-dihydroxyterephthalic acid as an organic linker bridging to the Cu(II)ions. Like HKUST-1 MOF, each copper centre of Cu-CPO-27 also coordinates to a water molecule. 3-morpholinoquinoxalin-2(1H) was synthesized from morpholine with 74% yield using a 5 mol % catalyst (Cu-CPO-27) in DMA at 80 °C for 16 h under an oxygen atmosphere.

3.5 C–B Bond Formation

Organoboranes are versatile synthetic intermediates for carbon–carbon and carbon–heteroatom bond formations. Among many methods, transition-metal-catalyzed borylation of C–H bonds is the most efficient and atom economical for direct formation of organoboron compounds. Lin group has developed a cobalt-based heterogeneous catalyst using a terpyridine-based metal-organic layer (TPY-MOL) for borylation of benzylic position [9]. Metal-organic layers (MOLs) are category of 2D materials of few nanometers in thickness. MOLs provide a new platform for designing a new class of 2D single-site solid catalysts without any diffusional constraints. TPY-MOL is based on $Hf_6(\mu_3\text{-}O)_4(\mu_3\text{-}OH)_4(HCO_2)_6$ as a metal node and 4'-(4-carboxyphenyl)-[2,2':6',2"-terpyridine]-5,5"-dicarboxylate (TPY) as the bridging linkers (**Figure 10**). Treatment of TPY ligands in TPY-MOL with $CoCl_2$ and $FeBr_2$ afforded highly effective recyclable MOL catalysts for selective benzylic borylation (4.6:1).

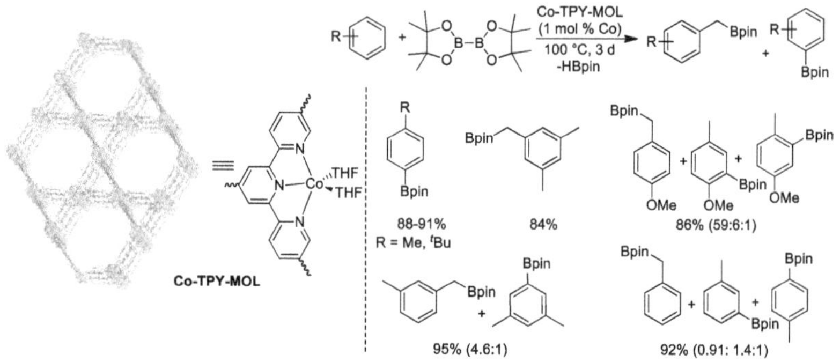

Figure 10. TPY-MOL containing terpyridine-coordinated cobalt catalyst for borylation of arenes. Examples of borylated products with their yields.

Cobalt-catalyst (pyrim-MOF-Co) coordinated with the redox non-innocent pyridylimine-functionalized metal–organic frameworks have exhibited high activity for activation of arene and benzylic C–H bond [50]. Pyrim-MOF is constructed from Zr-cluster nodes bridged to pyridylimine-functionalized dicarboxylate linkers (pyrim-TPDC) to give a UiO-type framework of $Zr_6O_4(OH)_4(pyrim\text{-}TPDC)_6$. Metalation of Pyrim-MOF with cobalt dichloride followed by its treatment with $NaEt_3BH$ afforded active catalysts Pyrim-MOF-Co. The catalysts showed chemoselective borylation of the secondary or tertiary $C(sp^3)$–H bond in the presence of $C(sp^2)$–H and primary $C(sp^3)$–H bonds. Borylation of arenes was carried out using B_2pin_2 or HBpin (pin = pinacolate) and 0.5 mol % of cobalt catalysts at 110 °C, either neat arene or heptane as solvent was used (**Figure 11**). Substituted arenes with electron-withdrawing and electron-donating groups have been successfully borylated using the pyrim-MOF-Co catalyst. Under similar reaction conditions, pyrim-MOF-Co catalysts also catalyze regioselective borylation of secondary or tertiary $C(sp^3)$–H bonds at the benzylic position over arene groups with selectivity up to 96%.

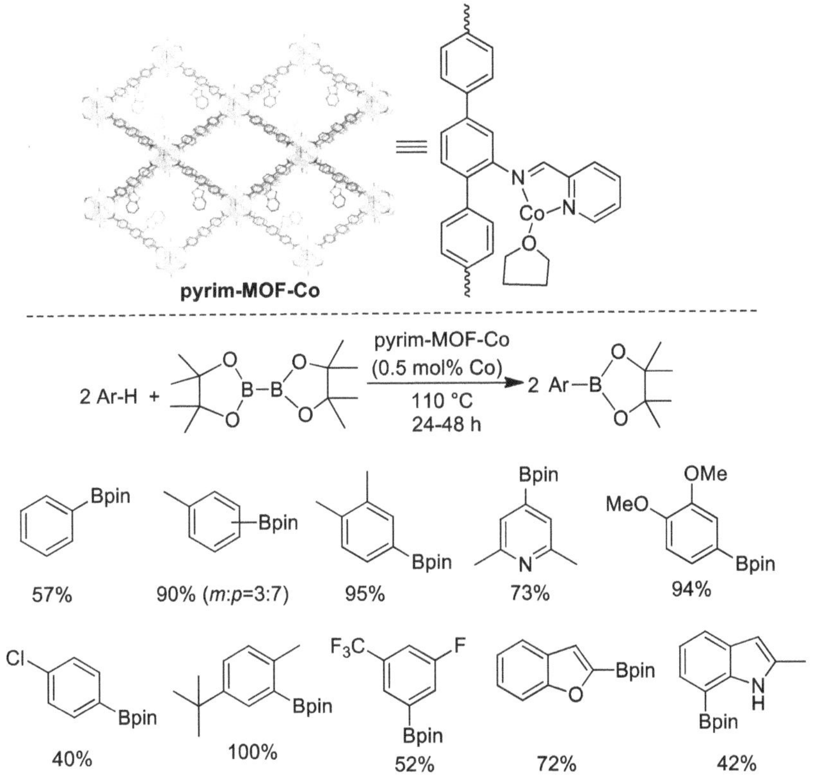

Figure 11. Borylation of arene and benzylic C–H bond catalyzed by pyrim-MOF-Co (0.5 mol % Co) using B_2pin_2 or HBpin.

4. Oxidation of Alkanes

Saturated hydrocarbons or alkanes are one of the main constituents of natural gas [51]. The oxygen-containing compounds produced via oxidation of these hydrocarbons using an oxidizing agent are an important chemical feedstock in chemical industries [52].

4.1 Cycloalkane Oxidation

The oxidation of cycloalkanes into the corresponding ketones or alcohols is considered as very important industrial process. Cyclohexanone and cyclohexanol obtained after cyclohexane oxidation have important commercial values as these are used as intermediates in the production of nylon fibres [53,54]. Most of the industrial processes for the production of cycloalcohols or cycloketones involve homogeneous catalysts. But using heterogeneous catalysts have their own advantages such as easy recovery, robustness and lower operational cost. In this regard, Metal-organic frameworks (MOFs) have been found to be excellent heterogeneous catalysts and solid supports to carry out the oxidation of cyclohexane because of their structural tunability [55-57].

In 2009, Corrado Di Nicole et al. developed a Cu-MOF having molecular formula $[Cu_3(\mu_3\text{-}OH)(\mu\text{-}pz)_3(CH_2\text{=}C(CH_3)COO)_2]$ derived from copper(II) methacrylate and pyrazole (pz) [58]. Under the optimized reaction condition, this MOF oxidized cyclohexane and cyclopentanes to their corresponding cycloalcohols and cycloketones with aqueous H_2O_2 in MeCN at 25 °C. The highest total oxidation product of 32% was obtained in the case of oxidation of cyclohexane when the molar ratio of H_2O_2/Cu-MOF catalyst is 1000, and addition of HNO_3 in HNO_3/Cu-MOF catalyst molar ratio of 10 (**Figure 12**). The addition of HNO_3 was necessary as trinuclear Cu-complexes do not exhibit catalytic activity in the absence of acid. Protonation at the ligand creates unsaturation at the metal centre, thus stabilizing the peroxo species and preventing the decomposition of H_2O_2. Moreover, in presence of simple salts such as copper(II) acrylate and copper(II) methacrylate, the oxidation of cyclohexane under similar reaction conditions, gave a conversion of only 3.3% and 4.3%, respectively, and in the absence of the metal catalyst, no oxidation product was obtained. Under similar reaction conditions, with H_2O_2/Cu-MOF catalyst molar ratio of 500, the oxidation of cyclopentane to give cyclopentanol and cyclopentanone gave the highest total oxidation product of 29%.

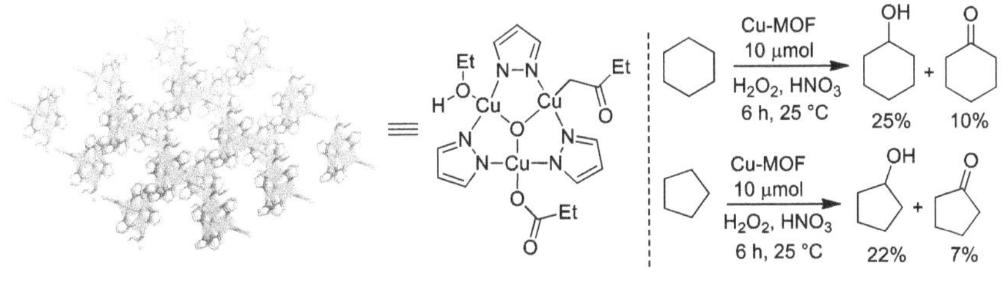

Figure 12. Cu-MOF and its application in catalytic oxidation of cycloalkanes.

4.2 Methane Oxidation

Natural gas is primarily used for domestic cooking/heating, conversion to syn-gas via reforming, and transportation fuels in the form of compressed natural gas (CNG). Methane is the major component of natural gas, and the transportation of natural gas over long distances is economically challenging. The direct oxidation of methane to value-added products such as methanol is highly lucrative in the petrochemical industry due to the low-cost feedstocks [59]. In nature, methane is converted to methanol by the enzyme methane monooxygenase [60,61]. In 2017, Johannes and his group reported Cu-NU-1000 MOF using $Zr_6(\mu_3\text{-}O)_4(\mu_3\text{-}OH)_4(OH)_4(OH_2)_4$-[1,3,6,8-tetrakis($p$-benzoate)pyrene]$_2$ as a support for stabilising Cu-oxo clusters [62]. For testing the catalytic activity towards the oxidation of methane to methanol, the Cu-NU-1000 MOF was activated by treatment with O_2 at 200 °C followed by exposing the MOF to a flow of CH_4 at 150 °C. Further, for the product desorption, 50/50 or 10/90 mixture of H_2O/He was purged to it. The isolated oxidised products contained methanol, dimethyl ether and CO_2 as detected by mass spectrometry. A maximum selectivity of 45-61% was observed in the production of methanol and dimethyl ether, whereas the remaining was CO_2 (**Figure 13**).

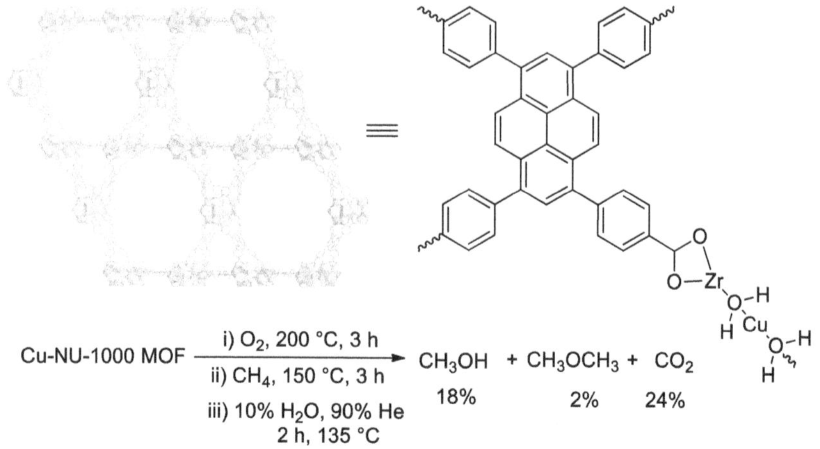

Cu-NU-1000 MOF $\xrightarrow{\text{i) O}_2\text{, 200 °C, 3 h}}$ CH$_3$OH + CH$_3$OCH$_3$ + CO$_2$

i) O$_2$, 200 °C, 3 h
ii) CH$_4$, 150 °C, 3 h
iii) 10% H$_2$O, 90% He
2 h, 135 °C

CH$_3$OH + CH$_3$OCH$_3$ + CO$_2$
18%   2%   24%

Figure 13. Oxidation reaction of methane using Cu-NU-1000 MOF.

In 2019, Zebao Rui and co-workers synthesized a MOF derived IrO$_2$/CuO catalyst for the oxidation of methane to methanol with high activity and selectivity in mild conditions [63]. For the synthesis of the catalyst, a methanolic solution of trimesic acid, Cu(NO$_3$)$_2$, polyvinylpyrrolidone (PVP) and a certain amount of Ir NPs were mixed to obtain Ir encapsulated Cu-MOF that was further calcined air to attain hybrid IrO$_2$/CuO MOF. Under optimised reaction conditions of 3 bar CH$_4$ and 1 bar air in water at 150 °C for 3 h, yielded 872 μmolg$_{cat}^{-1}$ of methanol by preventing over oxidation. However, they also reported that with increasing methane pressure in the reaction, methanol production increases, thus indicating CH$_4$ activation as the rate-determining step. The methanol yield reached as high as 1937 μmolg$_{cat}^{-1}$ at elevated CH$_4$ feeding pressure up to 20 bar.

4.3 Oxidation of Alcohols

Oxidation of alcohols to ketones and aldehydes has emerged as an important reaction in the field of organic chemistry [64]. Over oxidation of alcohols can also yield carboxylic acids which are also important chemical feedstocks in the industry. These reactions are prompted in the presence of an oxidant and a catalyst. In 2007, a Pd based MOF, [Pd(pymo)$_2$]$_n$.3H$_2$O (pymo = hydroxypyrimidinolate) has been reported by H. Garcia and co-workers for aerobic alcohol oxidation in cinnamyl alcohol [65]. This MOF was synthesized in aqueous medium and hence insensitive towards moisture. The refluxing of Pd(2-hydroxypyrimidine)$_2$Cl$_2$ in an aqueous solution adjusted with 1 M NaOH to pH 6.0 for 5 d yielded the Pd-MOF. The catalytic activity of the MOF was tested for the oxidation of cinnamyl alcohol to cinnamyl aldehyde using atmospheric air as the oxidant. This resulted 99% conversion and 74% selectivity in 20 h at 90 °C. The crystal structure of the MOF was intact after the reaction conditions, allowing the reusability of the catalyst.

In another instance, a team led by Yingwei Li developed an Au/MIL-101(CD/PVP) catalyst that showed high selectivity and conversion for a variety of benzyl alcohols [66]. The Au-catalyst, derived from HAuCl$_4$ was deposited in MIL-101 (Cr$_3$(F,OH)O[(O$_2$C)-C$_6$H$_4$-(CO$_2$)]$_3$) framework using polyvinylpyrrolidone as a protecting agent. This was reported to be the first example of MOF supported Au catalyst that was synthesized using a solution-based synthetic strategy for the deposition of gold NPs. At 1 mol% Au, the alcohols were oxidised under 1 atm of oxygen, when stirred at 80 °C in toluene in base-free condition (**Figure 14**). It has been observed that benzyl alcohols containing electron-donating groups such as CH$_3$, OCH$_3$ and OH were easily oxidised in comparison to those having electron-withdrawing groups (such as NO$_2$, Cl). The catalyst also promoted the oxidation of reluctant aliphatic alcohols. A TOF of 29300 h^{-1} was achieved for the aerobic oxidation of 1-phenylethanol to give acetophenone with a selectivity of >99%.

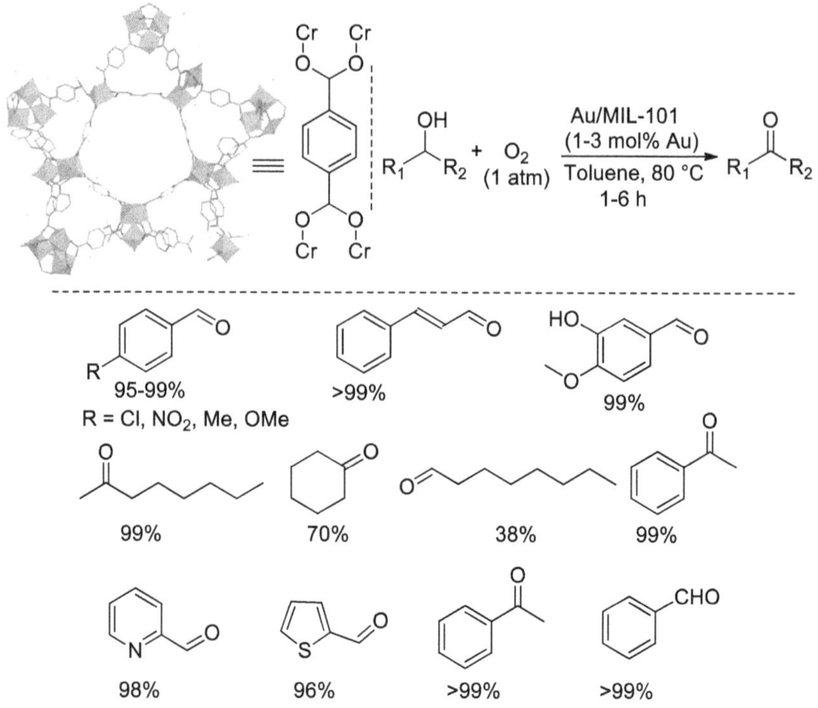

Figure 14. Oxidation of alcohols using Au/MIL-101 MOF.

4.4 Oxidation of Cyclohexene

The oxidation of cyclohexene is an important organic reaction. Depending upon the oxidant used, we can obtain numerous types of value-added chemicals via oxidation of cyclohexenes. Among these products, 2-cyclohexene-1-one obtained by the allylic oxidation of cyclohexene, is a very important intermediate as it is used in medication, pesticides, spices etc. [67]. In 2009, hydrothermally synthesized Cu based MOF [Cu(H$_2$btec)(bipy)]$_\infty$ (btec = 1,2,4,5-benzenetetracarboxylic acid; bipy = 2,2'- bipyridine) was reported by Spodine and co-workers for the oxidation of cyclohexene with *tert*-butyl hydroperoxide as the oxidant [68]. The highest TOF of 79 h^{-1} was obtained with 0.01 mol % of Cu at 75 °C giving 64.5% conversion in case of oxidation of cyclohexene with high chemoselectivity to give the corresponding epoxide (**Figure 15**). In the absence of MOF catalyst, cyclohexanone was obtained as the major product under similar reaction conditions.

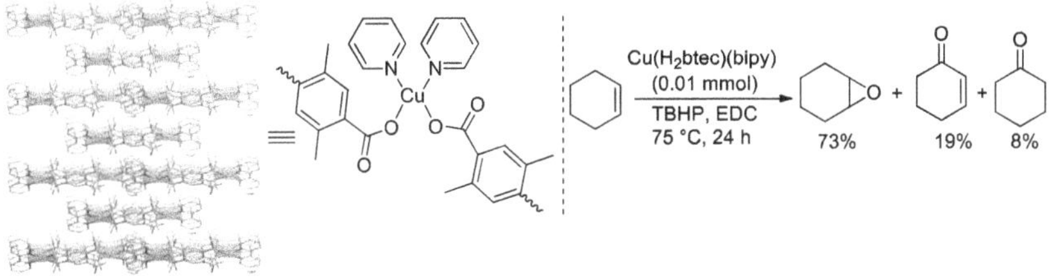

Figure 15. Oxidation of cyclohexene using [Cu(H₂btec)(bipy)]∞ MOF.

In 2015, a mixed metal Co/Ni-MOF-74 catalyst having three-dimensional honeycomb-like network was reported by Z. Li and his group for aerobic cyclohexene oxidation to cyclohexene oxide, 2-cyclohexen-1-ol, 2-cyclohexen-1-one and cyclohexene hydroperoxide [69]. The Ni-MOF-74 was synthesized using 2,5-dihydroxyterephthalic acid and nickel precursor. The Ni^{2+} was then further substituted by Co^{2+} via metal-exchange method. The Co^{2+} were incorporated in the synthesized Ni-MOF-74 in such a way that the Co^{2+} were more accessible to the substrates, thus increasing the catalytic activity of MOF. Under optimized reaction conditions, the catalyst showed high selectivity in oxidation of cyclohexene at 80 °C in O_2 under atmospheric pressure. The use of catalyst increased the selectivity, to form 2-cyclohexen-1-ol and 2-cyclohexen-1-one as the main products, while in the absence of catalyst, auto-oxidation of cyclohexene in air yielded cyclohexene hydroperoxide as the major product. The highest selectivity of 93.4% with 54.7% conversion was observed when 61% of Ni^{2+} was exchanged with Co^{2+}.

4.5 Oxidation of CO

CO is a very toxic gas released from various industries, automobiles, etc. CO poisoning is harmful to humans. It combines with the haemoglobin in our body to form carboxyhaemoglobin, thereby reducing the capability of haemoglobin to carry oxygen [70]. In the United States, CO poisoning has become very common, resulting in almost 50,000 cases per annum [71]. Much attention has been given to prevent the release of CO in the environment and in this regard, CO oxidation has emerged as one of the important processes.

In 2010, Chang-jun Liu and his group reported the oxidation of CO using $Cu_3(BTC)_2$ MOF supported Pd-catalyst [72]. $Cu_3(BTC)_2$ MOF catalyst was amorphized by treating it with dielectric-barrier discharge

(DBD) plasma to give $Cu_3(BTC)_2$-P catalyst. Further, $Cu_3(BTC)_2$-P was again modified to 1% $Pd/Cu_3(BTC)_2$-P by impregnating with an aqueous solution of $Pd(NO_3)_2$. They tested the catalytic activity of the catalyst for oxidation of CO, under atmospheric pressure in a quartz-tube fixed-bed reactor. It was purged with argon at 150 °C followed by purging with a gaseous mixture of carbon monoxide, oxygen and argon at 50 °C. After analysis of effluent with GC, they found that the catalysts gave 100% conversion at 220 °C to give CO_2 (**Figure 16**). Moreover, without Pd-loading, $Cu_3(BTC)_2$ and $Cu_3(BTC)_2$-P gave no conversion below 190 °C and 210 °C, they gave 100% conversion at 240 °C and 231 °C respectively.

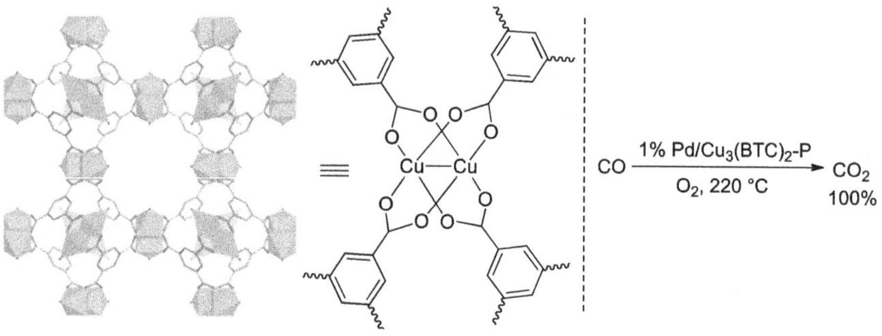

Figure 16. Oxidation of CO catalysed by 1% $Pd/Cu_3(BTC)_2$-P MOF.

In 2017, EL-Shall and co-workers developed a very efficient catalyst (Pd@Ce-MOF) for the oxidation of CO as well as for the capturing of CO_2 at the lowest temperature (273 K) [73]. The catalyst has been synthesised by the incorporation of Pd nanoparticles within a Ce-based MOF, namely $\{[Ce(BTC)(H_2O)]\cdot DMF\}_n$ (where BTC is benzene-1,3,5-tricarboxylate) by microwave irradiation method. Due to the presence of f-orbital in cerium, it has a high affinity for oxygen, and it could easily store oxygen. Moreover, Pd-nanoparticle enhances the catalytic activity towards the oxidation of CO, hence making Pd@Ce-MOF the ideal candidate for the oxidation of CO. At 3% Pd-loading, 100% conversion for CO oxidation was observed at 190 °C. While at 5% and 7% Pd-loading, full conversion was observed at 92 °C and 148 °C respectively. This was due to the aggregation of Pd at Ce-surface that might have resulted in the decrease of the surface area of MOF. Furthermore, without the Pd-loading, below 200 °C, the catalyst showed no activity in the oxidation of CO.

4.6 Oxidation of Aromatic Hydrocarbons

Catalytic oxidation of aromatic hydrocarbons has been of great importance in terms of industrial utilization as they can be converted into valuable products. Phthalic and maleic anhydrides obtained after oxidation of aromatic hydrocarbons such as benzene, naphthalene and xylenes are used in the synthesis of polyesters or resins and also employed as plasticisers [74]. Toluene is harmful to both human health and environment, and its catalytic oxidation to less toxic and useful aldehydes and alcohols is an effective process [75]. Peedikakkal and his group synthesized some mixed-metal MOFs of 1,3,5-benzenetricarboxylate (BTC) namely, M-Zn-BTC where M = Cu(II), Co(II) and Fe(II) as well as Cu-BTC MOF [76]. Firstly, Zn-BTC and Cu-BTC MOFs were synthesized using BTC and respective metal precursor as $Zn(NO_3)_2 \cdot 6H_2O$ and $Cu(NO_3)_2 \cdot 3H_2O$ in DMF. Further, M-Zn-BTC was synthesized using the post-synthetic exchange method by soaking Zn-BTC crystals in a methanolic solution of Cu-, Co- or Fe- salts. The catalytic activity of the catalysts towards oxidation of toluene was investigated using hydrogen peroxide as an oxidant in a 10-point electrothermal sealed reactor at 70 °C. In 24 h, Cu-BTC MOF showed the highest conversion of up to 80% and selectivity of 40%, while Zn-BTC gave 22% conversion with 52% selectivity to yield *o*-cresol. However, Cu-Zn-BTC, Co-Zn-BTC and Fe-Zn-BTC gave conversions of 70, 38 and 30%, and selectivity of 40, 44 and 23%, respectively, for *o*-cresol. However, selectivity towards benzaldehyde formation (66%) increased upon the addition of Cu to Zn-BTC (**Figure 17**). Interestingly, the reaction did not proceed in the presence of BTC ligand alone without the MOF catalyst suggesting metal nodes as the active sites for the catalytic activity.

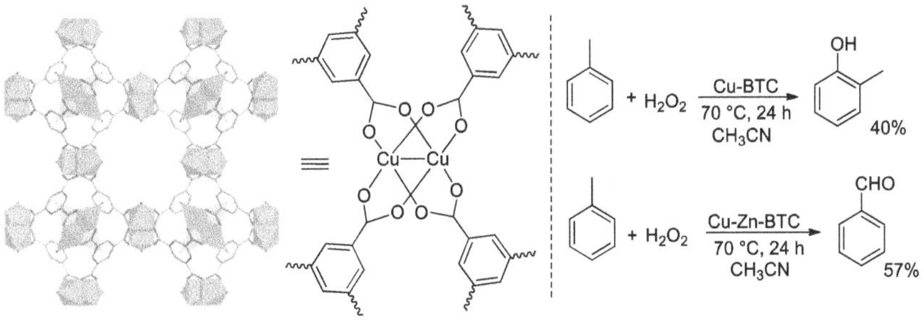

Figure 17. Cu-BTC and Cu-Zn-BTC MOF catalysed oxidation of toluene.

Ge and co-workers in 2020, synthesised MnO_x-CeO_2-MOF via *in situ* pyrolysis of Ce/Mn-MOF-74 as precursor. Ce/Mn-MOF-74 was hydrothermally synthesized using $MnCl_2 \cdot 4H_2O$, $Ce(NO_3)_3 \cdot 6H_2O$ and 2,5-dihydroxyterephthalic acid followed by calcination at 550 °C in air to form MnO_x-CeO_2-MOF catalyst [77]. The catalytic activity of the catalyst towards oxidation of toluene were investigated. Under optimized reaction conditions of 1000 ppm toluene, 20% O_2 and WHSV 60,000 mL/g_{cat}h, MnO_x-CeO_2-MOF gave 90% conversion at 220 °C whereas, MnO_x-MOF gave 90% conversion at 230 °C. Thus, they concluded that the incorporation of Ce enhanced the low-temperature reducibility of catalysts.

5. Tandem Reactions

A tandem catalysis is a "one-pot" process involving multiple catalytic transformations via different mechanisms using one or more catalysts followed by a single workup stage [78]. Tandem catalysis eliminates intermediate workup and isolation of the intermediates, thus reduces reaction steps to afford the desired products. In case of homogenous tandem catalyst systems, designing a tandem catalytic system is difficult due to incompatibility of different catalysts in solution and also the detrimental intermolecular catalyst-catalyst interactions that lead to catalysts' decomposition. Due to their modular nature, rapid mass transport through open channels, and isolation of active catalytic sites, MOFs provide a suitable solid support to assemble multiple catalysts together for efficient tandem catalytic reactions [79].

5.1 MOFs having Acidic Active Sites

Different types of MOFs were synthesised having coordinatively unsaturated open metal sites (OMSs) on secondary building units [80]. In these, the metals are perhaps surrounded by ligands such as water or other solvent molecules which can be easily removed by heating or keeping in vacuum. Thus, the finally obtained OMSs become active Lewis acidic sites for Lewis-acid catalysed tandem reactions. Hou and his group, in 2018 demonstrated the first example of MOF catalyzed tandem benzene acylation–Nazarov cyclization reaction. They synthesised two metal cluster-based MOF having molecular formula $\{(H_3O)_2[Co_4(pbcd)_2(\mu_3\text{-}OH)_2] \cdot CH_3CH_2OH \cdot 4H_2O\}_n$ (pbcd=9,9'-(propan-1,3-diyl)bis(9H-carbazole-3,6-dicarboxylicacid) and

$\{[Co_{5.25}(mcd)_2(HCO_2)(\mu_2\text{-}O)_{0.5}(\mu_3\text{-}OH)_{0.5}(H_2O)_4]\cdot6H_2O\cdot5DMF\}_n$ (mcd=9,9'-methylenebis(9H-carbazole-3,6-dicarboxylicacid). They investigated their catalytic activity by catalysing a tandem acylation–Nazarov cyclization reactions of 1,3-dimethoxybenzene with α,β-unsaturated carboxylic acids for the synthesis of cyclopentenone[b]benzene products [81]. One of the main reason behind the formation of cyclopentenone[b]benzene products is attributed not only to the availability of active sites but also to the high density of Lewis acidic sites within the catalyst frameworks and also to the synergy among the Co(II) clusters. At 30 mol % catalyst loading, the highest conversion of >99% was obtained at 90 °C after 5 h to yield cyclization product (5,7-dimethoxy-2,3-dimethyl-2,3-dihydro-1H-inden-1-one) with 81% selectivity when the MOF was used after dehydrating it at 363 K by vacuum baking to remove water and the solvent, thus facilitating the faster diffusion of the substrate under similar reaction conditions.

Recently in 2020, H. Garcia and his co-workers synthesised nitro functionalized chromium terephthalate MIL-101(Cr)-NO_2 MOF for the formation of a variety of benzimidazoles via tandem reaction of catalysed condensation on acidic sites followed by oxidative dehydrogenation reaction [82]. The hydrothermal reaction between 2-nitroterephthalic acid and $CrCl_3$ in demineralized water yielded MIL-101(Cr)-NO_2 MOF. For testing the catalytic activity of the catalyst towards cyclocondensation of *o*-phenylendiamine with benzaldehyde to form 2-phenylbenzimidazole, 0.1 mmol of the MOF catalyst were added to a solution of substrates (0.5 mmol each) in acetonitrile and heated the reaction mixture at 70 °C for 4 h (**Figure 18**). They observed that the controlled experiments in argon atmosphere gave a conversion of 94% with a selectivity of 10% for 2-phenylbenzimidazole while in oxygen atmosphere, the conversion was 100% with a selectivity of 90%. The increased selectivity in oxygen suggested that the oxygen promoted the imidazole formation which was also supported by the proposed mechanism.

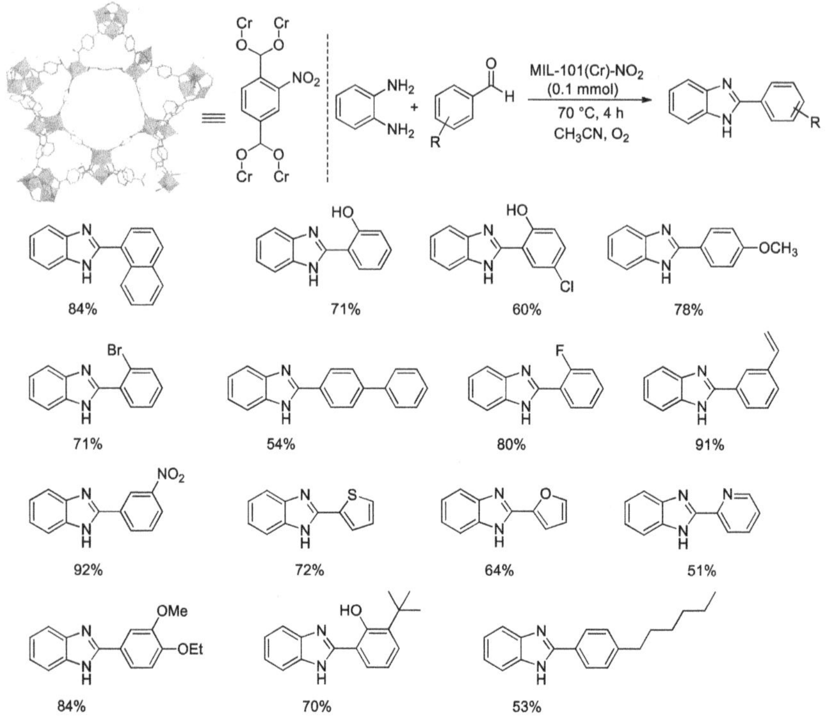

Figure 18. MIL-101(Cr)-NO$_2$ MOF and its application in tandem cyclocondensation reactions.

5.2 MOFs having Basic Active Sites

Many MOFs having active basic sites have been prepared using different types of basic N-containing ligands which can be functionalised in the nodes or the organic linkers [83, 84]. These types of modifications can lead to the formation of MOFs with various types of characteristics and the basic functionalised MOFs can undergo a variety of reactions including Knoevenagel condensation, Michael addition, etc. [83].

Tandem or cascade reactions have also been carried out using MOFs having basic sites. In 2015, Seth M. Cohen and his group prepared a tandem catalyst, Zn(II)-based IRMOF-9-Irdcppy-NH$_2$ (IRMOF = isoreticular metal-organic framework) that contains both -NH$_2$ as well as organometallic Ir(I) groups which are integrated into the MOF by using the pre- as well as post-synthetic method [85]. The MOF is capable of acting as a catalyst for both Knoevenagel condensation and allylic N-alkylation reaction. The basic amino group of the catalyst was responsible for catalysing Knoevenagel condensation while the Ir(I) species catalyses the allylic N-alkylation reaction. Firstly, IRMOF-9-dcppy-NH$_2$ was synthesized using 2,2′-diamino[1,1′-biphenyl]-4,4′-dicarboxylic acid, 6-(4-carboxyphenyl)nicotinic acid and Zn(NO$_3$)$_2$·6H$_2$O in DMF. Further, a solution of [Ir(COD)(OCH$_3$)]$_2$ (COD = 1,5-cyclooctadiene) in chloroform was added to it and heated at 55 °C for 24 h to obtain IRMOF-9-Irdcppy-NH$_2$ MOF (**Figure 19**). At 0.015 mmol of Ir, this catalyst was able to catalyse tandem Knoevenagel condensation reaction and allylic N-alkylation reaction between indoline-7-carboxaldehyde, diallyl carbonate and malonitrile in CDCl$_3$ at 55 °C for 36 h to yield 2-((1-allylindolin-7-yl)methylene)malononitrile with 95% conversion. However, the homogenous catalyst, [Ir(COD)(OCH$_3$)]$_2$ (5 mol % of Ir) gave only 35% conversion after 5 d in similar reaction conditions.

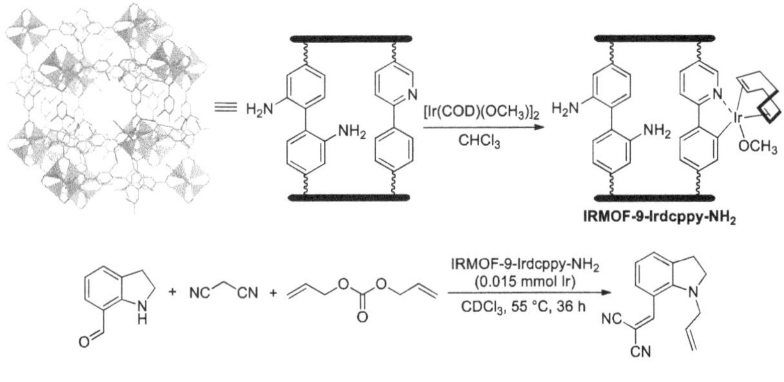

Figure 19. Synthesis of IRMOF-9-Irdcppy-NH$_2$ MOF and its application in tandem Knoevenagel condensation reaction and allylic N-alkylation reaction.

5.3 MOFs having Both Acid-Base Bifunctional Sites

MOFs have the ability to tune their structure and can undergo modifications by changing their functionalities which makes them suitable for having bifunctional sites and thereby catalysing various types of reactions including tandem or cascade reactions [86]. There have been many examples where acidic and basic active sites have been incorporated in the metal nodes or the organic linkers of MOFs directly or by post-synthetic modifications.

In 2012, Zhou and his co-workers, synthesised PCN-124 MOF from copper nitrate and ligands having various functional groups such as carboxylate, pyridine and amide [5,5′-((pyridine-3,5-dicarbonyl)bis(azanediyl))diisophthalate] (PDAI) [87]. Here, the Cu(II) centres serve as the Lewis acid centres while the pyridine and amide groups are the Lewis basic centres and there is a cooperative action between these two centres which ultimately makes PCN-124 an efficient catalyst to perform tandem deacetalisation–Knoevenagel condensation reactions. At 0.5 mol % catalyst loading, deacetalization of dimethoxymethylbenzene in DMSO-d_6 yielded benzaldehyde after 12 h. Further, Knoevenagel condensation of benzaldehyde with malononitrile gave benzylidene malononitrile with 100% conversion.

In 2019, Su and coworkers opted for a dynamic installation strategy for the construction of various types of Zr- MOFs with different functionality from the *proto*-MOF LIFM-28 having a formula of [(Zr$_6$(μ_3-O)$_4$(μ_3-OH)$_4$(H$_2$O)$_4$(OH)$_4$L$_4$]·solvents where L is 2,2′-bis-(trifluoromethyl)-4,4′-biphenyldicarboxylate [88]. BPYDC (2,2′-bipyridine-5,5′-dicarboxylic acid) and NH$_2$-TPDC (2-amino-terphenyl4,4′-dicarboxylic acid) were inserted into single *proto*-MOF LIFM-28 to get LIFM-80 and insertion of BPYDC led to the formation of LIFM-28-BPYDC which were further modified and metalated with Cu(I) to give LIFM-80(Cu) and LIFM-28-BPYDC(Cu). The open Cu(I)-chelating sites were supposedly worked as Lewis acidic sites for the aerobic oxidation of the benzyl alcohol to benzaldehyde while the amine groups(-NH$_2$) showing Brønsted basicity are responsible for promoting the Knoevenagel condensation by converting the intermediate benzaldehyde to benzylidene malononitrile (**Figure 20**). LIFM-80(Cu) catalyzed tandem reaction by oxidising phenylmethanol to benzaldehyde, and then acetal reaction of the aldehydes to form (dimethoxymethyl)benzene with 99% conversion after 6 h.

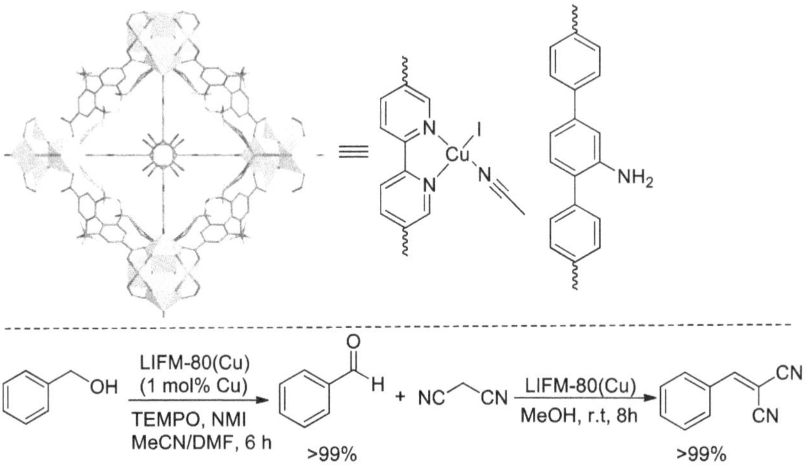

Figure 20. LIFM-80(Cu) MOF having both acid-base bifunctional sites assisting in tandem catalysis.

6. Asymmetric Catalysis

6.1 Early Examples

Kim and co-workers were the first to report a chiral MOF catalyst and demonstrated its catalytic activity in the stereospecific transesterification reactions in 2000 [89]. The chiral MOF was synthesized by using enantiopure metal-organic clusters (Zinc) as secondary building blocks (SBUs) and the organic linker was derived from a tartaric acid derivative. This report was a breakthrough as the polymeric structure of those polymeric chiral catalysts reported earlier was not retained after the reaction conditions [1, 90]. This chiral Zinc MOF (*D*-POST-1) is constructed from (4S,5S)-2,2-dimethyl-5-(pyridin-4-ylcarbamoyl)-1,3-dioxolane-4-carboxylate (**1L**) linker and trinuclear zinc cluster. The zinc cluster has connected with the six carboxylates group of chiral ligands (**1L**) and a bridging oxo oxygen to form SBU and these SBUs are then interconnected through coordinate covalent bonds between the zinc ions and pyridyl groups of **1L**, to afford a two-dimensional (2D) MOF (*D*-POST-1). Out of the six coordinated ligands, half of the pyridine groups did not participate in the network structure. This accessible basic site was utilized for catalysis in the asymmetric transesterification reactions. Later, the next year Lin and his team reported the first truly asymmetric catalytic transformation occurring at the metal centre of chiral MOF that was derived from a lanthanum group metals and a phosphonate-substituted diethoxy-BINOL ligand [91]. The MOF was an active catalyst for asymmetric cyanosilylation of aldehydes, although poor enantioselectivity was observed (5% ee). Since then, a significant progress in asymmetric catalysis using chiral MOFs has been made.

6.2 Asymmetric Hydrogenation

Catalytic asymmetric hydrogenation of prochiral alkenes, ketones, and imines is one of the most efficient methods to produce optically active compounds. Among many chiral catalysts, Noyori's 2,2'-bis-(diphenylphosphino)-1,1'-binaphthyl (BINAP)-based Ru and Rh homogeneous catalysts are widely used for the hydrogenation of a range of substrates with high enantioselectivity [92-95]. Lin and co-workers in 2003 heterogenized the Noyori's Ru-BINAP-DPEN (DPEN: 1,2-diphenylethylenediamine) catalytic moieties within a porous Zr phosphonate framework [96]. The chiral zirconium phosphonate (Zr-Ru-L_1 MOF) was synthesized by solvothermal reaction of a chiral phosphonic acid-substituted Ru-BINAP-DPEN and Zr(O^{t}Bu)$_4$. The chiral Zr phosphonate was an active and recyclable heterogeneous catalyst for asymmetric hydrogenation of aromatic ketones giving the corresponding chiral secondary alcohols with ee up to 99%. Importantly, the enantioselectivity of Zr phosphonates was superior to their homogeneous counterpart, Ru-BINAP-DPEN (**Figure 21**). A seminal Zr phosphonate framework (Zr-Ru-L_2 MOF) bearing Ru(BINAP)(DMF)$_2$Cl$_2$ moiety, reported by the same group, was also highly active and enantioselective in the hydrogenation of a wide range of β-keto esters with *ee* values of up to 95% [97].

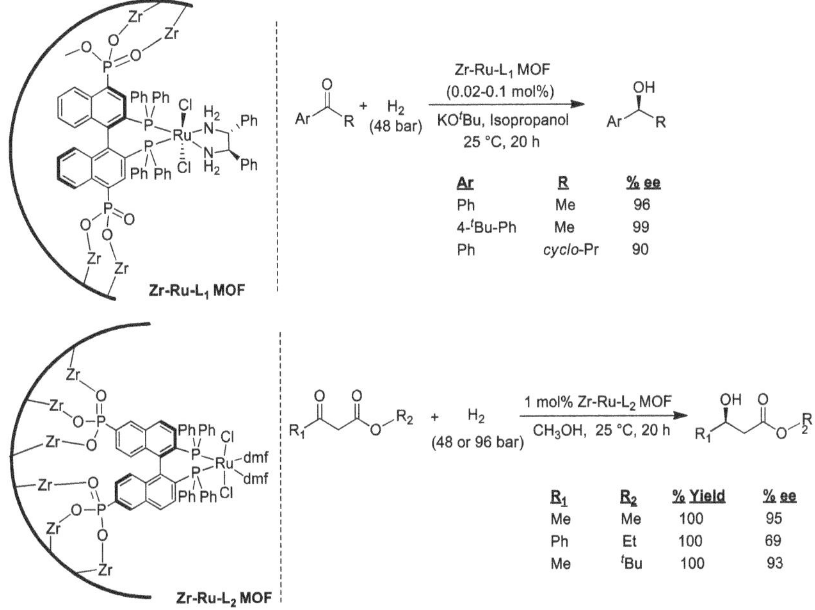

Figure 21. MOF-supported Ru catalyst for asymmetric hydrogenation of aromatic ketones and β-keto esters.

In 2014, Lin and coworkers reported the development of MOF-supported Ru-BINAP catalyst via post-synthetic metalation of BINAP-functionalized MOF [13]. A solvothermal reaction between 4,4′-bis(4-carboxyphenylethynyl)BINAP and $ZrCl_4$, in presence of trifluoroacetic acid (TFA) in DMF yielded the chiral BINAP-MOF. The MOF was built from $Zr_6O_4(OH)_4(O_2CR)_{12}$ cluster and BINAP-derived dicarboxylate linker to adopt a UiO topology (**Figure 22**). The post-synthetic metalation of BINAP-MOF with Ru-precursor afforded highly enantioselective single-site chiral metal catalyst for asymmetric hydrogenation of β-keto esters and substituted alkenes.

Recently, Liu, Cui and coworkers synthesized several highly robust chiral Zr-MOF with a **flu** or **ith** topology built from Zr-oxo-hydroxo cluster and enantiopure 1,1′-biphenol-derived tetracarboxylate linker [98]. The sequential post-synthetic metalation of **flu** MOFs, featuring dihydroxyl groups of biphenol in open and large cages, with $P(NMe_2)_3$ and $[Ir(COD)Cl]_2$, afforded heterogeneous chiral catalysts for enantioselective hydrogenation of α-dehydroamino acid esters with ee up to 98%.

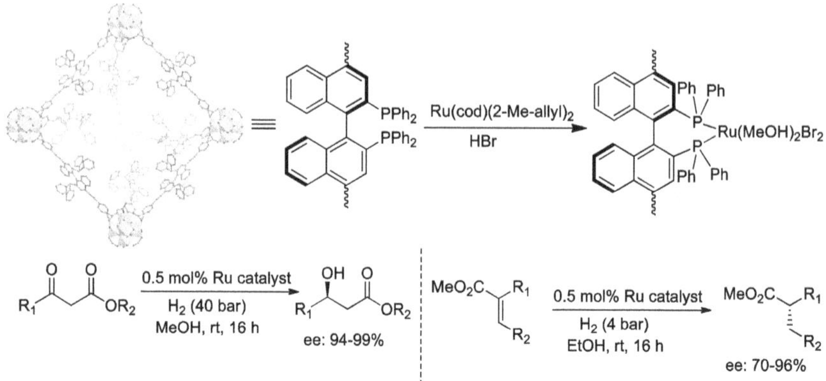

R1	R2	% ee
Me	Me	97
Et	Et	94
Me	tBu	>99

R1	R2	% ee
NHAc	H	85
NHAc	Ph	70
CH2CO2Me	H	91

Figure 22. Synthesis of Ru-catalyst via post-synthetic metalation of chiral BINAP-functionalized MOF for asymmetric hydrogenation of β-keto ester and substituted alkenes.

At 0.5 mol% Ir-loading, the flu-MOF-Ir catalysts efficiently hydrogenated a wide range of α-dehydroamino acid esters under 5 bar of H_2 in dichloromethane as the solvent at room temperature to obtain the corresponding hydrogenated products in high yields and enantiomeric excesses up to 98% (**Figure 23**). In contrast to flu-MOF-Ir, the analogous iridium complexes of **ith** Zr-MOFs could not be prepared as the **ith** Zr-MOFs did not react with $P(NMe_2)_3$ due to the steric crowding around the dihydroxyl groups in small cages that prevents coordination of biphenol to the phosphine.

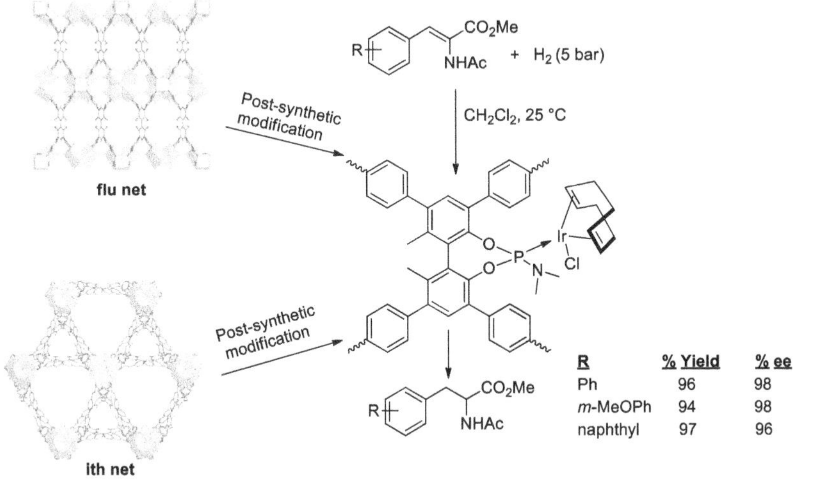

Figure 23. 1,1′-Biphenol-functionalized Zr-MOFs with a flu or ith topology as the chiral supported ligands to develop iridium catalyst for asymmetric hydrogenation reaction of α-dehydroamino acid esters.

6.3 Asymmetric Cyanosilylations of Carbonyl Compounds

Asymmetric cyanosilylation of carbonyls is important for synthesizing chiral cyanohydrins, which are useful intermediates to prepare various optically active compounds [91, 99-102]. In 1994, Fujita, Ogura and co-workers first reported a crystalline two-dimensional square network material constituted of cadmium(II) and 4,4'-bipyridine, which was active in shape selectively cyanosilylation of aldehydes [1]. Duan and coworkers fabricated two homochiral MOFs having one-dimensional channels and coordinatively unsaturated metal sites using Ce^{+3} nodes and methylenediisophthalate as the linkers in the presence of chiral pyrrolidine-2-yl-imidazole integrated within the frameworks [103]. The handedness of the MOF could be controlled by changing the chiral template; D- or L-pyrrolidine-2-yl-imidazole. Both Homochiral Ce-MOFs displayed excellent activity and enantioselectivity in cyanosilylation of aldehydes at room temperature to obtain optically active cyanohydrins with ee up to 98% (**Figure 24**).

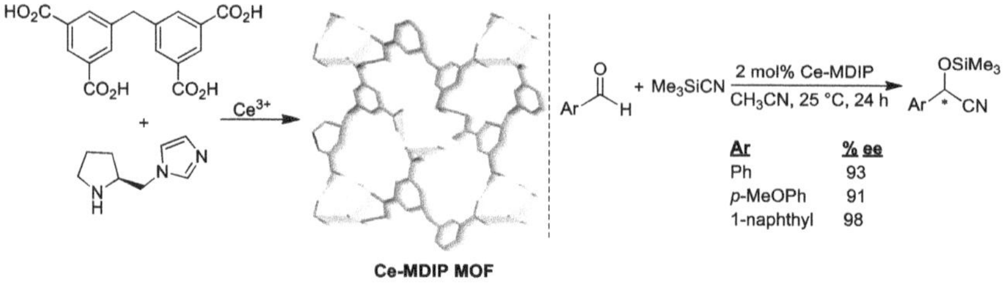

Ce-MDIP MOF

Figure 24. Cyanosilylation of aldehydes catalyzed by Ce-MDIP.

Cui and his team in 2014 reported a homochiral Zn-MOF built from Zn_4O node and enantiopure 2,2′-dihydroxy-1,1′-biphenyl as the chiral linker [104]. Upon lithiation of one hydroxyl group, the lithiated MOF (Li-Zn MOF) became an active heterogeneous catalyst for asymmetric cyanation of aldehydes with up to >99% ee (**Figure 25**).

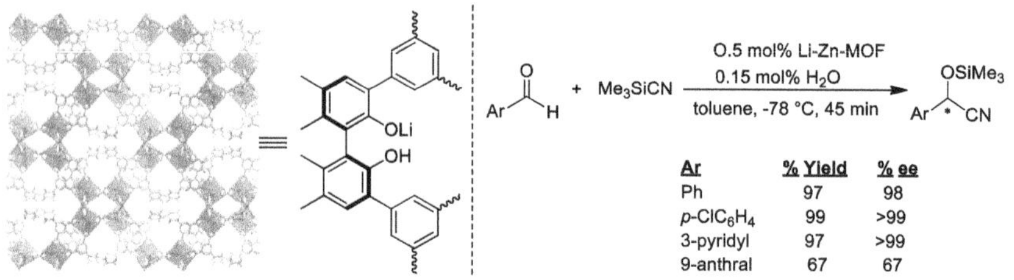

Figure 25. Asymmetric cyanosilylation of aldehydes catalyzed by Li-Zn MOF.

6.4 Asymmetric Aldol Reaction

Asymmetric aldol reactions using chiral organocatalysts are common for the enantioselective C–C bond formations. Several MOF-supported chiral organocatalysts had been developed in recent years for asymmetric aldol reactions [105]. Prof. Kimon Kim and his co-workers in 2009, demonstrated a post-synthetic modification of MIL-101-MOF (MIL, Matérial Institut Lavoisier) by attaching L-proline derived chiral ligands, [(*S*)-N-(pyridin-3-yl)-pyrrolidine-2-carboxamide] and [(*S*)-N-(pyridin-4-yl)-pyrrolidine-2-carboxamide], to its open coordination sites affording chiral MOFs for asymmetric aldol reactions [106]. The resultant chiral MOFs showed higher catalytic activities than the chiral ligand itself. At 10 mol% catalyst loading, these MOFs showed moderate to good enantioselectivity in aldol reaction between various aldehydes and ketones at room temperature producing the aldol products with ee up to 80% (**Figure 26**).

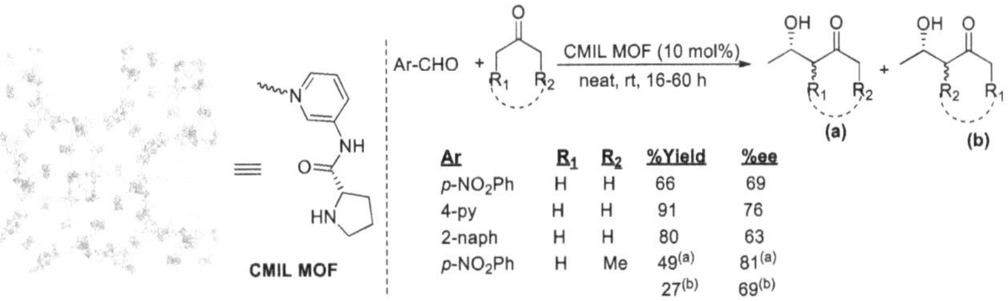

Ar	R_1	R_2	%Yield	%ee
p-NO$_2$Ph	H	H	66	69
4-py	H	H	91	76
2-naph	H	H	80	63
p-NO$_2$Ph	H	Me	49[a]	81[a]
			27[b]	69[b]

Figure 26. CMIL MOF catalyzed asymmetric aldol reactions between different aldehydes and ketones.

6.5 Asymmetric Diels-Alder Reaction

The [4 + 2] cycloaddition between a diene and a dienophile, known as the Diels–Alder reaction, is a useful C–C bond forming method in synthetic organic chemistry. In asymmetric Diels–Alder reactions, a chiral MOF facilitates the encapsulation of the reactants within its pores and organizes them in a constricted chiral reaction space containing catalytic sites. In 2011, Jeong and co-workers developed a Cu-MOF, (*S*)-KUMOF-1 having formula of Cu$_2$(*S*)-L)$_2$(H$_2$O)$_2$ [L = 2,2′-dihydroxy-6,6′-dimethyl(1,1′biphenyl)-4,4′-dicarboxylate] [107]. The MOF has a non-interpenetrating NbO type framework proving a large pore of 2 × 2 × 2 nm³, and two Cu(II) ions are held by four carboxylate groups of chiral biphenyl linkers to form a paddle-wheel SBU. After reaction of (*S*)-KUMOF-1 with Ti(O-iPr)$_4$, the resultant chiral MOFs exhibited excellent enantioselectivity for hetero Diels-Alder reaction between diene and aldehyde, resulting in cyclic ether product with ee up to 55% (**Figure 27**). The homogeneous control of this catalyst yielded the product in a much lower yield (58%) and ee (18%) compared to those with Ti/(*S*)-KUMOF-1 under identical reaction conditions, highlighting the beneficial effect of MOF frameworks in the improvement of enantioselectivity and yield of the products.

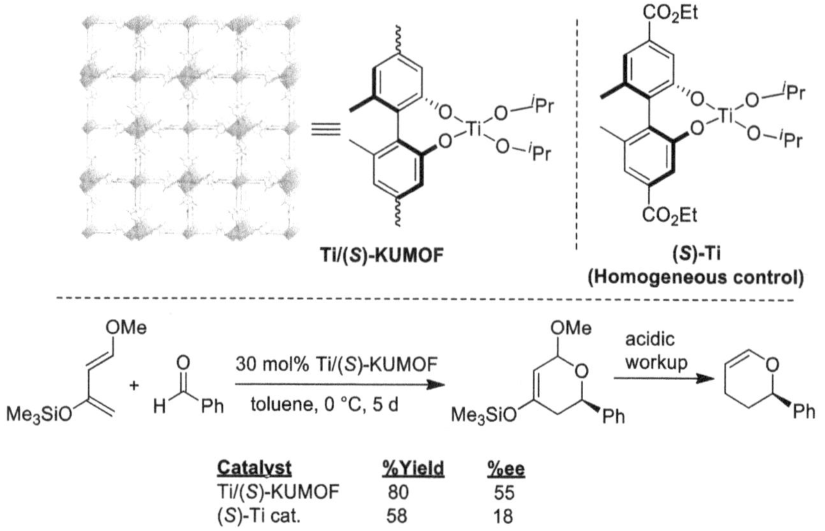

Catalyst	%Yield	%ee
Ti/(S)-KUMOF	80	55
(S)-Ti cat.	58	18

Figure 27. Ti/(S)-KUMOF for catalytic asymmetric hetero Diels-Alder reaction and comparison of its catalytic activity with the homogeneous control [(S)-Ti].

In 2016, Tanaka and his co-workers reported a homochiral porous MOF for asymmetric Diels-Alder reaction of isoprene and N-ethyl maleimide [108]. The MOF was synthesized using $Cu(NO_3)_2.3H_2O$ and (R)-2,2'-dihydroxy-1,1'-binaphthyl-4,4'-dibenzoic acid as the chiral ligand in DMF and water. Under optimised reaction conditions, ees up to 75% were obtained when the reactions were carried out at 0 °C in ethyl acetate. In the same year, Cui and co-workers developed a highly enantioselective Cr(salen)-based MOF catalyst for Nazarov cyclization [109].

6.6 Asymmetric Organozinc Addition to Carbonyls

Organozinc addition to aldehydes and ketones using chiral catalysts is an important method to prepare optically active alcohols. Lin and co-workers first developed a chiral BINOL-ligated titanium catalyst supported by a Cd-MOF for asymmetric $ZnEt_2$ additions to aromatic aldehydes [110]. The homochiral porous MOF having a formula of $[Cd_3Cl_6L_3] \cdot 4DMF \cdot 6MeOH \cdot 3H_2O$ was constructed from (R)-6,6'-dichloro-2,2'-dihydroxy1,1'-binaphthyl-4,4'-bipyridine linker (L) and $[Cd(\mu\text{-}Cl)_2]_n$ SBUs and has a non-interpenetrating 3D network with very large chiral channels of ~1.6 × 1.8 nm cross-section (**Figure 28**).

The reaction of dihydroxy groups in chiral BINOL within the Cd-MOF and Ti(OiPr)$_4$ formed (BINOLate)Ti(O^{i}Pr)$_2$ species, which was an active catalyst for the addition of ZnEt$_2$ to aromatic aldehydes to afford optically active secondary alcohols with ee up to 93%.

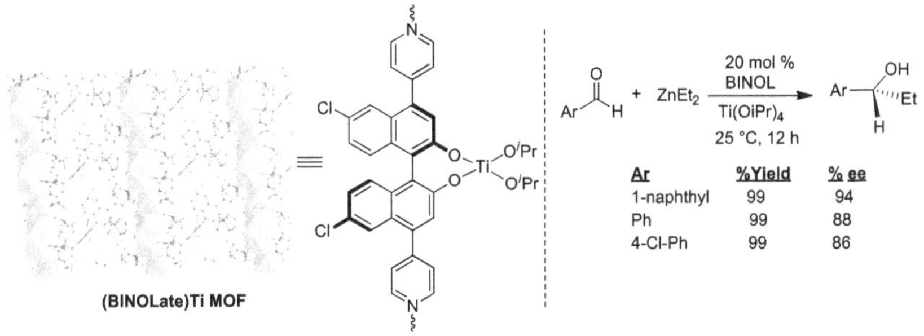

Figure 28. (BINOLate)Ti MOF and its application in asymmetric diethyl zinc addition reactions.

The same group later developed a series of isoreticular chiral MOFs bearing chiral 1,1′-bi-2-naphthol moiety at their linkers, which have the same structures but different sizes of pores [111]. These MOFs were constructed from copper paddle-wheels nodes and BINOL-derived tetracarboxylic acid bridging ligands. Upon metalation with Ti(O^{i}Pr)$_4$, the resultant chiral MOF became an efficient catalyst for diethylzinc and alkynylzinc addition to carbonyls to afford a range of chiral secondary alcohols. The pore sizes of these MOFs were tuned to demonstrate optimizing the reaction rate and enantioselectivity of the catalysts. The MOF with a smaller linker and narrow channels showed almost zero or no enantioselectivity, while in contrast, MOFs having larger pores showed good enantioselectivity of up to 91%.

6.7 Epoxidations and Epoxide Ring-opening Reactions

Epoxidation of alkenes transforms C=C double bond into epoxides or cyclic ethers by using oxidising agents such as peracids. However, using MOF as a catalyst can be fruitful in terms of selectivity, stability and activity of the catalyst as they possess properties like high surface area, high thermal stability, tunable porosity, recyclability etc [112-114]. MOF provides a unique platform for synthesizing single-site solid catalyst for various organic transformations with high activity and high selectivity. Moreover, encapsulation of molecular catalysts inside the pores of the MOF results in catalytic site isolation thus preventing the intermolecular decomposition and deactivation of the catalyst. The first example of MOF-catalyst for asymmetric epoxidation of alkene was reported by Hupp and co-workers in 2006, in which the active site, (*R,R*)-1,2-cyclohexanediamino-N,N'-bis(3-*tert*-butyl-5-(4-pyridyl)salicylidene)Mn^{III}Cl was dopped in a Zn-MOF having biphenyldicarboxylate as the second linker and pore sizes of 0.62 x 1.57 nm. This MOF was a highly effective solid catalyst for enantioselective olefin epoxidation affording chiral epoxides with ee up to 82% and TON up to 1400. The solid catalyst could be recycled up to 3 times without diminishing the enantioselectivity [112]. Later in 2010, Lin and co-workers synthesized a series of isoreticular MOFs having primitive cubic network topology but with different pore sizes built from [Zn$_4$(μ_4-O)(O$_2$CR)$_6$] SBUs and systematically elongated dicarboxylate linkers bearing chiral Mn-Salen catalytic subunits (**Figure 29**) [113]. At 0.1 mol% catalyst loading, these MOFs showed high activity and enantioselectivity in the epoxidation of olefins with ee up to 92%. The authors also concluded that the catalytic activity of MOFs with larger pore sizes is limited by the intrinsic reactivity of the catalytic building blocks, although the diffusion of bulky substrates can be a rate-limiting factor for the catalysis.

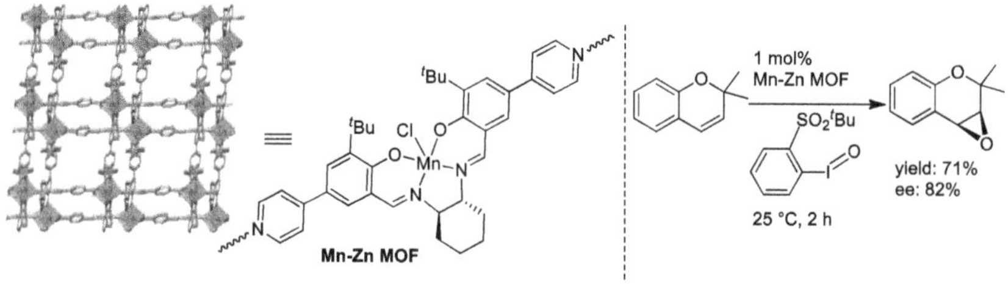

Figure 29. Mn/Zn MOF-catalyzed enantioselective epoxidation of 2,2-dimethylchromene.

In 2017, Yong Cui and his team demonstrated a synthetic strategy to incorporate different synergistic functionalities into one crystalline network to develop the MOF that can be used as an excellent platform to engineer heterogeneous catalysts containing multiple and cooperative active sites [115]. Following this strategy, the group developed a series of MOFs bearing chiral salen-derived bridging linkers and different metal centres like Mn, Co, Fe, V, Ru, Cu, Cr. They performed a variety of asymmetric sequential alkene epoxidation and epoxide ring-opening reactions using those chiral MOFs to exhibit moderate to good enantioselectivities (11-96% ee). They also demonstrated that the increasing catalyst loading value would increase the conversion rate and enantioselectivity, which was also higher than its homogeneous counterparts.

6.8 Asymmetric Cyclopropanations

The generation of cyclopropane rings in any enantioenriched forms via chemical reaction processes is commonly known as asymmetric cyclopropanation reactions. The high ring strain present in the three-membered ring makes the synthesis process challenging. In 2011, Lin and co-workers developed a chiral MOF of the primitive cubic unit (pcu) topology having a formula of $[Zn_4(\mu_4\text{-}O)[(Ru(L)(py)_2Cl]_3]\cdot(DEF)_{19}\cdot(DMF)_5\cdot(H_2O)_{17}$, where L is $(R,R)\text{-}(-)\text{-}$N,N′-(methyl-3-carboxyl-5-tert-butylsalicylidene)-1,2-cyclohexanediamine [116]. The SBU of the synthesized MOF was constructed with Zn metal cluster, whereas the Ruthenium centre is the catalyst's active site. The Ru^{II} MOF obtained by reducing catalytically inactive Ru^{III} MOF using LiEt₃BH has an open channel dimension of 1.4×1.0 nm² and a cavity size of 1.7 nm. At a 3 mol% Ru- loading and in the presence of NaBH(OMe)₃, the MOF was a highly active and enantioselective catalyst for asymmetric cyclopropanation of various substituted alkenes (ee up to 91%) at room temperature (**Figure 30**). The addition of NaBH(OMe)₃ was necessary to prevent the oxidation of Ru^{II} species that prevent catalyst deactivation.

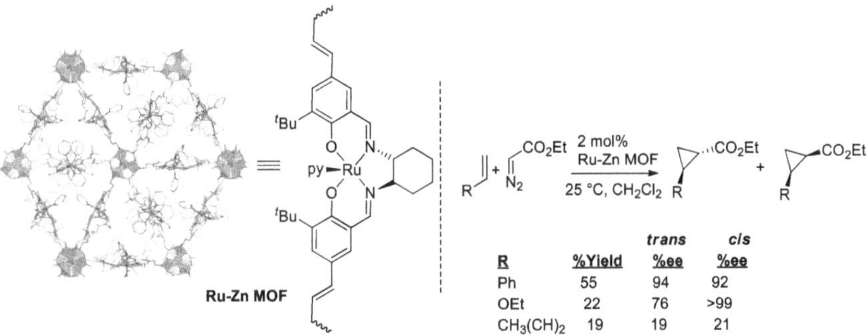

Figure 30. Ru-Zn MOF for catalytic asymmetric cyclopropanation reactions of terminal olefins.

6.9 Cycloaddition

When two or more unsaturated molecules or two unsaturated bonds in the same molecules combine to form a cyclic adduct, it is known as cycloaddition reaction. Chiral cyclic compounds can be obtained in different asymmetric reaction pathways such as reductive cyclization, Alder-enecycloisomerization, CO_2 cycloaddition reactions, alkene oxidation, Pauson-Khand cycloaddition and 1,3-dipolar cycloaddition. In addition, asymmetric cycloaddition is one of the best routes to obtain optically active cyclic carbonates or cyclic urethanes using CO_2 [117]. Prof. Chunying Duan and co-workers in 2012 designed a homochiral metal-organic silver framework for 1,3-dipolar cycloaddition reactions [118]. They incorporated three pyridine–imine bidentate chelators into a triphenylamine fragment which upon treatment with $AgBF_4$ in methanol in presence of cinchonine or cinchonidine as a chiral source, afforded the desired silver-based chiral MOFs. The cycloaddition reaction between 2-(benzylidene-amino)acetate (MBA) and N-methylmaleimide (NMM), resulted high conversion (90%) and enantioselectivity (90%) after 24 h in presence of 3 mol% catalyst to give the *endo* adduct, (1*S*,2*R*,4*S*,5*R*)-Methyl-4-phenyl-7-methyl-6,8-dioxo-3,7-diazabicyclo[3.3.0]octane-2-carboxylate (**Figure 31**). However, the 1,3-dipolar cycloaddition reaction did not proceed in the presence of only the chiral ligand, (+)-cinchonine or (−)-cinchonidin without the MOF-catalyst. Authors concluded that the high enantioselectivity of the chiral Ag-TPHA MOF-catalyst is originated presumably due to the restricted movement of the substrates within the confined pores of the MOF.

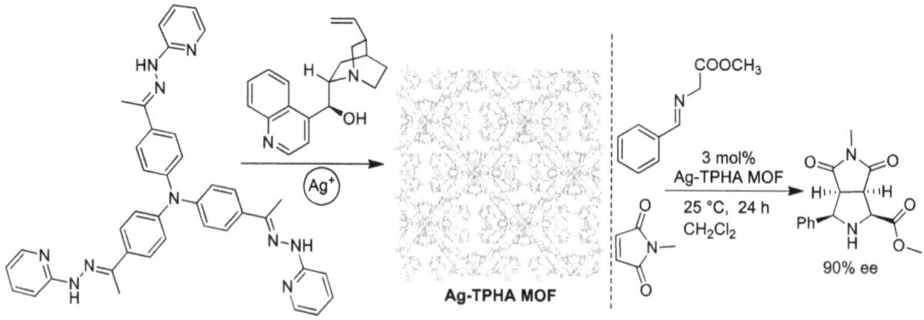

Figure 31. Ag-TPHA MOF and its application in asymmetric 1,3-dipolar cycloaddition.

Lin and co-workers in 2015 reported Rh-catalysts supported by the chiral BINAP-based UiO MOF, discussed in section 2.2, for asymmetric cyclization reactions of 1,6-enynes [119]. The chiral BINAP-MOF was metalated with $Rh(nbd)_2BF_4$ or $[Rh(nbd)Cl]_2/AgSbF_6$ to obtain a highly enantioselective catalyst for reductive cyclization and Alderene cycloisomerization of 1,6-enynes, respectively (**Figure 32**).

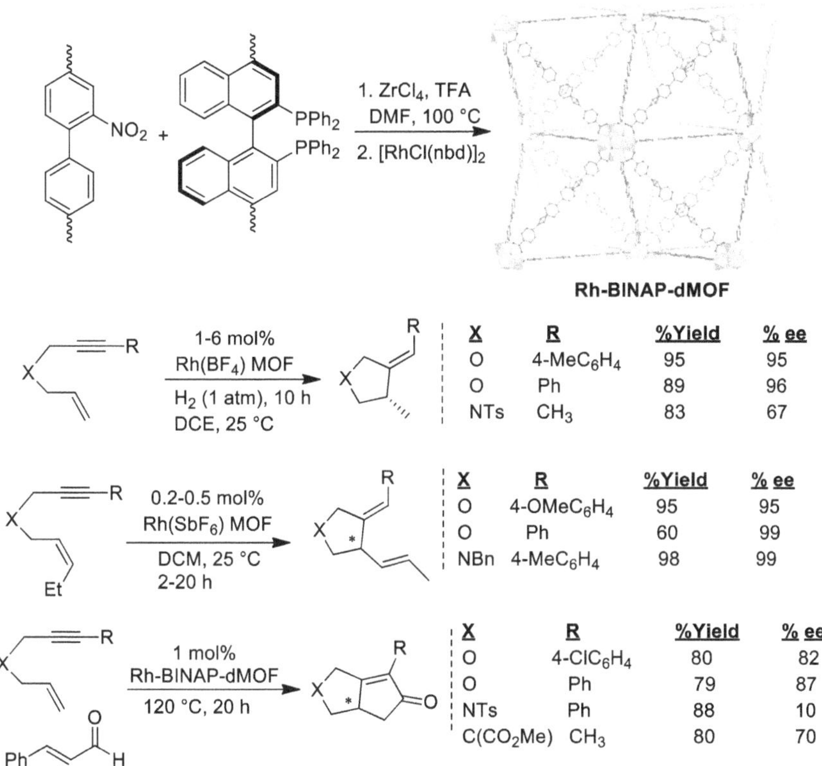

First reaction (1-6 mol% Rh(BF₄) MOF, H₂ (1 atm), 10 h, DCE, 25 °C):

X	R	%Yield	% ee
O	4-MeC₆H₄	95	95
O	Ph	89	96
NTs	CH₃	83	67

Second reaction (0.2-0.5 mol% Rh(SbF₆) MOF, DCM, 25 °C, 2-20 h):

X	R	%Yield	% ee
O	4-OMeC₆H₄	95	95
O	Ph	60	99
NBn	4-MeC₆H₄	98	99

Third reaction (1 mol% Rh-BINAP-dMOF, 120 °C, 20 h):

X	R	%Yield	% ee
O	4-ClC₆H₄	80	82
O	Ph	79	87
NTs	Ph	88	10
C(CO₂Me)	CH₃	80	70

Figure 32. Synthesis of Rh-BINAP-mix-MOF and application of Rh-BINAP MOF in asymmetric cyclisation reactions.

The authors also reported the synthesis of Rh-BINAP-mix-MOF following a mixed-ligand strategy. The mixed BINAP-MOF was built from 4,4'-((7,7'-bis(diphenylphosphino)-[1,1'-binaphthalene]-4,4'-diyl)bis(ethyne-2,1-diyl))dibenzoic acid and unfunctionalized dicarboxylate linkers (4,4'-((2-nitro-[1,1'-biphenyl]-4,4'-diyl)bis(ethyne-2,1-diyl))dibenzoic acid) in 1:4 ratio (**Figure 12**). At 1-6 mol% Rh-loading, and 1 atm H_2, $Rh(BF_4)$/BINAP-MOF was the active catalyst for asymmetric reductive cyclization of various substituted (electron-donating groups like Me, *t*Bu) and unsubstituted 1,6-enynes in room temperature after 10 h, (ee up to 96%). In addition, Rh(SbF$_6$)/BINAP-MOF was highly active in Alder-ene cycloisomerization of a range of substituted 1,6-enynes (*p*-Me, *p*-OMe substitution) to obtain 1,4-diene products in excellent yields and ee up to 99%. However, at 1 mol% Rh-loading, Rh-BINAP-mix-MOF was the most active catalyst for the Pauson-Khand reaction of 1,6-enyne and *trans*-cinnamyl aldehyde at 120 °C in 20 h with high conversion (up to 91%) and enantioselectivity (up to 87% ee). The inclusion of an unfunctionalized linker within the MOF created extra void space which made diffusion of sterically demanding bicyclic rhodacycle intermediates produced during the reaction cycle feasible, thus accelerating the reaction rate and increasing the selectivity. Under the similar reaction condition, Rh-BINAP-mix-MOF catalyzed Pauson-Khand reaction of various substituted 1,6-enynes and *trans*-cinnamyl aldehyde to give the cyclic product with good yield and ee up to 87%.

7. Conclusion

The design of MOF-based catalysts holds promises in developing highly active and selective heterogeneous catalysts for various applications due to its numerous advantages such as crystallinity, high porosity, isoreticular and tunable structures and high surface area. A range of catalytical reactions, including reductions, oxidations, tandem and asymmetric catalysis using MOF catalysts, has been discussed in this chapter. The chapter has covered recent developments in preparing MOF catalysts using several synthetic approaches, such as encapsulating metal or metal nanoparticles within the pores and incorporating metal centres at SBUs or organic linkers. Well defined single-site metal catalysts can easily be prepared at the nodes and linkers of the MOFs via active-site isolation that prevents intermolecular decomposition.

References

1. Fujita, M., Kwon, Y. J.,Washizu, S., and Ogura, K. (1994) Preparation, Clathration Ability, and Catalysis of a Two-Dimensional Square Network Material Composed of Cadmium(II) and 4,4'-Bipyridine. *Journal of the American Chemical Society*, **116**(3), 1151-1152.

2. Hönicke, I. M., Senkovska, I., Bon, V., Baburin, I. A., Bönisch, N., Raschke, S., Evans, J. D., and Kaskel, S. (2018) Balancing Mechanical Stability and Ultrahigh Porosity in Crystalline Framework Materials. *Angewandte Chemie International Edition*, **57**(42), 13780-13783.

3. Eddaoudi, M., Kim, J., Rosi, N., Vodak, D., Wachter, J., O'Keeffe, M., and Yaghi Omar, M. (2002) Systematic Design of Pore Size and Functionality in Isoreticular MOFs and Their Application in Methane Storage. *Science*, **295**(5554), 469-472.

4. Phan, A., Czaja, A. U., Gándara, F., Knobler, C. B., and Yaghi, O. M. (2011) Metal–Organic Frameworks of Vanadium as Catalysts for Conversion of Methane to Acetic Acid. *Inorganic Chemistry*, **50**(16), 7388-7390.

5. Schlichte, K., Kratzke, T., and Kaskel, S. (2004) Improved synthesis, thermal stability and catalytic properties of the metal-organic framework compound $Cu_3(BTC)_2$. *Microporous and Mesoporous Materials*, **73**(1), 81-88.

6. Alaerts, L., Séguin, E., Poelman, H., Thibault-Starzyk, F., Jacobs, P. A., and De Vos, D. E. (2006) Probing the Lewis Acidity and Catalytic Activity of the Metal–Organic Framework $[Cu_3(BTC)_2]$ (BTC=Benzene-1,3,5-tricarboxylate). *Chemistry – A European Journal*, **12**(28), 7353-7363.

7. Ravon, U., Savonnet, M., Aguado, S., Domine, M. E., Janneau, E., and Farrusseng, D. (2010) Engineering of coordination polymers for shape selective alkylation of large aromatics and the role of defects. *Microporous and Mesoporous Materials*, **129**(3), 319-329.

8. Ji, P., Manna, K., Lin, Z., Feng, X., Urban, A., Song, Y., and Lin, W. (2017) Single-Site Cobalt Catalysts at New $Zr_{12}(\mu_3\text{-}O)_8(\mu_3\text{-}OH)_8(\mu_2\text{-}OH)_6$ Metal–Organic Framework Nodes for Highly Active Hydrogenation of Nitroarenes, Nitriles, and Isocyanides. *Journal of the American Chemical Society*, **139**(20), 7004-7011.

9. Lin, Z., Thacker, N. C., Sawano, T., Drake, T., Ji, P., Lan, G., Cao, L., Liu, S., Wang, C., and Lin, W. (2018) Metal–organic layers stabilize earth-abundant metal–terpyridine diradical complexes for catalytic C–H activation. *Chemical Science*, **9**(1), 143-151.

10. Li, Z., Schweitzer, N. M., League, A. B., Bernales, V., Peters, A. W., Getsoian, A. B., Wang, T. C., Miller, J. T., Vjunov, A., Fulton, J. L., Lercher, J. A., Cramer, C. J., Gagliardi, L., Hupp, J. T., and Farha, O. K. (2016) Sintering-Resistant Single-Site Nickel Catalyst Supported by Metal–Organic Framework. *Journal of the American Chemical Society*, **138**(6), 1977-1982.

11. Hwang, Y. K., Hong, D.-Y., Chang, J.-S., Jhung, S. H., Seo, Y.-K., Kim, J., Vimont, A., Daturi, M., Serre, C., and Férey, G. (2008) Amine Grafting

on Coordinatively Unsaturated Metal Centers of MOFs: Consequences for Catalysis and Metal Encapsulation. *Angewandte Chemie International Edition*, **47**(22), 4144-4148.

12. Akiyama, G., Matsuda, R., Sato, H., Takata, M., and Kitagawa, S. (2011) Cellulose Hydrolysis by a New Porous Coordination Polymer Decorated with Sulfonic Acid Functional Groups. *Advanced Materials*, **23**(29), 3294-3297.

13. Falkowski, J. M., Sawano, T., Zhang, T., Tsun, G., Chen, Y., Lockard, J. V., and Lin, W. (2014) Privileged Phosphine-Based Metal–Organic Frameworks for Broad-Scope Asymmetric Catalysis. *Journal of the American Chemical Society*, **136**(14), 5213-5216.

14. Manna, K., Zhang, T., and Lin, W. (2014) Postsynthetic Metalation of Bipyridyl-Containing Metal–Organic Frameworks for Highly Efficient Catalytic Organic Transformations. *Journal of the American Chemical Society*, **136**(18), 6566-6569.

15. Bogaerts, T., Van Yperen-De Deyne, A., Liu, Y.-Y., Lynen, F., Van Speybroeck, V., and Van Der Voort, P. (2013) Mn-salen@MIL101(Al): a heterogeneous, enantioselective catalyst synthesized using a 'bottle around the ship' approach. *Chemical Communications*, **49**(73), 8021-8023.

16. Jiang, H.-L., Liu, B., Akita, T., Haruta, M., Sakurai, H., and Xu, Q. (2009) Au@ZIF-8: CO Oxidation over Gold Nanoparticles Deposited to Metal–Organic Framework. *Journal of the American Chemical Society*, **131**(32), 11302-11303.

17. Pi, Y., Feng, X., Song, Y., Xu, Z., Li, Z., and Lin, W. (2020) Metal–Organic Frameworks Integrate Cu Photosensitizers and Secondary Building Unit-Supported Fe Catalysts for Photocatalytic Hydrogen Evolution. *Journal of the American Chemical Society*, **142**(23), 10302-10307.

18. Feng, X., Pi, Y., Song, Y., Brzezinski, C., Xu, Z., Li, Z., and Lin, W. (2020) Metal–Organic Frameworks Significantly Enhance Photocatalytic Hydrogen Evolution and CO_2 Reduction with Earth-Abundant Copper Photosensitizers. *Journal of the American Chemical Society*, **142**(2), 690-695.

19. Béchamp Reduction (2009). In: *Comprehensive Organic Name Reactions and Reagents*, Wang, Z.(ed.), Wiley, pp. 284-287.

20. Formenti, D., Ferretti, F., Scharnagl, F. K., and Beller, M. (2019) Reduction of Nitro Compounds Using 3d-Non-Noble Metal Catalysts. *Chemical Reviews*, **119**(4), 2611-2680.

21. Antil, N., Kumar, A., Akhtar, N., Newar, R., Begum, W., Dwivedi, A., and Manna, K. (2021) Aluminum Metal–Organic Framework-Ligated Single-Site Nickel(II)-Hydride for Heterogeneous Chemoselective Catalysis. *ACS Catalysis*, **11**(7), 3943-3957.

22. Zaheer, M., and Kempe, R. (2015) Catalytic Hydrogenolysis of Aryl Ethers: A Key Step in Lignin Valorization to Valuable Chemicals. *ACS Catalysis*, **5**(3), 1675-1684.

23. Zakzeski, J., Bruijnincx, P. C. A., Jongerius, A. L., and Weckhuysen, B. M. (2010) The Catalytic Valorization of Lignin for the Production of Renewable Chemicals. *Chemical Reviews*, **110**(6), 3552-3599.

24. Azadi, P., Inderwildi, O. R., Farnood, R., and King, D. A. (2013) Liquid fuels, hydrogen and chemicals from lignin: A critical review. *Renewable and Sustainable Energy Reviews*, **21**(C), 506-523.

25. Song, Y., Feng, X., Chen, J. S., Brzezinski, C., Xu, Z., and Lin, W. (2020) Multistep Engineering of Synergistic Catalysts in a Metal–Organic Framework for Tandem C–O Bond Cleavage. *Journal of the American Chemical Society*, **142**(10), 4872-4882.

26. Song, Y., Li, Z., Ji, P., Kaufmann, M., Feng, X., Chen, J. S., Wang, C., and Lin, W. (2019) Metal–Organic Framework Nodes Support Single-Site Nickel(II) Hydride Catalysts for the Hydrogenolysis of Aryl Ethers. *ACS Catalysis*, **9**(2), 1578-1583.

27. Shiramizu, M., and Toste, F. D. (2012) Deoxygenation of Biomass-Derived Feedstocks: Oxorhenium-Catalyzed Deoxydehydration of Sugars and Sugar Alcohols. *Angewandte Chemie International Edition*, **51**(32), 8082-8086.

28. Schwob, T., Kunnas, P., de Jonge, N., Papp, C., Steinrück, H. P., and Kempe, R. (2019) General and selective deoxygenation by hydrogen using a reusable earth-abundant metal catalyst. *Science Advances*, **5**(11), eaav3680.

29. Clive, D. L. J., and Wang, J. (2002) A Tin Hydride Designed To Facilitate Removal of Tin Species from Products of Stannane-Mediated Radical Reactions. *The Journal of Organic Chemistry*, **67**(4), 1192-1198.

30. Srikrishna, A., Sattigeri, J. A., Viswajanani, R., and Yelamaggad, C. V.(1995) A Simple and Convenient One Step Method for the Reductive Deoxygenation of Aryl Ketones to Hydrocarbons. *Synlett*, **1995**(1), 93-94.

31. Wei, T., Lan, Y., and Xu, J. (2005) Mildly and highly selective reductive deoxygenation of para-di- and monoalkylaminophenyl ketones by borane. *Heteroatom Chemistry*, **16**(1), 2-5.

32. Clemmensen, E. (1913) Reduktion von Ketonen und Aldehyden zu den entsprechenden Kohlenwasserstoffen unter Anwendung von amalgamiertem Zink und Salzsäure. *Berichte der deutschen chemischen Gesellschaft*, **46**(2), 1837-1843.

33. Ansell, L. L., Rangarajan, T., Burgess, W. M., Eisenbraun, E. J., Keen, G. W., and Hamming, M. C. (1976) The synthesis of 1,2,3,7,8,9-hexahydrodibenzo[def, mno]chrysene and the use of hydriodic acid-red phosphorus in the deoxygenation of ketones. *Organic Preparations and Procedures International*, **8**(3), 133-140.

34. Huang, M. (1949) Reduction of Steroid Ketones and other Carbonyl Compounds by Modified Wolff--Kishner Method. *Journal of the American Chemical Society*, **71**(10), 3301-3303.

35. Antil, N., Kumar, A., Akhtar, N., Newar, R., Begum, W., and Manna, K. (2021) Metal–Organic Framework-Confined Single-Site Base-Metal Catalyst for Chemoselective Hydrodeoxygenation of Carbonyls and Alcohols. *Inorganic Chemistry*, **60**(12), 9029-9039.

36. Stanislaus, A., and Cooper, B. H. (1994) Aromatic Hydrogenation Catalysis: A Review. *Catalysis Reviews*, **36**(1), 75-123.

37. Källström, S., and Leino, R. (2008) Synthesis of pharmaceutically active compounds containing a disubstituted piperidine framework. *Bioorganic & Medicinal Chemistry*, **16**(2), 601-635.

38. Chida, N., and Sato, T. (2014) Synthesis of natural products containing cyclohexane units utilizing the Ferrier carbocyclization reaction. *Chemical record (New York, N.Y.)*, **14**(4), 592-605.

39. Ji, P., Song, Y., Drake, T., Veroneau, S. S., Lin, Z., Pan, X., and Lin, W. (2018) Titanium(III)-Oxo Clusters in a Metal–Organic Framework Support Single-Site Co(II)-Hydride Catalysts for Arene Hydrogenation. *Journal of the American Chemical Society*, **140**(1), 433-440.

40. Manna, K., Zhang, T., Carboni, M., Abney, C. W., and Lin, W. (2014) Salicylaldimine-Based Metal–Organic Framework Enabling Highly Active Olefin Hydrogenation with Iron and Cobalt Catalysts. *Journal of the American Chemical Society*, **136**(38), 13182-13185.

41. Thacker, N. C., Lin, Z., Zhang, T., Gilhula, J. C., Abney, C. W., and Lin, W. (2016) Robust and Porous β-Diketiminate-Functionalized Metal–Organic Frameworks for Earth-Abundant-Metal-Catalyzed C–H Amination and Hydrogenation. *Journal of the American Chemical Society*, **138**(10), 3501-3509.

42. Chen, L., Rangan, S., Li, J., Jiang, H., and Li, Y. (2014) A molecular Pd(II) complex incorporated into a MOF as a highly active single-site heterogeneous catalyst for C–Cl bond activation. *Green Chemistry*, **16**(8), 3978-3985.

43. Kaslow, C. E., and Stayner, R. D. (1945) Ozonolysis of Styryl Derivatives of Nitrogen Heterocycles. *Journal of the American Chemical Society*, **67**(10), 1716-1717.

44. Dang, H. T., Lieu, T. N., Truong, T., and Phan, N. T. S. (2016) Direct alkenylation of 2-substituted azaarenes with carbonyls via C-H bond activation using iron-based metal-organic framework Fe$_3$O(BPDC)$_3$ as an efficient heterogeneous catalyst. *Journal of Molecular Catalysis A: Chemical*, **420**, 237-245.

45. Phan, N. T. S., Nguyen, C. K., Nguyen, T. T., and Truong, T. (2014) Towards applications of metal–organic frameworks in catalysis: C–H direct activation of benzoxazole with aryl boronic acids using

Ni$_2$(BDC)$_2$(DABCO) as an efficient heterogeneous catalyst. *Catalysis Science & Technology*, **4**(2), 369-377.

46. Le, H. T. N., Nguyen, T. T., Vu, P. H. L., Truong, T., and Phan, N. T. S. (2014) Ligand-free direct C-arylation of heterocycles with aryl halides over a metal-organic framework Cu$_2$(BPDC)$_2$(BPY) as an efficient and robust heterogeneous catalyst. *Journal of Molecular Catalysis A: Chemical*, **391**, 74-82.

47. Tran, N. T. T., Tran, Q. H., and Truong, T. (2014) Removable bidentate directing group assisted-recyclable metal–organic frameworks-catalyzed direct oxidative amination of Sp^2 C–H bonds. *Journal of Catalysis*, **320**, 9-15.

48. Xu, F., Kang, W.-F., Wang, X.-N., Kou, H.-D., Jin, Z., and Liu, C.-S. (2017) Synergic effect of copper-based metal–organic frameworks for highly efficient C–H activation of amidines. *RSC Advances*, **7**(81), 51658-51662.

49. Hoang, T. T., To, T. A., Cao, V. T. T., Nguyen, A. T., Nguyen, T. T., and Phan, N. T. S. (2017) Direct oxidative C-H amination of quinoxalinones under copper-organic framework catalysis. *Catalysis Communications*, **101**, 20-25.

50. Newar, R., Begum, W., Antil, N., Shukla, S., Kumar, A., Akhtar, N., Balendra, and Manna, K. (2020) Single-Site Cobalt-Catalyst Ligated with Pyridylimine-Functionalized Metal–Organic Frameworks for Arene and Benzylic Borylation. *Inorganic Chemistry*, **59**(15), 10473-10481.

51. Shul'pin, G. B., and Shul'pina, L. S. (2021) Oxidation of Organic Compounds with Peroxides Catalyzed by Polynuclear Metal Compounds. *Catalysts*, **11**(2), 186.

52. Das, B., Al-Hunaiti, A., Haukka, M., Demeshko, S., Meyer, S., Shteinman, A. A., Meyer, F., Repo, T., and Nordlander, E. (2015) Catalytic Oxidation of Alkanes and Alkenes by H$_2$O$_2$ with a μ-Oxido Diiron(III) Complex as Catalyst/Catalyst Precursor. *European Journal of Inorganic Chemistry*, **2015**(21), 3590-3601.

53. Khirsariya, P., and Mewada, R. (2018) Review of a Cyclohexane Oxidation Reaction Using Heterogenous Catalyst. *International Journal of Engineering Research and Development*, **2**(4), 3911-3914.

54. Andrade, M. A., and Martins, L. M. D. R. S. (2020) Sustainability in Catalytic Cyclohexane Oxidation: The Contribution of Porous Support Materials. *Catalysts*, **10**(1), 2.

55. Lee, J., Farha, O. K., Roberts, J., Scheidt, K. A., Nguyen, S. T., and Hupp, J. T. (2009) Metal–organic framework materials as catalysts. *Chemical Society Reviews*, **38**(5), 1450-1459.

56. Dhakshinamoorthy, A., Alvaro, M., and Garcia, H. (2012) Commercial metal–organic frameworks as heterogeneous catalysts. *Chemical Communications*, **48**(92), 11275-11288.

57. Leus, K., Liu, Y.-Y., and Van Der Voort, P. (2014) Metal-Organic Frameworks as Selective or Chiral Oxidation Catalysts. *Catalysis Reviews*, **56**(1), 1-56.

58. Di Nicola, C., Karabach, Y. Y., Kirillov, A. M., Monari, M., Pandolfo, L., Pettinari, C., and Pombeiro, A. J. L. (2007) Supramolecular Assemblies of Trinuclear Triangular Copper(II) Secondary Building Units through Hydrogen Bonds. Generation of Different Metal–Organic Frameworks, Valuable Catalysts for Peroxidative Oxidation of Alkanes. *Inorganic Chemistry*, **46**(1), 221-230.

59. Simons, M. C., Prinslow, S. D., Babucci, M., Hoffman, A. S., Hong, J., Vitillo, J. G., Bare, S. R., Gates, B. C., Lu, C. C., Gagliardi, L., and Bhan, A. (2021) Beyond Radical Rebound: Methane Oxidation to Methanol Catalyzed by Iron Species in Metal–Organic Framework Nodes. *Journal of the American Chemical Society*, **143**(31), 12165-12174.

60. In Yeub, H., Seung Hwan, L., Yoo Seong, C., Si Jae, P., Jeong Geol, N., In Seop, C., Choongik, K., Hyun Cheol, K., Yong Hwan, K., and Jin Won, L. (2014) Biocatalytic Conversion of Methane to Methanol as a Key Step for Development of Methane-Based Biorefineries. *Journal of Microbiology and Biotechnology*, **24**(12), 1597-1605.

61. Blanchette, C. D., Knipe, J. M., Stolaroff, J. K., DeOtte, J. R., Oakdale, J. S., Maiti, A., Lenhardt, J. M.; Sirajuddin, S., Rosenzweig, A. C., and Baker, S. E. (2016) Printable enzyme-embedded materials for methane to methanol conversion. *Nature Communications*, **7**(1), 11900.

62. Ikuno, T., Zheng, J., Vjunov, A., Sanchez-Sanchez, M., Ortuño, M. A., Pahls, D. R., Fulton, J. L., Camaioni, D. M., Li, Z., Ray, D., Mehdi, B. L., Browning, N. D., Farha, O. K., Hupp, J. T., Cramer, C. J., Gagliardi, L., and Lercher, J. A. (2017) Methane Oxidation to Methanol Catalyzed by Cu-Oxo Clusters Stabilized in NU-1000 Metal–Organic Framework. *Journal of the American Chemical Society*, **139**(30), 10294-10301.

63. Yang, L., Huang, J., Ma, R., You, R., Zeng, H., and Rui, Z. (2019) Metal–Organic Framework-Derived IrO_2/CuO Catalyst for Selective Oxidation of Methane to Methanol. *ACS Energy Letters*, **4**(12), 2945-2951.

64. Bai, C., Li, A., Yao, X., Liu, H., and Li, Y. (2016) Efficient and selective aerobic oxidation of alcohols catalysed by MOF-derived Co catalysts. *Green Chemistry*, **18**(4), 1061-1069.

65. Llabrés i Xamena, F. X., Abad, A., Corma, A., and Garcia, H. (2007) MOFs as catalysts: Activity, reusability and shape-selectivity of a Pd-containing MOF. *Journal of Catalysis*, **250**(2), 294-298.

66. Liu, H., Liu, Y., Li, Y., Tang, Z., and Jiang, H. (2010) Metal–Organic Framework Supported Gold Nanoparticles as a Highly Active Heterogeneous Catalyst for Aerobic Oxidation of Alcohols. *The Journal of Physical Chemistry C*, **114**(31), 13362-13369.

67. Yin, C., Yang, Z., Li, B., Zhang, F., Wang, J., and Ou, E. (2009) Allylic Oxidation of Cyclohexene with Molecular Oxygen Using Cobalt Resinate as Catalyst. *Catalysis Letters*, **131**(3), 440-443.

68. Brown, K., Zolezzi, S., Aguirre, P., Venegas-Yazigi, D., Paredes-García, V., Baggio, R., Novak, M. A., and Spodine, E. (2009) [Cu(H$_2$btec)(bipy)]∞: a novel metal organic framework (MOF) as heterogeneous catalyst for the oxidation of olefins. *Dalton Transactions*, (8), 1422-1427.

69. Sun, D., Sun, F., Deng, X., and Li, Z. (2015) Mixed-Metal Strategy on Metal–Organic Frameworks (MOFs) for Functionalities Expansion: Co Substitution Induces Aerobic Oxidation of Cyclohexene over Inactive Ni-MOF-74. *Inorganic Chemistry*, **54**(17), 8639-8643.

70. Blumenthal, I. (2001) Carbon monoxide poisoning. *Journal of the Royal Society of Medicine*, **94**(6), 270-272.

71. Rose, J. J., Wang, L., Xu, Q., McTiernan, C. F., Shiva, S., Tejero, J., and Gladwin, M. T. (2017) Carbon Monoxide Poisoning: Pathogenesis, Management, and Future Directions of Therapy. *American Journal of respiratory and critical care medicine*, **195**(5), 596-606.

72. Ye, J.-y., and Liu, C.-j. (2011) Cu$_3$(BTC)$_2$: CO oxidation over MOF based catalysts. *Chemical Communications*, **47**(7), 2167-2169.

73. Lin, A., Ibrahim, A. A., Arab, P., El-Kaderi, H. M., and El-Shall, M. S. (2017) Palladium Nanoparticles Supported on Ce-Metal–Organic Framework for Efficient CO Oxidation and Low-Temperature CO$_2$ Capture. *ACS Applied Materials & Interfaces*, **9**(21), 17961-17968.

74. (1983) Chapter XXI: The Oxidation of Aromatic Hydrocarbons. In: *Studies in Surface Science and Catalysis*, Golodets, G.I. (ed.), Elsevier, Netherlands, pp. 650-742.

75. Murindababisha, D., Yusuf, A., Sun, Y., Wang, C., Ren, Y., Lv, J., Xiao, H., Chen, G. Z., and He, J. (2021) Current progress on catalytic oxidation of toluene: a review. *Environmental Science and Pollution Research*, **28**(44), 62030-62060.

76. Peedikakkal, A. M. P., Jimoh, A. A., Shaikh, M. N., and Ali, B. E. (2017) Mixed-Metal Metal–Organic Frameworks as Catalysts for Liquid-Phase Oxidation of Toluene and Cycloalkanes. *Arabian Journal for Science and Engineering*, **42**(10), 4383-4390.

77. Sun, H., Yu, X., Ma, X., Yang, X., Lin, M., and Ge, M. (2020) MnO$_x$-CeO$_2$ catalyst derived from metal-organic frameworks for toluene oxidation. *Catalysis Today*, **355**, 580-586.

78. Wasilke, J.-C., Obrey, S. J., Baker, R. T., and Bazan, G. C. (2005) Concurrent Tandem Catalysis. *Chemical Reviews*, **105**(3), 1001-1020.

79. Huang, Y.-B., Liang, J., Wang, X.-S., and Cao, R. (2017) Multifunctional metal–organic framework catalysts: synergistic catalysis and tandem reactions. *Chemical Society Reviews*, **46**(1), 126-157.

80. Kökçam-Demir, Ü., Goldman, A., Esrafili, L., Gharib, M., Morsali, A., Weingart, O., and Janiak, C. (2020) Coordinatively unsaturated

metal sites (open metal sites) in metal–organic frameworks: design and applications. *Chemical Society Reviews*, **49**(9), 2751-2798.

81. Liu, M., Gao, K., Fan, Y., Guo, X., Wu, J., Meng, X., and Hou, H. (2018) Co-Cluster-Based Metal–Organic Frameworks as Selective Catalysts for Benzene Tandem Acylation–Nazarov Cyclization to Benzocyclopentanone. *Chemistry – A European Journal*, **24**(6), 1416-1424.

82. Vallés-García, C., Cabrero-Antonino, M., Navalón, S., Álvaro, M., Dhakshinamoorthy, A., and García, H. (2020) Nitro functionalized chromium terephthalate metal-organic framework as multifunctional solid acid for the synthesis of benzimidazoles. *Journal of Colloid and Interface Science*, **560**, 885-893.

83. Zhu, L., Liu, X.-Q., Jiang, H.-L., and Sun, L.-B. (2017) Metal–Organic Frameworks for Heterogeneous Basic Catalysis. *Chemical Reviews*, **117**(12), 8129-8176.

84. Karmakar, A., and Pombeiro, A. J. L. (2019) Recent advances in amide functionalized metal organic frameworks for heterogeneous catalytic applications. *Coordination Chemistry Reviews*, **395**, 86-129.

85. Dau, P. V., and Cohen, S. M. (2015) A Bifunctional, Site-Isolated Metal–Organic Framework-Based Tandem Catalyst. *Inorganic Chemistry*, **54**(7), 3134-3138.

86. Zhang, Y., Huang, C., and Mi, L. (2020) Metal–organic frameworks as acid- and/or base-functionalized catalysts for tandem reactions. *Dalton Transactions*, **49**(42), 14723-14730.

87. Park, J., Li, J.-R., Chen, Y.-P., Yu, J., Yakovenko, A. A., Wang, Z. U., Sun, L.-B., Balbuena, P. B., and Zhou, H.-C. (2012) A versatile metal–organic framework for carbon dioxide capture and cooperative catalysis. *Chemical Communications*, **48**(80), 9995-9997.

88. Cao, C.-C., Chen, C.-X., Wei, Z.-W., Qiu, Q.-F., Zhu, N.-X., Xiong, Y.-Y., Jiang, J.-J., Wang, D., and Su, C.-Y. (2019) Catalysis through Dynamic Spacer Installation of Multivariate Functionalities in Metal–Organic Frameworks. *Journal of the American Chemical Society*, **141**(6), 2589-2593.

89. Seo, J. S., Whang, D., Lee, H., Jun, S. I., Oh, J., Jeon, Y. J., and Kim, K. (2000) A homochiral metal–organic porous material for enantioselective separation and catalysis. *Nature*, **404**(6781), 982-986.

90. Evans, D. A., Woerpel, K. A., and Scott, M. J. (1992) 'Bis(oxazolines)' as Ligands for Self-Assembling Chiral Coordination Polymers—Structure of a Copper(I) Catalyst for the Enantioselective Cyclopropanation of Olefins. *Angewandte Chemie International Edition in English*, **31**(4), 430-432.

91. Li, Z., Liu, Y., Kang, X., and Cui, Y. (2018) Chiral Metal–Organic Framework Decorated with TEMPO Radicals for Sequential Oxidation/Asymmetric Cyanation Catalysis. *Inorganic Chemistry*, **57**(16), 9786-9789.

92. Kitamura, M., Ohkuma, T., Inoue, S., Sayo, N., Kumobayashi, H., Akutagawa, S., Ohta, T., Takaya, H., and Noyori, R. (1988) Homogeneous asymmetric hydrogenation of functionalized ketones. *Journal of the American Chemical Society*, **110**(2), 629-631.

93. Noyori, R., and Takaya, H. (1990) BINAP: an efficient chiral element for asymmetric catalysis. *Accounts of Chemical Research*, **23**(10), 345-350.

94. Ohkuma, T., Ooka, H., Yamakawa, M., Ikariya, T., and Noyori, R. (1996) Stereoselective Hydrogenation of Simple Ketones Catalyzed by Ruthenium(II) Complexes. *The Journal of Organic Chemistry*, **61**(15), 4872-4873.

95. Noyori, R., and Ohkuma, T. (2001) Asymmetric Catalysis by Architectural and Functional Molecular Engineering: Practical Chemo- and Stereoselective Hydrogenation of Ketones. *Angewandte Chemie International Edition*, **40**(1), 40-73.

96. Hu, A., Ngo, H. L., and Lin, W. (2003) Chiral Porous Hybrid Solids for Practical Heterogeneous Asymmetric Hydrogenation of Aromatic Ketones. *Journal of the American Chemical Society*, **125**(38), 11490-11491.

97. Hu, A., Ngo, H. L., and Lin, W. (2003) Chiral, porous, hybrid solids for highly enantioselective heterogeneous asymmetric hydrogenation of beta-keto esters. *Angewandte Chemie (International ed. in English)*, **42**(48), 6000-6003.

98. Jiang, H., Zhang, W., Kang, X., Cao, Z., Chen, X., Liu, Y., and Cui, Y. (2020) Topology-Based Functionalization of Robust Chiral Zr-Based Metal–Organic Frameworks for Catalytic Enantioselective Hydrogenation. *Journal of the American Chemical Society*, **142**(21), 9642-9652.

99. Gregory, R. J. H. (1999) Cyanohydrins in Nature and the Laboratory: Biology, Preparations, and Synthetic Applications. *Chemical Reviews*, **99**(12), 3649-3682.

100. Du, X., Li, Z., Liu, Y., Yang, S., and Cui, Y. (2015) Chiral porous metal–organic frameworks containing μ-oxo-bis[Ti(salan)] units for asymmetric cyanation of aldehydes. *Dalton Transactions*, **44**(29), 12999-13002.

101. Xi, W., Liu, Y., Xia, Q., Li, Z., and Cui, Y. (2015) Direct and Post-Synthesis Incorporation of Chiral Metallosalen Catalysts into Metal–Organic Frameworks for Asymmetric Organic Transformations. *Chemistry – A European Journal*, **21**(36), 12581-12585.

102. Li, J., Ren, Y., Qi, C., and Jiang, H. (2017) The first porphyrin–salen based chiral metal–organic framework for asymmetric cyanosilylation of aldehydes. *Chemical Communications*, **53**(58), 8223-8226.

103. Dang, D., Wu, P., He, C., Xie, Z., and Duan, C. (2010) Homochiral Metal–Organic Frameworks for Heterogeneous Asymmetric Catalysis. *Journal of the American Chemical Society*, **132**(41), 14321-14323.

104. Mo, K., Yang, Y., and Cui, Y. (2014) A Homochiral Metal–Organic Framework as an Effective Asymmetric Catalyst for Cyanohydrin Synthesis. *Journal of the American Chemical Society*, **136**(5), 1746-1749.

105. Demuynck, A. L. W., Goesten, M. G., Ramos-Fernandez, E. V., Dusselier, M., Vanderleyden, J., Kapteijn, F., Gascon, J., and Sels, B. F. (2014) Induced Chirality in a Metal–Organic Framework by Postsynthetic Modification for Highly Selective Asymmetric Aldol Reactions. *ChemCatChem*, **6**(8), 2211-2214.

106. Banerjee, M., Das, S., Yoon, M., Choi, H. J., Hyun, M. H., Park, S. M., Seo, G., and Kim, K. (2009) Postsynthetic Modification Switches an Achiral Framework to Catalytically Active Homochiral Metal–Organic Porous Materials. *Journal of the American Chemical Society*, **131**(22), 7524-7525.

107. Jeong, K. S., Go, Y. B., Shin, S. M., Lee, S. J., Kim, J., Yaghi, O. M., and Jeong, N. (2011) Asymmetric catalytic reactions by NbO-type chiral metal–organic frameworks. *Chemical Science*, **2**(5), 877-882.

108. Tanaka, K., Nagase, S., Anami, T., Wierzbicki, M., and Urbanczyk-Lipkowska, Z. (2016) Enantioselective Diels–Alder reaction in the confined space of homochiral metal–organic frameworks. *RSC Advances*, **6**(112), 111436-111439.

109. Xia, Q., Liu, Y., Li, Z., Gong, W., and Cui, Y. (2016) A Cr(salen)-based metal–organic framework as a versatile catalyst for efficient asymmetric transformations. *Chemical Communications*, **52**(89), 13167-13170.

110. Wu, C.-D., Hu, A., Zhang, L., and Lin, W. (2005) A Homochiral Porous Metal–Organic Framework for Highly Enantioselective Heterogeneous Asymmetric Catalysis. *Journal of the American Chemical Society*, **127**(25), 8940-8941.

111. Ma, L., Falkowski, J. M., Abney, C., and Lin, W. (2010) A series of isoreticular chiral metal–organic frameworks as a tunable platform for asymmetric catalysis. *Nature Chemistry*, **2**(10), 838-846.

112. Cho, S.-H., Ma, B., Nguyen, S. T., Hupp, J. T., and Albrecht-Schmitt, T. E. (2006) A metal–organic framework material that functions as an enantioselective catalyst for olefin epoxidation. *Chemical Communications*, (24), 2563-2565.

113. Song, F., Wang, C., Falkowski, J. M., Ma, L., and Lin, W. (2010) Isoreticular Chiral Metal–Organic Frameworks for Asymmetric Alkene

Epoxidation: Tuning Catalytic Activity by Controlling Framework Catenation and Varying Open Channel Sizes. *Journal of the American Chemical Society*, **132**(43), 15390-15398.

114. Song, F., Zhang, T., Wang, C., and Lin, W. (2012) Chiral porous metal-organic frameworks with dual active sites for sequential asymmetric catalysis. *Proceedings of the Royal Society A: Mathematical, Physical and Engineering Sciences*, **468**(2143), 2035-2052.

115. Xia, Q., Li, Z., Tan, C., Liu, Y., Gong, W., and Cui, Y. (2017) Multivariate Metal–Organic Frameworks as Multifunctional Heterogeneous Asymmetric Catalysts for Sequential Reactions. *Journal of the American Chemical Society*, **139**(24), 8259-8266.

116. Falkowski, J. M., Liu, S., Wang, C., and Lin, W. (2012) Chiral metal–organic frameworks with tunable open channels as single-site asymmetric cyclopropanation catalysts. *Chemical Communications*, **48**(52), 6508-6510.

117. Sakakura, T., and Kohno, K. (2009) The synthesis of organic carbonates from carbon dioxide. *Chemical Communications*, (11), 1312-1330.

118. Jing, X., He, C., Dong, D., Yang, L., and Duan, C. (2012) Homochiral Crystallization of Metal–Organic Silver Frameworks: Asymmetric [3+2] Cycloaddition of an Azomethine Ylide. *Angewandte Chemie International Edition*, **51**(40), 10127-10131.

119. Sawano, T., Thacker, N. C., Lin, Z., McIsaac, A. R., and Lin, W. (2015) Robust, Chiral, and Porous BINAP-Based Metal–Organic Frameworks for Highly Enantioselective Cyclization Reactions. *Journal of the American Chemical Society*, **137**(38), 12241-12248.

7

Proton Conducting Metal-Organic Frameworks (MOFs) and their Applications

Sakharam B. Tayade and Sanjog S. Nagarkar
Department of Chemistry, Indian Institute of Technology (IIT)
Bombay, Maharashtra, India

1. Introduction

Energy requirement all over the world is growing at a fast pace.[1] The environmental problems associated with conventional energy sources are among the most pressing issues faced by the world today.[2] Energy demand has risen to unprecedented magnitudes owing to rapid urbanization, population growth, and industrialization.[3] The traditional stocks of fossil fuels, such as coal, gas, and oil, are being used to meet this expanding need for energy by burning them in combustion engines. In addition to energy, these also produce greenhouse gasses and other hazardous by-products which contribute to air, water, and land pollution, with an adverse effect on flora and fauna.[4] Besides, these fossil fuel reserves are diminishing at a seemingly faster rate alongside a significant increase in their cost. Thus, it is necessary to explore alternative energy sources, which are abundant, sustainable, safe, and clean.

The electrochemical energy conversion and storage systems are of particular interest in this regard, as they typically convert the green and sustainable energy sources, for example, solar, wind, and tidal, into a more '*usable*' form of energy that is electricity. The typical electrochemical energy transformation and storage technologies are Batteries, Fuel Cells (FCs), Supercapacitors, etc. Figure 1 shows the range of energy stored and power available from these technologies as compared to the established technologies.[5] To compete, if not fully replace the fossil fuel technologies, both energy and power efficiencies of available from electrochemical energy conversion and storage needs to be improved. While electrochemical energy conversion and storage mechanisms of the above-mentioned electrochemical systems are different, there are similarities like the electron and ion transport are disconnected and the energy-providing processes happen at the phase boundary of the electrode/electrolyte interface. In this chapter, we limit our discussion to electrolytes only.

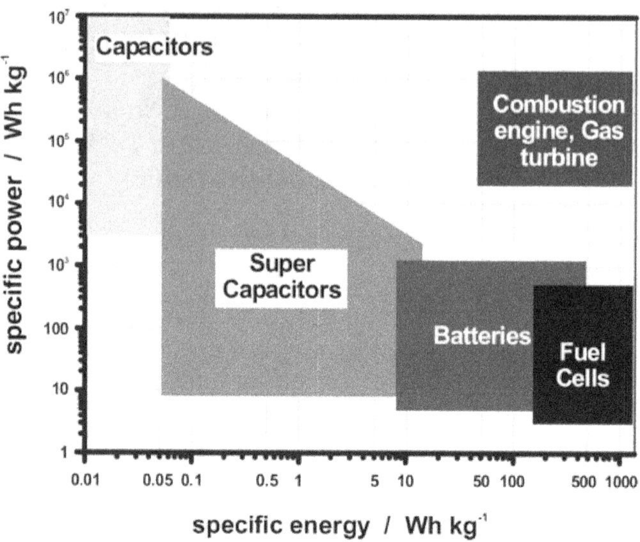

Figure 1. Fuel cells, Batteries, and Capacitors are compared to traditional systems in terms of energy storage and power capacity. "Reproduced with permission from Ref. [5]. Copyright 2004 American Chemical Society."

Proton (H^+), as a charge carrier, is most attractive due to its size and the extraordinary mobility (36.2×10^{-8} m^2 s^{-1} V^{-1}) in H_2O.[6] Great interest is being paid to the FCs, a proton-based energy conversion system because it can furnish both heat and electricity.[7] FC is an electrochemical device that efficiently converts chemical energy into electrical energy with environmental friendly by-products.[8] Typically, Hydrogen (H_2) gas dissociates at the anode to protons (H^+) and electrons (e^-) and migrates to the cathode via electrolyte and external circuit, respectively. At the cathode, H^+ and e^- combine with Oxygen (O_2) gas (from the air) to produce water and electricity generated as long as the fuel is supplied. As a consequence, FCs are considered green and efficient energy generation technologies.[9] In addition to FCs, the recent emergence of proton-based electrochemical storage systems namely protonic supercapacitors[10] and proton batteries[11] as a stand-alone hybrid system to reach high power density and quick charge-discharge rates have attracted significant attention. While all these systems employ proton transporting electrolytes the conductivity, operating temperatures, and mechanical properties requirements are significantly different.

All-solid-state electrochemical storage and conversion systems have attracted notable scientific and industrial interest owing to high power and energy densities, increased safety, and newer applications.[12] For example, it allows multiple FCs or batteries to be stacked in a single package to increase the energy density or flexible energy storage and conversion cells for wearable devices. The ionic conductivity, electrically insulating nature and mechanical properties of electrolytes play a very important part in determining the overall performance and safety of electrochemical devices. Thus the development of solid-state electrolytes (SSEs) is extensively pursued in the chemistry and materials domain.[5]

2. Metal-Organic Frameworks (MOFs) as Solid Electrolyte

Metal-organic frameworks (MOFs), also known as Porous Coordination Polymers (PCPs) are a crystalline class of compounds composed of multidimensional (0D, 1D, 2D, and 3D) network structures prepared by metal ions or metal clusters united through organic ligands/linkers (Fig. 2).[13] For the sake of consistency, we used MOF as an abbreviation throughout this chapter. These framework structures are diverse and usually possess a highly ordered nanoporous structure. As a result of the high surface area and tunable pore size, these materials have demonstrated exceptional properties and explored for a range of applications such as gas storage and separation,[14] catalysis,[15] magnetism,[16] biomedical,[17] sensing,[18] energy storage and conversion, etc.[19]

In 1979, Kanda *et al.* reported the first example of proton conductivity (PC) in MOFs in 2D dtoa copper(II) complexes[20]. However, because of the amorphous nature of these complexes, the proton transport phenomenon was uncertain. This was followed by the report in 2003 on conductivity by Kitagawa *et al.* in isomorphic 2D MOF comprised of Cu(II) dimers and bridging H_2dtoa ligand.[21] This H_2dtoaCu(II) complex exhibited a PC of 2.2×10^{-6} S·cm^{-1}. The PC in this MOF was ascribed to the water molecules. While the conductivity value in this MOF was not very high, it opened an avenue for the MOFs-based solid-state proton conductors.

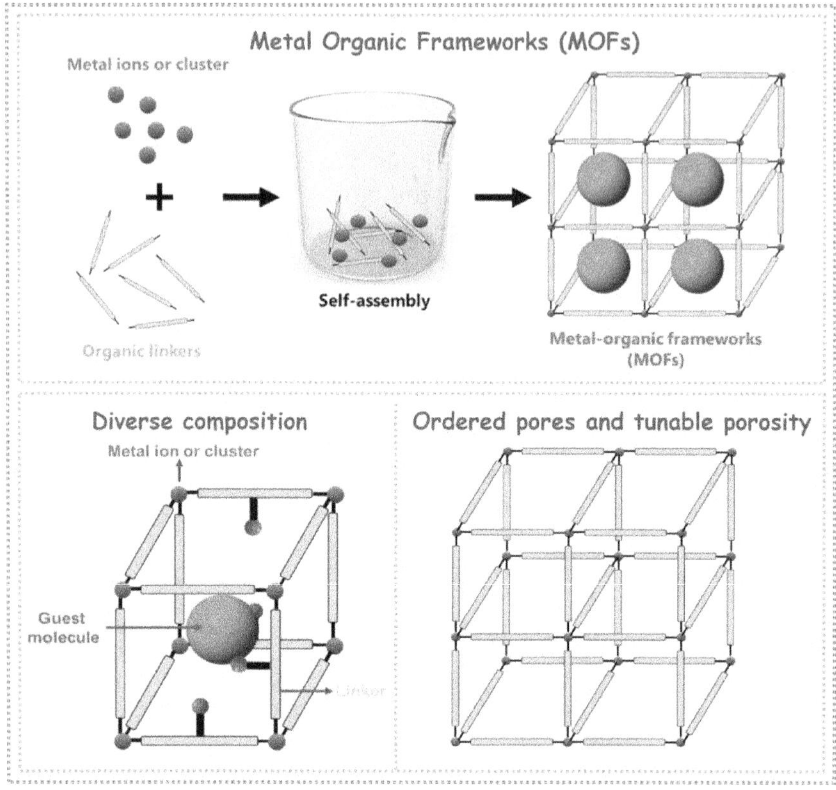

Figure 2. Self-assembly of inorganic (metal ions) and organic (ligands) building blocks forming Metal-organic Frameworks (MOFs) structure. The chemical diversity offered by building blocks, guests, and the physical diversity in terms of ordered porous structure with variable pore size, shape, etc.

3. Fundamentals of Proton (H⁺) Conduction

3.1 Measurement of Proton Conductivity

The PC of the MOFs is measured by the AC impedance spectroscopy technique. The MOF is placed between two electrodes (2-probe) and is held at constant temperature and relative humidity (RH) conditions. Pelletized MOF samples are the most commonly used form to calculate the sample's bulk conductivity (Fig. 3a). The grain boundary and inhomogeneity of temperature and/or humidity often result in lower conductivity values in the bulk sample. Additionally, the anisotropic nature of MOF crystal also offers differential conductivity along the crystallographic axis. Thus, to carefully analyze such effects and contributions, conductivities are also measured in the form of thin films or single crystals along the selected directions (Fig. 3b and c).

The PC value for the MOFs is determined using the following equation:

$$\sigma_{(T)} = \frac{L}{RA} \qquad \text{Eqn. (1)}$$

where, $\sigma_{(T)}$ = ionic conductivity ($S \cdot cm^{-1}$) of the sample at measured temperature, L = thickness of the sample in cm, R = resistance of the sample in Ω, and A = area of the cross-section of the sample in cm^2. The resistance value (R) is obtained from the fitting of the Nyquist plot obtained from the AC impedance measurement. The conductivity values are typically reported at measurement temperature and RH values.

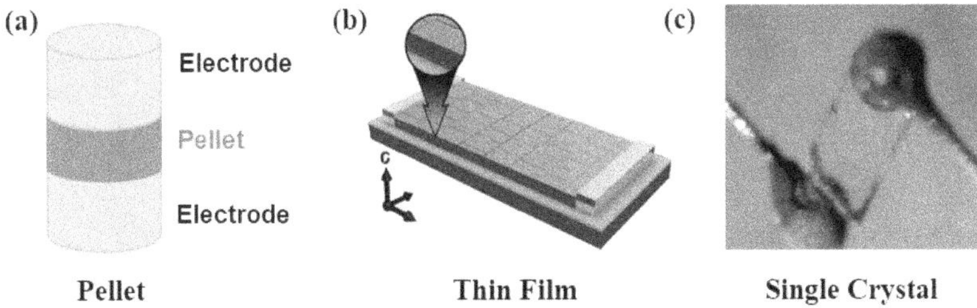

Figure 3. Measurement of solid-state PC of MOFs in the form of (a) Pellet; (b) Thin Film and (c) Single crystal. "Reproduced with permission from Ref. [22, 23]. Copyright 2017 and 2013 American Chemical Society."

The Arrhenius equation determines the activation energy (E_a) for the proton transport.

$$\sigma = \frac{\sigma_0}{T}\exp\left[-\frac{E_a}{kT}\right] \qquad \text{Eqn. (2)}$$

where, σ = conductivity, σ_0 = pre-exponential factor, E_a = activation energy, k = Boltzmann constant, and T = temperature.

3.2 Mechanism of Proton Conduction

Proton transfer is thought to take place via two mechanisms: the Grötthuss and the Vehicle mechanisms. The value of E_a is usually used to determine the operating mechanism.

Grötthuss mechanism of Proton transport: In 1804, Grötthuss hypothesized a mechanism for proton transfer through networks of H-bonds, such as those seen in the water, where the protons (H+) *"hop"* by swapping H-bonds with covalent from one water molecule to the nearby water molecules (Fig. 4a).[24] The hopping requires less energy and the E_a for the Grötthuss type of mechanism is observed in the range of 0.1-0.4 eV.[25]

Vehicle mechanism of proton transport: As the name implies the protons are transported by proton-carrying species (vehicle) via self-diffusion. For example, as shown in Figure 4b, the excess protons travel with the support of H_3O^+ ions as vehicles.[27] As the proton transport involves self-diffusion of the vehicle molecule in the media the E_a is > 0.4 eV.

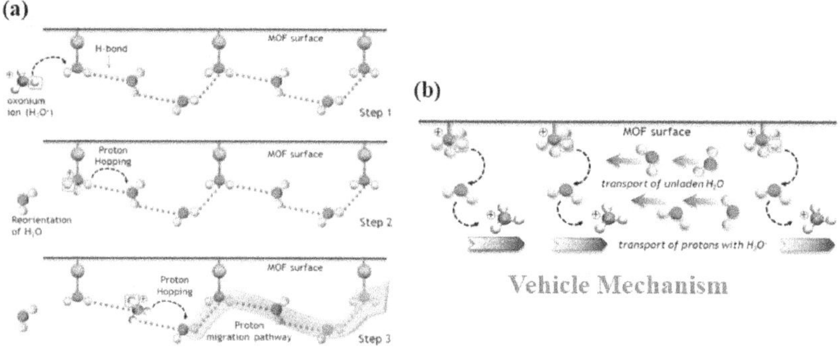

Figure 4. Schematic illustration of water-mediated (a) Grötthuss and (b) Vehicle mechanism of proton transport in MOFs. "Reproduced with permission from Ref. [26]. Copyright 2021 American Chemical Society."

3.3 Factors Governing Proton Transport

Concentration and mobility of mobile ions: As evident from equation 3, the ionic conductivity at a given temperature (T) is governed by three main factors:

$$\sigma_{(T)} = \sum n_i q_i \mu_i \quad \text{Eqn. (3)}$$

where, n_i = number of charge carriers, μ_i = mobility of charge carriers, and q_i = a charge on the mobile species. For proton (H^+) the charge on the ion (q_i) is fixed to +1, so typically to increase the PC of the material, efforts are made to increase the number of charge carriers by loading protic guests or increasing the concentration of acidic protons in the structure. A strategy to increase proton mobility by amorphization or defect creation is recently presented but is far less explored.[28] Additionally, the humidity is expected to decrease the local viscosity, in turn, increases the mobility of the proton on top of creating continuous proton hopping pathways. But no detailed experimental evidence is available to decipher the exact role of local viscosity change on proton conduction.

Transport number ($t_\pm$)**:** In addition to high ionic conductivity, one of the important performance indexes for ionic conductors/electrolytes is transport number (or transference number) ($t_\pm$).[12] It is the fraction of the total current carried by the ion (cation or anion) of interest (in the present case H^+). It is calculated using equation 4.

$$t_+ = \frac{I_+}{I_{total}} \qquad \text{Eqn. (4)}$$

I_+ and I_{total} are current carried by protons and the total current respectively. The higher the transport number (~ 1; also known as single ion conduction), the higher the efficiency. In general, for solid electrolytes, the transport number is close to 1, but in the case of liquid electrolyte impregnated porous MOF systems; the transport number is ~ 0.5, due to the movement of both cations and anions. However, the transport number measurements are rare in MOF literature.

4. Proton Transport in MOFs Under Humid and Anhydrous Conditions

The growth in the domain of proton-conducting MOFs-based electrolytes is primarily aimed at using it as a solid-state electrolyte in FC. In the literature, MOFs based conductors have been categorized into two distinct types: one work under humid conditions (require the presence of water vapors) similar to benchmark electrolyte Nafion which typically work up to ~80 °C, and the other works under anhydrous condition (does not require the presence of water vapors) similar to $CsHSO_4$ which typically work ~100-250 °C (intermediate temperature range). While solid oxide-based electrolytes are known to work at an even higher temperature the MOF thermal stability and guest retention limit the operational temperature range.

4.1 Proton Conduction Under the Humid Condition

Under humid conditions, the water molecules adsorbed on the MOF surface create an H-bonded network structure with functionalities present in MOFs, thus creating a facile protons transport pathway, as discussed earlier. To retain high PC ($>10^{-2}$ S·cm^{-1}), retention of the H-bonded structure created by ad/absorbed water molecules is crucial, thus typically the PC is measured between 20-100 °C (similar to Nafion).

Kitagawa *et al.* reported the preparation and PC in a 2D anionic MOF, $(NH_4)_2(adp)[Zn_2(ox)_3]\cdot3H_2O$.[29] To integrate proton carriers within this MOF, three strategies were employed: functionalization of the framework with acidic groups, usage of acidic molecules to fill pore spaces, and utilization of protonated counter ions. Ligand Adp, NH_4^+ ions, H_2O molecules, and Ox of the framework form 2D H-bonds networks within the pores (Fig. 5). The PC measurement under humid conditions revealed the conductivity value of 8×10^{-3} S·cm^{-1} (25 °C; 85% RH; $E_a = 0.63$ eV). The PC of this MOF is attributed to the H-bond among the NH_4^+ counter ions, Adp, and H_2O guest molecules. The inclusion of acidic groups via the judicious design of the framework structure leads to high PC in MOF.

Preservation of MOF structures under humid conditions remains a crucial factor to retain their intrinsic functionalities and properties. Shimizu *et. al* synthesized PCMOF-5 with ligands having coordinating phosphonate groups with the formula of [La(H$_5$L)(H$_2$O)$_4$].[30] As can be seen in Fig. 6, three out of four phosphonate groups coordinate to produce a 3D open framework structure, while the uncoordinated phosphonic acid group points towards the channel. The uncoordinated phosphonate moieties form a continuous 1D H-bond chain with the water molecules in the channel. Under the 98% RH, PCMOF-5 exhibits a conductivity of 2.5×10^{-3} S·cm^{-1} at 60.1 °C with E_a of 0.16 eV, indicating the Grötthuss mechanism through the well-defined 1D H-bond arrays.

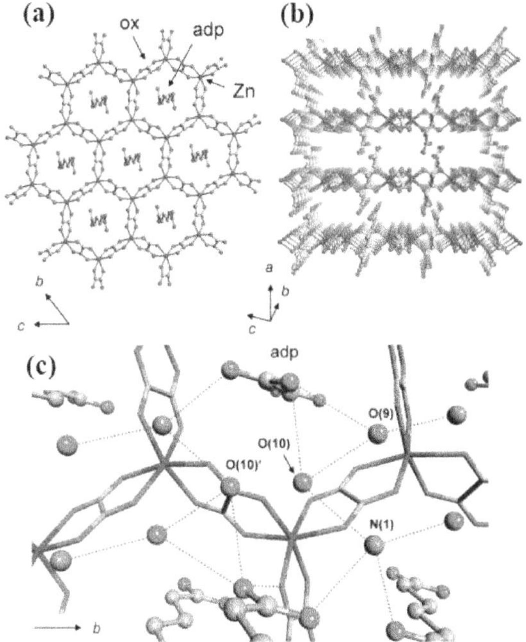

Figure 5. (a) Honeycomb-like layered framework of (NH$_4$)$_2$(adp)[Zn$_2$(ox)$_3$]·3H$_2$O 2D MOF. (b) Structure of guest-free framework viewed along the *b*-axis. (c) H-bonding interactions between water, NH$_4$$^+$ cation, and adp. (Colours scheme: red = oxygen, green = nitrogen, gray = carbon, and blue = zinc atoms, respectively). "Reproduced with permission from Ref. [29]. Copyright 2009 American Chemical Society."

The PC of a MOF can be improved by increasing the fraction of water molecules in the metal coordination layer with respect to metal ions. In this connection, Banerjee *et al.* synthesized Ca-based MOF in presence of solvents such as H$_2$O, DMA, DMSO, and DMF employing BTC as a coordinating ligand.[31] The MOF with the high content of water coordinated with the metal center (Ca-BTC-H_2O MOF) displayed the maximum conductivity of 1.2×10^{-4} S·cm⁻¹ and the lowest E_a of 0.18 eV among the synthesized MOFs.

Further, Das *et al.* using the template promoted method synthesized three Co(II) based high proton-conducting MOFs by employing bpy as the ligand. These MOFs involve a common links of [Co(bpy) $(H_2O)_4]^{2+}$ unit annulled by templated polycarboxylates, such as fdc, btc and btec in 1-Co-fdc, 2-Co-tri, and 3-Co-tetra MOFs, respectively.[32] These MOFs adopt 1D architectures and crystallized waters in the lattices. Among these three MOFs, Co-tri MOF was found to be a superb H⁺ conductor (10^{-1} S · cm⁻¹). In these MOFs, water molecules and –COOH groups act as conducting media for the transfer of a proton.

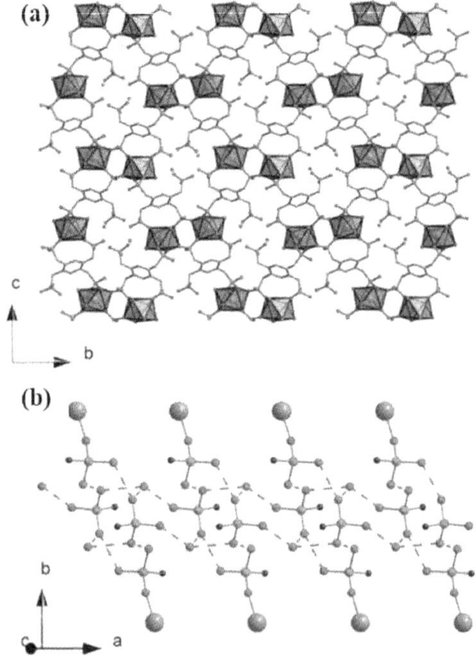

Figure 6. (a) PCMOF-5 crystal structure featuring non-coordinated water molecules and dangling phosphonic acid groups. (b) The phosphonic acid groups and unbound water molecules form a 1D H-bond configuration. "Reproduced with permission from Ref. [30]. Copyright 2013 American Chemical Society."

Maji and co-workers reported a new 3D functional MOF with PTC ligand and and K^I metal node formulated as, $\{[K_8(PTC)_2(H_2O)_{1.5}]\cdot 4H_2O\}_n$.[33] This MOF consists of a 3D pillared-layer framework in which perylene moieties are organized in a special end-to-end off-slipped and zigzag order is controlled by K^I-carboxylate bonding in the framework. At room temperature (RT) and 98% RH, the PC value of MOF was 1×10^{-3} S·cm^{-1}. The H-bonding network formed via connection of the guests and coordinated H_2O molecules serves as a proton transfer path.

4.2 Proton Conduction Under the Anhydrous Condition

In the absence of water, the mobile protic guest species such as Im, Tz, histamine, or inorganic acids (H_2SO_4 or H_3PO_4) existing in the open channels of MOFs act as a potential source of H^+ increasing charge concentration and form an H-bonded network to transport the proton. The working temperature range is governed by the nature of the guest species and framework stability which can be sub-zero to ~200 °C (Intermediate temperature range).

In an early report, Kitagawa and co-workers reported PC under the anhydrous condition in MOFs.[34] They loaded Im molecules into Al-based porous MOFs ($[Al(\mu_2\text{-OH})(1,4\text{-ndc})]_n$) and $[Al(\mu_2\text{-OH})(1,4\text{-bdc})]_n$. The PC of the guest-free form of ($[Al(\mu 2\text{-OH})(1,4\text{-ndc})]_n$) was found to be very low (10^{-13} S·cm^{-1}). The conductivity was raised to 5.5×10^{-8} S·cm^{-1} on the introduction of Im as a protic conducting species at RT which was then found to increase to 2.2×10^{-5} S·cm^{-1} at 120 °C. The enhancement in the conductivity is assigned to the rotation of the Im molecules within the porous framework. As stated before, a higher carrier concentration improves ionic conductivity as per equation 3. While guest free MOF based on bdc linker displayed less PC of ~10^{-10} S·cm^{-1} (RT) and on Im loading it increased to 1×10^{-7} S·cm^{-1} (120 °C).

Another anhydrous PC study was reported by Shimizu *et al.* using β-PCMOF2, based on Zn(II) metal center.[35] The 3D honeycomb structure has 1D channels lined with −SO_3H groups of ligands. On the introduction of increasing amounts of Tz into the channels (β-PCMOF2(Tz)x, x = 0.3, 0.45, and 0.6) high conductivity of 2×10^{-4} S·cm^{-1}, 5×10^{-4} S·cm^{-1}, and 4×10^{-4} S·cm^{-1}, at 150 °C and under anhydrous condition were recorded. The group also tested the FC performance of MOF as the electrolyte which we will discuss in the latter part.

The PC under the anhydrous condition in the *'plastic crystal'*, which is a MOF made from Zn(II), hydrogen phosphates, and protonated Imidazole (ImH$_2$), was reported by Kitagawa *et al.*[36] The MOF contained an anionic 1D chain of Zn^{2+}-phosphate with ImH$_2$ situated in the inter-chain space resulting in [Zn(HPO$_4$)(H$_2$PO$_4$)$_2$](ImH$_2$)$_2$ MOF (Fig. 7a-c). On heating, the compound showed a solid-solid phase transition at 70 °C followed by melting at 160 °C, this is known as plastic crystal behavior. The solid-solid phase change is attended by a sudden and sharp increase in anhydrous ionic conductivity by 4 orders from 10^{-8} S·cm^{-1} at 55 °C to 10^{-4} S·cm^{-1} at 130 °C (Fig. 7d). The increased dynamics of the imidazolium cations within the chains were reasoned to abrupt enchantment in the PC with increasing temperature. Interestingly, on subsequent cooling and re-crystallization, no abrupt PC change was observed. The maximum conductivity value obtained was almost the same for both as-synthesized and MOF recrystallized from melt quenched glass (Fig. 7e).

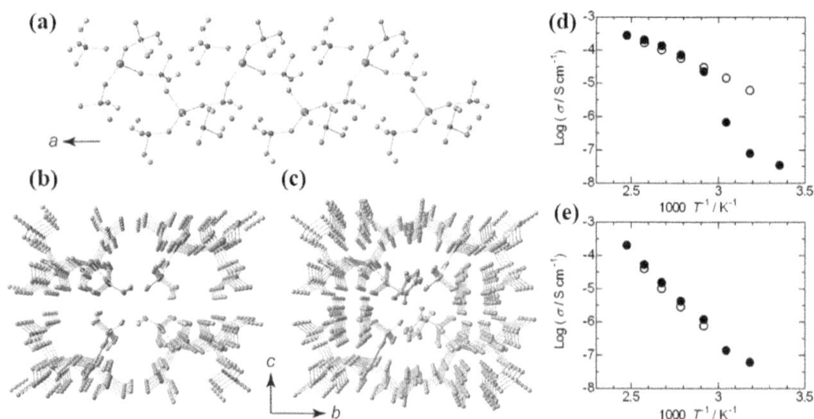

Figure 7. (a) Structure of anionic 1D chain of [Zn(HPO$_4$)(H$_2$PO$_4$)$_2$]$^{2-}$ MOF and (b) 3D arrangement. (c) Framework structure with ImH$_2^+$ ions present between the 1D chains (H atoms are omitted). (d) and (e) Arrhenius plots under moisture-free conditions for as-synthesized and recrystallized Zn-MOF respectively. (Heating cycle = filled circles; cooling cycle = open circle). "Reproduced with permission from Ref. [36]. Copyright 2012 American Chemical Society."

4.3 Simultaneous Humid and Anhydrous Conditions Proton Conduction

PC in both humid and anhydrous conditions is highly desirable to expand the working capabilities of solid electrolytes under a variety of environments and for a large number of possible uses. A very first example of MOF electrolyte, with high PC under humid as well as anhydrous conditions, was reported by Ghosh and co-workers.[37] The Zn(II) MOF; $\{[(Me_2NH_2)_3(SO_4)]_2[Zn_2(ox)_3]\}_n$, where the anionic framework $[Zn_2(ox)_3]_n^{2-}$ is interlocked with a sulfate (SO_4^{2-}) and dimethylammonium (Me_2NH_2) ions containing supramolecular net $[(Me_2NH_2)_3SO_4]_n^+$ (Fig. 8). In humid and anhydrous circumstances, high conductivity values of 4.2×10^{-2} S·cm^{-1} (98% RH) and 1×10^{-4} S·cm^{-1} (150 °C, N_2 atmosphere) were recorded.

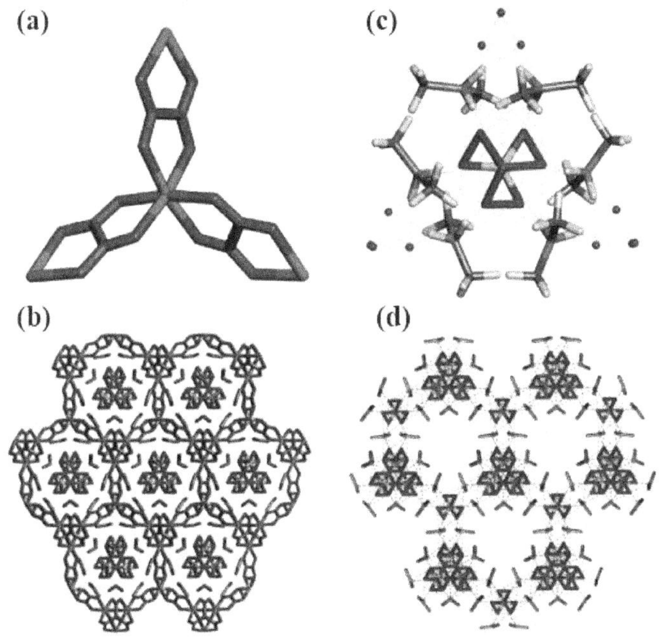

Figure 8. (a) The tris-chelated Zn forming $[Zn_2(ox)_3]_2^-$ anionic structure (b) Bulk neutral structure of $\{[(Me_2NH_2)_3(SO_4)]_2[Zn_2(ox)_3]\}_n$. (c) Multiple H-bonding associations among dimethyl ammonium and sulfate ions. d) $[(Me_2NH_2)_3SO_4]_n^+$ supramolecular network in 3D. "Reproduced with permission from Ref. [37]. Copyright 2014 Wiley-VCH."

5. Design Strategies for Efficient Proton Conduction in MOFs

The reticular chemistry knowledge along with the theoretically infinite combinations of metal ions and ligands offers a high degree of designability and structural predictability. Additionally, the host-guest chemistry and post-synthetic modification (PSM) of the MOF framework offer a wide range of opportunities to modulate both structure and functions.

MOFs have received a lot of interest in the last decade as solid-state proton conducting materials due to simple design and synthesis, repeated organization of voids, and host-guest chemistry.[38] PC in MOFs originates from the presence of guest molecules in the pores or channels, such as H_2O, H_3PO_4, H_2SO_4, ammonium ions, Im, histamine, and Tz.[39, 40] These guest molecules serve as carriers for the transfer of protons within the framework. Furthermore, hydrophilic and protic moieties in the frameworks, such as $-NH_2$, $-O-H$, $-C=O$, $-SO_3H$, $-COOH$, and others, play an important role in proton transfer *via* H-bonding interactions. Herein, we discuss the variety of ways to impart or enhance the PC in MOFs (Fig. 9).

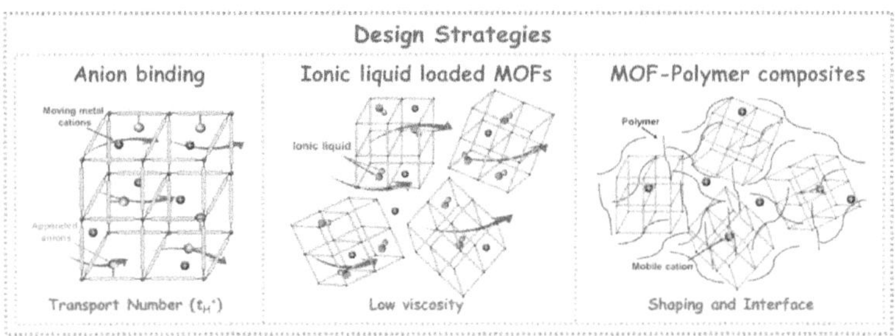

Figure 9. Various design strategies employed to synthesize efficient solid-state proton-conducting MOF materials. Anion binding via ligand or open metal site modification for single ion transport. Ionic liquid loading in porous MOF matrix for retaining low viscosity and hence high conductivity even in solid-state. MOF Polymer composite for shaping the solid electrolyte and interface engineering with other functional materials.

5.1 Encapsulation of Functional Guest Molecules

To achieve the goal of high PC in MOFs, the general criterion is to enhance the proton concentration inside the framework. Typically, the inclusion of the acidic guest molecules within the pore void has been investigated for high PC. Nonetheless, acidic groups, whether organic or inorganic, usually cause coordination bond breakage and structural degradation. As a result, incorporating them into pores is difficult, because high structural stability is required for non-volatile acid guest impregnation. The use of various inorganic acids has been extensively studied, as they have been shown to improve PC in many cases. The PC can be tuned depending on the nature of the doped acids and their concentration. Furthermore, the pK_a of the embodied acids is a critical parameter for providing an adequate number of protons when constructing an H-bonding route for an efficient PC. Lower is the pK_a value, higher will be proton dissociation and hence higher PC. Using this strategy, Fedin *et al.* impregnated acid moieties H_3PO_4 or H_2SO_4 in highly robust and porous MIL MOF (Fig. 10).[41] MIL-101 MOF was loaded with H_2SO_4 or H_3PO_4 and dehydrated to yield solid compounds H_2SO_4@MIL-101 and H_3PO_4@MIL-101. At 150 °C and 0.13% RH, H_2SO_4 loaded and H_3PO_4 loaded MOFs exhibited conductivity values of 1×10^{-2} S·cm^{-1} and 3×10^{-3} S·cm^{-1} respectively.

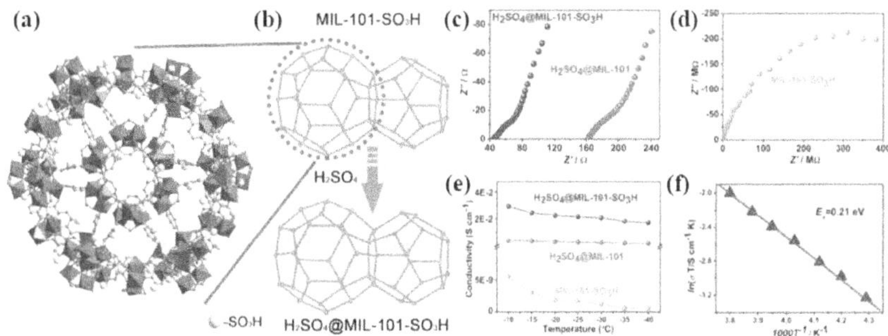

Figure 10. (a) Porous framework of (Cr)MIL-101 with dangling –SO₃H groups (yellow spheres). (b) Schematic representation of acid loading in porous MOF host to synthesize H_2SO_4 loaded MIL-101-SO₃H. (c) Nyquist plots, red and blue for H_2SO_4 loaded MIL-101 and H_2SO_4 loaded MIL-101-SO₃H, respectively. (d) Nyquist plot of MIL-101-SO₃H (-40 °C). (e) Temperature-dependent variation of conductivities for acid-loaded MOFs (f) Arrhenius plot and E_a of H_2SO_4 loaded MIL-101-SO₃H. "Reproduced with permission from Ref. [42]. Copyright 2017 American Chemical Society."

In a similar vein, the synergistic high PC was observed further in MIL-101-SO₃H, H_2SO_4 loaded MIL-101, and H_2SO_4 loaded MIL-101-SO₃H by Lan *et al.* [42] Herein they utilized the acidic functional moieties present within the framework and in addition to the non-volatile acid molecule encapsulated in open framework structure. The 3M H_2SO_4 loaded MIL-101-SO_3H and H_2SO_4 loaded MIL-101 showed the highest conductivities of 1.82×10^{-1} and 6.09×10^{-1} S·cm^{-1} at 70 °C, while un-loaded MIL-101-SO_3H showed significantly lower conduction under the same condition. The high conductivity in H_2SO_4-loaded MIL-101-SO_3H is due to the high dense Brønsted acidic –SO_3H moieties.

Considering the poor stability of MOFs towards acids, another unique strategy is to use ionic liquids (ILs) as a proton carrier in MOFs. The ILs are promising candidate materials as a safe electrolyte at moderate temperatures due to their non-volatility, high thermal stability, non-flammability, and low corrosivity.[43] Combining ILs and MOFs to build a unique soft-media-in-hard-matrix material is one viable technique. Xu *et al.* developed EIMS-HTFSA@-MIL as a solid-state electrolyte material by integrating the binary ionic liquid EIMS-HTFSA into a mesoporous MOF.[44] The EIMS-HTFSA impreg-nated samples were made considering the pore volume of 1.895 cm^3 g^{-1} of activated MIL-101, and with 25, 50, and 100% pore volume (theoretically) occupied by the IL (assigned as BIL25, BIL50 and BIL100, respectively. The thermal stability of EIMS-HTFSA@MIL compounds was found to be 320 °C, which is sufficient for a safe PEM operating at moderate temperatures. BIL100 sample exhibited H$^+$ conductivity of 2×10^{-4} S·cm^{-1} (140 °C) when compared to pristine MIL-101 MOF sample is around six orders of magnitude higher. In comparison to MIL-101, BIL25, and BIL50, the conductivity of BIL100 is observed to be increased six, five, and three orders of mag-nitude at 140 °C. These findings suggest that the amount of EIMS-HTFSA impregnated in the MOF has a substantial relationship with the PC of EIMS HTFSA@MIL. The increased conductivity is due to in-creasing formation of continuous PC channel within the MOF pores on IL impregnation and increased concentration of conducting spe-cies.

Another effective strategy for developing solid-state PC materials is to encapsulate *N*-heterocyclic compounds like Im, Tz, and histamine inside the unoccupied pores of MOF. Kitagawa *et al.* were the first to investigate the encapsulation of Im molecules into MOFs to produce remarkably conducting solid electrolytes, as previously men-tioned.[34]

5.2 Modifications at the Metal Node

Open metal sites (OMSs) in MOFs have been considered weaker spots, on account of the probability of getting the metal-oxygen bond hydrolyzed in the presence of water vapour.[45] Despite this, it allows MOF researchers to incorporate a variety of versatile and complicated functional moieties into MOFs, improving the stability and usefulness of modified MOFs. Under the right conditions, the auxiliary ligand's lability from the coordination sphere leads to the formation of OMSs.[46] Initially, Yamada *et al.* (2009), studied the significance of coordinated H_2O on PC of MOFs by using a di-hydrate form of 1D ferrous oxalate compound.[47] Notably, OMS acts as a Lewis acid site increasing the acidity of the coordinated water molecule leading to facile proton release than that of non-coordinated water molecules. Thus, the controlled coordination of protic molecules can be used as a tool to regulate the PC.

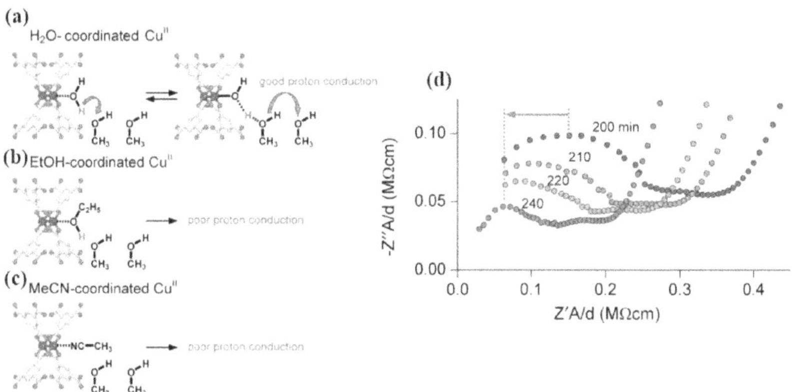

Figure 11. Qualitative illustration of proton transfer from Cu[II] centers coordinated with (a) H_2O, (b) EtOH, and (c) MeCN. (d) Impedance spectra of an H_2O-HKUST-1 sample after exposure to MeOH vapour at RT for a different time interval. "Reproduced with permission from Ref. [48]. Copyright 2012 American Chemical Society."

Further, Hupp and co-workers studied the effect of coordination of protic and aprotic molecules on the PC using HKUST-1, $[Cu_3(btc)_2(H_2O)_3]$ MOF, which has a Cu paddle-wheel with H_2O coordinated to OMS.[48] The activated HKUST-1 with OMSs subjected to solvent vapours of H_2O, MeOH, MeCN, and EtOH (Fig. 11). The coordination to OMS (Cu) leads to an increase in the acidity of water leading to enhancement in the conductivity (4.3×10^{-7} S·cm⁻¹; RT) compared to the MeOH exposed sample (1.5×10^{-5} S·cm⁻¹; RT).

Fe-MIL-88B is a classic MOF composed of 1,4-bdc ligand connecting iron(III) trimers.[49] This MOF, however, cannot operate as a proton conductor due to its limited temperature stability. A 1D channel in Fe-MIL-88B has significant numbers of OMS available for PSM. To overcome the low hydrothermal stability constraint of Fe-MIL-88B, Dong and Zang and co-workers employed the post-synthetic metal coordination strategy to synthesize Cr(III) analog (Cr-MIL-88B). The Cr-MIL-88B loaded with PSA and PESA, yielded Cr-MIL-88B-PSA and Cr-MIL-88B-PESA with pyridine functionality coordinated to OMS (Fig. 12). At 100 °C and 85 % RH, the PC reached maximum values of 1.58×10^{-1} S·cm⁻¹ and 4.50×10^{-2} S·cm⁻¹ for PSA and PESA loaded Cr-MIL-88B, respectively. Installation of acidic −SO_3H groups directed towards the center of the channel in a highly dense manner via PSM of Cr(III) OMS leads to improvement in PC values. Thus, metal node exchange in Fe-MIL-88B with Cr resulted in improved stability to water and site to anchor the acidic functionality for the introduction and modulation of PC.

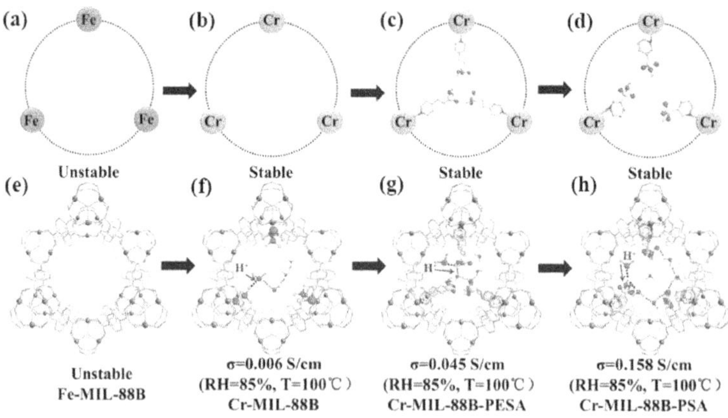

Figure 12. Schematic and corresponding crystal structures of (M)MIL-88B (M = Fe/Cr) structures with corresponding metal atoms and functional group attached to it viewed along the c-axis. (a) & (e) Fe-MIL-88B, (b) & (f) Cr-MIL-88B, (c) & (g) Cr-MIL-88B-PESA and (d) & (h) Cr-MIL-88B-PSA. "Reproduced with permission from Ref. [49]. Copyright 2020 American Chemical Society."

Chen *et al.* reported the PCs of NENU-3 MOF formulated as ([Cu$_{12}$(BTC)$_8$(H$_2$O)$_{12}$][HPW$_{12}$O$_{40}$])·Guest) with Im molecule coordinated to Cu(II) OMS (Im-Cu@(NENU-3a) and other with free Im encapsulated in pore space (Im@(NENU-3) (Fig. 13).[50] The ion conductivity measurement revealed two orders difference in both the samples with Im@(NENU-3) exhibiting PC of 1.82×10^{-2} S·cm^{-1} (70 °C; 90% RH). While under a similar environment, Im-Cu@(NENU-3a) showed a PC of 3.16×10^{-4} S·cm^{-1}. The difference was ascribed to the isolation of lattice water molecules by immobilized Im molecules and low loading in Im-Cu@(NENU-3a) which disrupts the proton transport route. On the other hand, plenty of unattached Im molecules (high loading amount) in Im@(NENU-3) MOF were said substantially enhance the sequential proton-transfer pathways via creating a facile H-bonded network with lattice water.

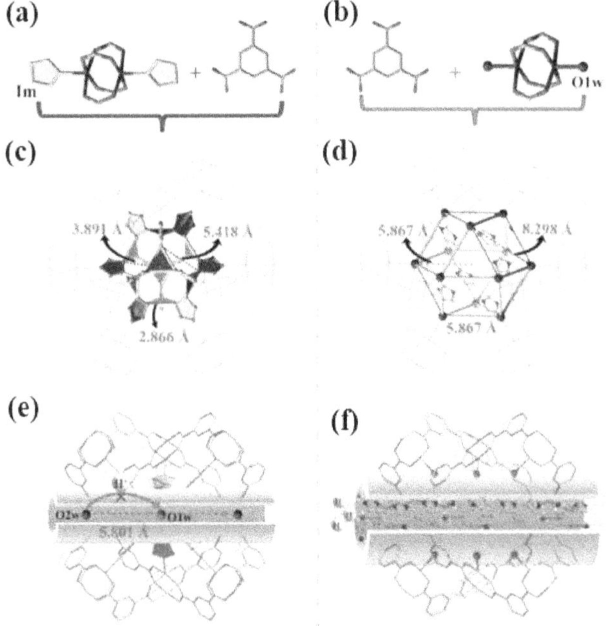

Figure 13. Comparison of structural features of Im-Cu@(NENU-3b) and NENU-3. (a) Cu-paddle wheel coordinated by Imidazolate, (c) cuboctahedra cage formed by BTC, and (e) proposed pathway of PC in Im-Cu@(NENU-3b). (b) Cu-paddle wheel coordinated with water. (d) cuboctahedral cage formed by BTC with free Im molecules in the pore. (f) PC pathway via self-diffusion of lattice water. "Reproduced with permission from Ref.[50]. Copyright 2017 American Chemical Society."

Zhou *et al.* reported contradicting observations in regards to Im arrangements and PC in MOFs.[51] They studied three MOFs composed of [1,1':3',1"-terphenyl]-4,4",5'-tricarboxylic acid ligand and $Fe_3O(CH_3CO_2)_6$ clusters as a parent Fe-MOF, blank, Im@Fe-MOF with embedded Im molecules inside the pores and Im-Fe-MOF which has chemically coordinated Im molecules (Fig. 14). The conductivities were found to be 2.56×10^{-5}, 8.41×10^{-5} and 2.06×10^{-3} S·cm^{-1} (25 °C, 98% RH with E_a = 0.385, 0.573, and 0.436 eV) for (Fe-MOF) (Im@Fe-MOF) (Im-Fe-MOF), respectively (Fig. 14e and f). These results indicate that the Im-Fe-MOF with bound Imidazoles strengthens the formation of a proton-conduction route and improves the PC, this is two orders higher than Fe-MOF and Im@Fe-MOF. It is worth mentioning that, improved conductivity was attained when Im molecules were encapsulated in the pores for Im@(NENU-3), but when Im molecules were coordinated with metal centers in the case of Im-Fe-MOF, even higher PC was obtained than mere cavity filled Im@Fe-MOF. Nonetheless, it should be noted that, in comparison to pristine NENU-3, the Im-Cu@(NENU-3a) displayed a higher PC value but is still less than that of Im-Fe-MOF, which implies that coordinated Im is not an essential criterion to realize increased conductivity.

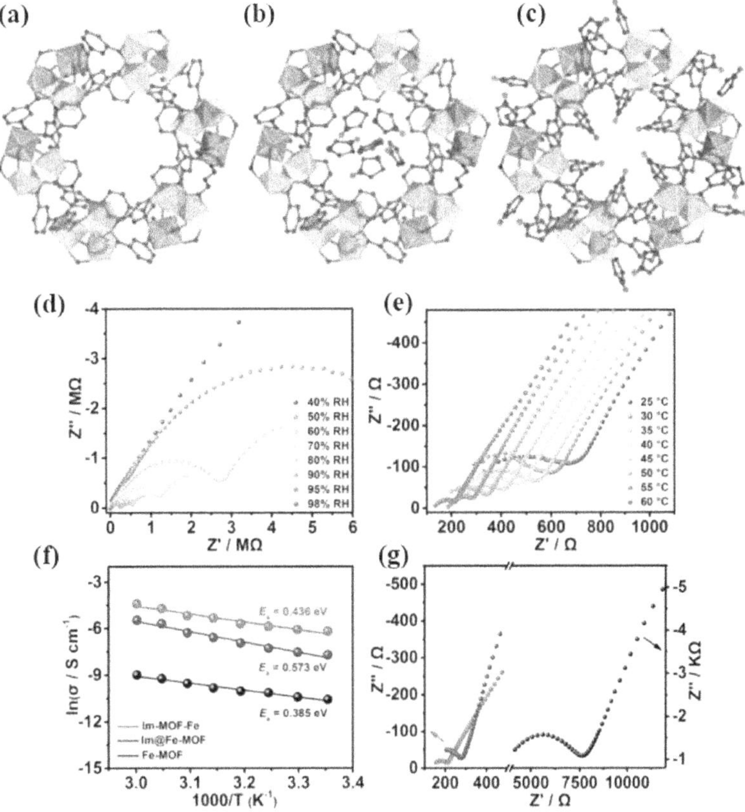

Figure 14. (a), (b) and (c) Structures of Fe-MOF, Im@Fe-MOF and Im-Fe-MOF, respectively. (d) Nyquist plot of Im–Fe–MOF (25 °C) exposed to various RH values. (e) Nyquist plot of Im–Fe–MOF (98% RH) with increasing temperatures. (f) Comparison of Arrhenius plots of PC for MOFs under 98% RH condition. (g) Nyquist plot comparison for MOFs (60 °C, 98% RH). "Reproduced with permission from Ref. [51]. Copyright 2017 American Chemical Society."

Shimizu *et al.* investigated PC in a set of MOFs, including PCMOF-5, which was made with a flexible 1,2,4,5 tetrakis(phosphonomethyl) benzene, and lanthanide ions.[52] The PC measurement reveals the conductivities of the La and Pr MOFs are approximately 10 folds greater than those of other members of the groups (10^{-3} S·cm^{-1} vs. 10^{-4} S·cm^{-1}). According to the author's findings, differences in conductivity can be linked-to crystal particle size owing to lattice strain effects and grain boundary differences when using different lanthanide ions.

Banerjee and colleagues noticed a similar trend in the series of alkaline earth metals, reporting three MOFs (M-SBBA) based on Ca(II), Sr(II), and Ba(II) as metal nodes with SBBA as an organic ligand.[53] PC measurement in these MOFs at 98% RH and 25 °C revealed that the Sr-SBBA MOF shows one order higher proton transport as compared to Ca-SBBA (10^{-5} S·cm^{-1} vs. 10^{-6} S·cm^{-1}) but surprisingly Sr-MOF also showed higher E_a (0.56 eV) than Ca-MOF (0.23 eV), while Ba-MOF disintegrated under RH. The difference in behavior was ascribed to the difference in framework structure and concentration of carrier (water).

5.3 Functionalization of Organic Ligands

Using predesigned chemical reactions, a variety of functional moieties anchored on the linker can be embedded in MOFs. Specifically, the non-coordinated functional moieties of the ligand interact strongly with guest molecules present in the framework. Inserting non-coordinated or partially coordinated acidic functional moieties in the ligand has become a frequent strategy to create proton-conducting MOFs. Li *et al.* studied proton transport in UiO-66 series MOFs with different functional groups $-SO_3H$, $-COOH$, $-NH_2$ and $-Br$.[54] They found the trends of PC values as, UiO-66-SO$_3$H≈UiO-66-(COOH)$_2$> UiO-66-NH$_2$> UiO-66> UiO-66-Br. The extremely acidic and strong hydrophilic nature of the functional groups particularly $-SO_3H$ and $-COOH$ boosts the conductivity performance (Fig. 15).

Further, Furukawa *et al.* have reported the synthesis of Fe$_4$(BDC)$_2$(NDC)(SO$_4$)$_4$(DMA)$_4$ termed, VNU-15(Fe) MOF.[55] The sulfate anions bridged the adjacent metal centers along with organic ligands forming a 3D structure while DMA cations were present in the channel with H-bond interactions between them. This resulted in a high PC value of 2.9×10^{-2} S·cm^{-1} (95 °C, 60% RH) (Fe). Thus the addition of bridging sulfate ligand against only BDC/NDC imparts anionic nature to MOF as against neutral structure (refer to MIL-53 structure) creating a facile proton transport pathway and yielding ~2.5 times higher PC than Nafion.

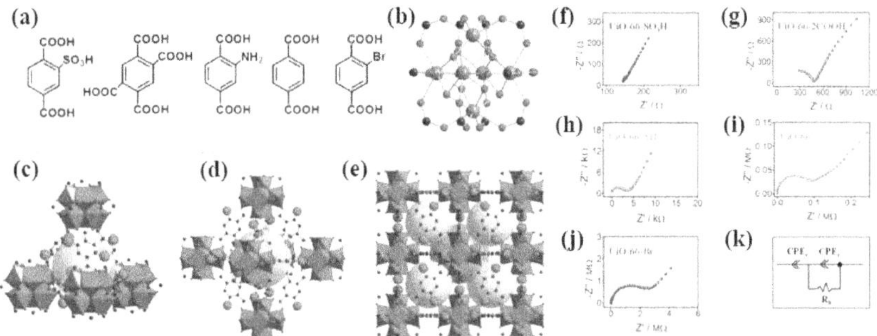

Figure 15. (a) Terephthalate linker with diverse functional groups. (b) Structure of Zr-cluster metal node. (c) tetrahedral and (d) octahedral cage structures. (e) 3D structure of UiO-66 MOF with cage-type pores in the framework and dangling functional groups. Color codes: blue = Zr, black = C, red = O. (f-j) are Nyquist plots of UiO-66 MOFs with respective functional groups. (k) The equivalent circuit was used to fit the data. "Reproduced with permission from Ref. [54]. Copyright 2015 American Chemical Society."

Shigematsu and co-workers reported another representative example of PC reliant on the functional group present on the organic ligand.[56] They chose the stable MIL-53 MOF with Al and Fe as metal ions. The ortho position of the $1,4$-H_2BDC ligand was modified using functional groups such as $-NH_2$, $-OH$, and $-COOH$ in this work. The results showed that MIL-$53(Fe)$-$(COOH)_2$ had the highest PC of 0.7×10^{-5} S·cm^{-1} among the modified MIL-MOFs. The proton transport capacity emerging out of the acidic functional moieties connected to the organic ligand is thought to be the fundamental factor controlling the PC of these MOFs. The order of discovered PC ($-NH_2 < -H < -OH < -COOH$) for the MIL-53 series is well related to the range of pK_a values of various groups of $1,4$-H_2BDC ($-NH_2 > -H > -OH > -COOH$).

5.4 Tuning Pore Space

The H-bonding interactions between the conductive medium and the framework surface are controlled by the hydrophilicity and hydrophobicity of the pore surface, which improves PC. The pore design and its environment can eventually be adorned with the right combination of ligands and functional groups. In this context, Park and co-workers reported an $Mg_2H_6(H_3O)(TTFTB)_3$) a mesoporous MOF (MIT-25) with cylindrical pores with sizes of 27 Å and 4.5 Å, respectively (Fig. 16).[57] The large mesopore is composed of two structurally integrated μ_2-H-bridged carboxylic acids per Mg^{2+} center. In contrast, the tiny pore has an additional pair of μ_2-H-bridged carboxylates that projects into the pore and form a helical channel. Water vapour sorption reveals a characteristic stepwise adsorption profile. Isotherm displays the first step which appeared below 40% RH, corresponding to a smaller hydrophilic pore. While the second step happens above 50% RH and was related to a less hydrophilic pore with a large free space. In this MOF, they studied proton transfer that is selective and reliant on vapour pressure. The PC increases from 1.58 × 10^{-5} S.cm^{-1} (at 25 °C) to 1.03 × 10^{-4} S·cm^{-1} (at 75 °C), with E_a of 0.36 eV at 40% RH. While, at 95% RH, the PC increases from 6.8 × 10^{-5} S·cm^{-1} (at 25 °C) to 5.1 × 10^{-4} S·cm^{-1} (at 75 °C), with E_a of 0.40 eV. According to H_2O vapour sorption and theoretical investigations, the small pore occupies guest H_2O at 40% RH, indicating that conductivity at low RH levels is mostly associated with the Grotthuss mechanism in hydrophilic small pores. The PC at higher RH values, on the other hand, comprises an average of both large and small pores due to the successive filling. Thus, this study highlights the importance of the design of hydrophilic pores for PC under humid conditions.

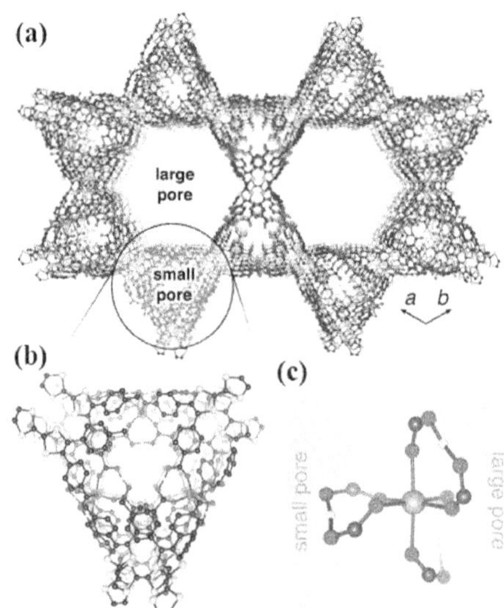

Figure 16. (a) MIT-25 MOF porous structure along the *c*-axis. (b) A small pore structure with one μ_2-H-bridged carboxylic acid pointing towards the pore. (c) The metal node coordination environment. "Reproduced with permission from Ref. [57]. Copyright 2018 American Chemical Society."

Further, Kitagawa *et al.* reported the metal-organic nanotube, with the formula $[Pt(dach)(bpy)Br]_4(SO_4)_4 \cdot 32H_2O$ obtained via oxidative polymerization of $[Pt(dach)(bpy)]_4(SO_4)_4$.[58] The MOF structure contains hydrophobic (A) and hydrophilic (B) 1D channels. Importantly the hydrophobic channel contains two types of water clusters, octamers and tetramers with weak interactions between pore walls expected to provide high PC. The sulfate anions present between the tubes balance the overall charge. A superprotonic PC of 1.7×10^{-2} S·cm^{-1} (55 °C, 95% RH) was recorded for MOF single-crystal along the hydrophobic channel direction. The efficient diffusion of water due to weak interactions with pore wall, acidity imparted by amine protons combined with the solid-liquid transition of water, markedly different from the bulk water, in the hydrophobic channel together contributed to superior PC performance. The situation is equivalent to that of carbon nanotubes (CNTs) or ion channels in ion transport proteins.

5.5 Post-synthetic Modifications (PSM) in MOFs

Synthetic chemists use bond breaking and making as a tool to realize molecules of interest with fascinating properties. In recent times, the PSM approach is being extensively used to construct a framework with outstanding PC ability. Usually, the multiple oxygen atoms of the linker serve as a donor atoms for metal coordination at the time of preparation of the MOF. As a result, proton donors such as –COOH, –SO_3H, and –PO_4H_3 that are totally or partially linked to the metal core have a limited role. The PSM reaction, which occurs after the MOF synthesis, has proven to be a useful strategy for overcoming this limitation. The structural stability of the MOF is required for successful PSM to prevent framework structure breakdown during the chemical reaction. Phang and co-workers introduced acidic –SO_3H sites in UiO-66-(SH)$_2$ using a post-synthesis oxidation process.[59] During the PSM, there was no decomposition or change in the structure of the framework was observed. A relatively low conductivity was obtained for UiO-66-(SH)$_2$ as compared to UiO-66(SO$_3$H)$_2$ (1.4×10^{-2} S·cm^{-1} 25 °C, 90% RH).

Das *et al.* reported a post-synthetic ring-opening method to introduce acidic –SO_3H functionality in UiO-66-NH$_2$ MOF.[60] Propane sultone and butane sultone undergo ring-opening reactions to yield covalently linked –SO_3H (Fig. 17a and c). The different alkyl chain lengths led to a large difference in conductivity at relatively high temperatures (1.64×10^{-1} and 4.6×10^{-3} S·cm⁻¹, 80 °C, 95% RH, respectively) as well as low temperatures (5.83×10^{-2} S·cm⁻¹, 11.5 °C; 5.9×10^{-4} S·cm⁻¹, 10.1 °C; both at 95% RH) (Fig. 17b and d). This work revealed the significance of the PSM approach in modulating and achieving high conductivity by changing the alkyl chain lengths via ring-opening reactions.

Not only ligands but the PSM strategy can also be used to modify the metal node to realize high PC. Very recently, Zhou *et al.* employing the PSM strategy successfully grafted organic acids, EDTA and OX into the MOF-808 framework.[61] MOF-808-OX has a PC of 1.94×10^{-4} S·cm⁻¹ (30 °C, 98% RH) which is ~150-times greater than the parent MOF-808.

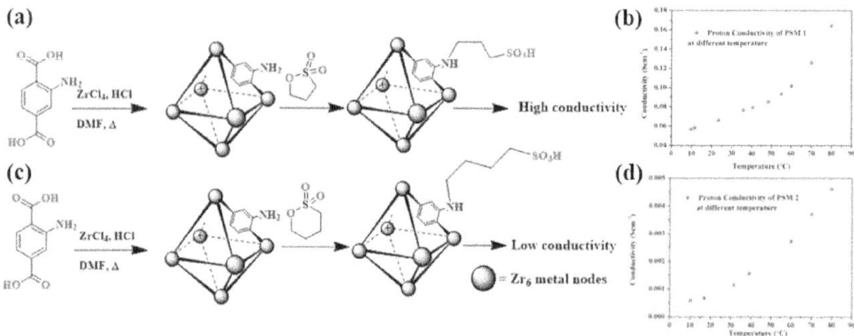

Figure 17. (a, c) Synthesis and PSM scheme of UiO-66-NH₂ MOF with respective sultone. (b, d) Ionic conductivity of PSM1 and PSM 2 as a function of temperature. "Reproduced with permission from Ref. [60]. Copyright 2019 American Chemical Society."

Another approach is the inclusion of counter ions in the framework via PSM, in which, the counter ions in MOFs act as a proton source and/or carrier. In this connection, Wei and co-workers investigated proton-conducting multi-layered MOF, formulated as (Me₂NH₂)[Eu(L)] (H₄L = 5-(phosphonomethyl)isophthalic acid).[62] The layered anionic framework [Eu(L)]⁻ with counter cations (Me₂NH₂)⁺ are present between the layers gave rise to a 3D structure. These counter cations interact with neighboring non-coordinated oxygen atoms from phosphonate groups to build strongly H-bonded networks (N–H···O) arranged parallel to the *c*-axis (Fig. 18). Along these chains, they found the facile PC for a single-crystal (1.25×10^{-3}

S·cm⁻¹, 150 °C), while, the pressed pellet yielded a PC of 3.76×10^{-3} S·cm⁻¹ (100 °C, 98% RH) (Fig. 18d).

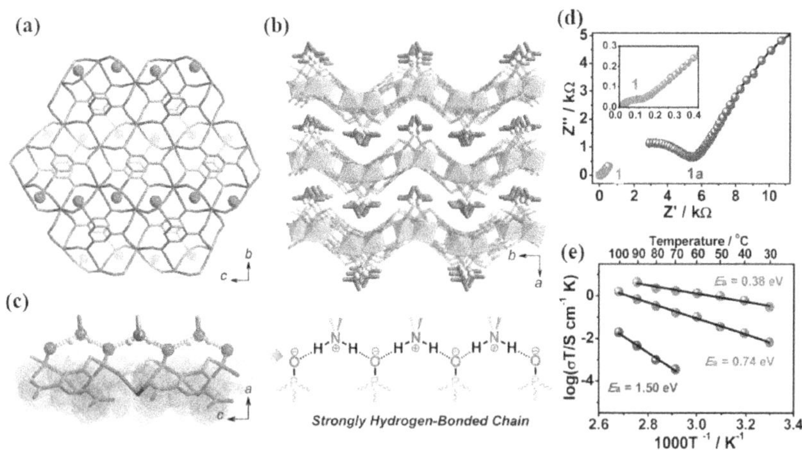

Figure 18. (a) Anionic host framework of the MOF $(Me_2NH_2)[Eu(L)]$. (b) The 2D anionic layered structure of MOF with the counter cations. (c) H-bonding interactions between O_7 atoms and $(Me_2NH_2)^+$ cations. (d) Nyquist plots of the single crystal (red) and compacted powder sample (blue) (98% RH, 100 °C). (e) Arrhenius plots for the single crystal (green) and powder sample (blue). "Reproduced with permission from Ref. [62]. Copyright 2017 American Chemical Society."

Furthermore, Biradha *et al.* reported discrete coordination metal complexes $\{[Zn(PIP)_2(H_2O)_4]\cdot NDS\}$ and $\{[Ni(PIP)_2(H_2O)_4]\cdot NDS\}$. They introduced NDS in the large rhomboidal channels of dimensions 27.59 Å × 20.27 Å (for Zn complex) and 27.40 Å × 20.21 Å (for Ni complex) in an *in-situ* reaction.[63] These complexes exhibited PC in the order of 10^{-5} to 10^{-3} S·cm⁻¹ (80 °C, 98% RH). PC has been aided in these complexes, owing to strong H-bonds between the coordinated H_2O molecules, Im moieties, and NDS anions.

5.6 Defects and Amorphization in MOFs

The deliberate introduction of structural defects or structural imperfections can be utilized to impart or promote the PC. The missing metal node or organic linkers in the framework act as defect sites and lower the energy barrier for the proton movement. These defects in the MOFs can be introduced during synthesis or post-synthetically. Kitagawa *et al.* studied the defective Zr-based MOF synthesized by

ZrCl$_4$ and 2-sulfoterephthalate having a hexanuclear Zr(IV) cluster [Zr$_6$O$_4$(OH)$_8$L$_{4.2}$.xH$_2$O] where the bulky ligand is expected to introduce the defects (Fig. 19).[64] These defects gave rise to the formation of 9- and 12-connected metal clusters as shown in Figure 19a. The tetrahedral pore (5 Å), an octahedral pore (6 Å), and a larger pore (12 × 25 Å) from the empty cluster are shown in Figures 19b and c. The PC for two batches of the samples was observed to be, 1.93 × 10^{-3} S·cm^{-1} and 1.82 × 10^{-3} S·cm^{-1} (65 °C, 95%). The lower value is assigned to defective metal clusters acting as proton-trapping sites and the calculations suggested that acid treatment of the MOF could improve the overall conductivity by saturating the trapping sites.

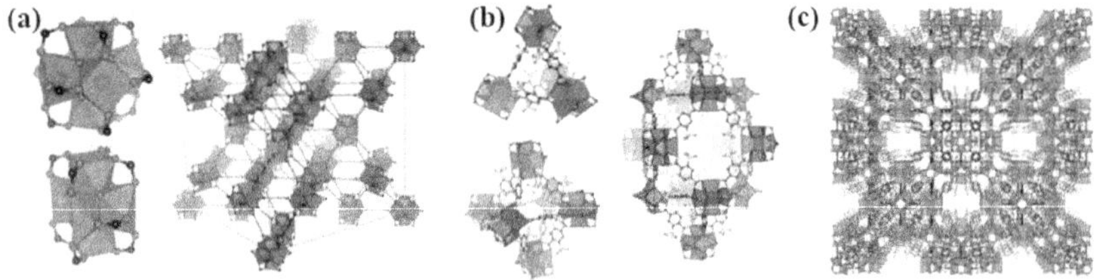

Figure 19. (a) The connectivity of the Zr clusters in (Zr$_6$O$_4$(OH)$_8$(O$_2$C–R–CO$_2$)$_{4.2}$·67H$_2$O) MOF in the unit cell is depicted. (b) Tetrahedral, octahedral, and large pore structure in the MOF. (c) Large continuous pores in cubic cells of zirconium 2- sulfoterephthalate. (Colour, black = C, red = O, yellow = S, green = Zr, green). "Reproduced with permission from Ref. [64]. Copyright 2015 American Chemical Society."

The MOF synthesized in presence of acetic acid and sulfoacetic acid did show higher values of 2.40 × 10^{-3} S·cm^{-1} (65 °C, 95% RH) and 5.62 × 10^{-3} S·cm^{-1} (95% RH). This study demonstrates that the surface properties of the MOF can be significantly altered through the creation of defective sites within MOFs that effectively act as the proton trapping sites in the defective clusters of the MOFs.

Kitagawa and co-workers synthesized 2D MOF formulated as [Zn-(H$_2$PO$_4$)$_2$(HTz)$_2$]$_n$ Zn-MOF (1) using Tz ligand for anhydrous PC.[65] Ligand HTz connects the adjacent octahedral metal centers and two axial positions are occupied by mono-dentate H$_2$PO$_4^-$ ions which serve as the dominant PC pathway for proton hopping. They enhanced the proton concentration and mobility by introducing defects and encapsulating the acidic and mobile H$_3$PO$_4$ species (Fig. 20). For this purpose, the excess amounts of H$_3$PO$_4$ mmol reacted were 4.0 for

Zn-MOF (2), 4.4 for Zn-MOF (3), 4.8 for Zn-MOF (4), and 5.2 mmol for Zn-MOF (5), leading to the incremental addition of H_3PO_4 as a defect in the framework. The Zn-MOF (5), exhibited four orders of magnitude higher PC of 4.6×10^{-3} S·cm^{-1} (150 °C), than that of the Zn-MOF (2). The defect sites in the structure, as well as the encapsulated mobile proton carriers (H_3PO_4, $H_2PO_4^-$, and H_2O), are critical in establishing an efficient PC path and increasing proton movement inside the framework. The facile proton conduction and nonporous characteristics of MOFs allowed them to be employed as an electrolyte in an FC (See section 6.1, for more details).

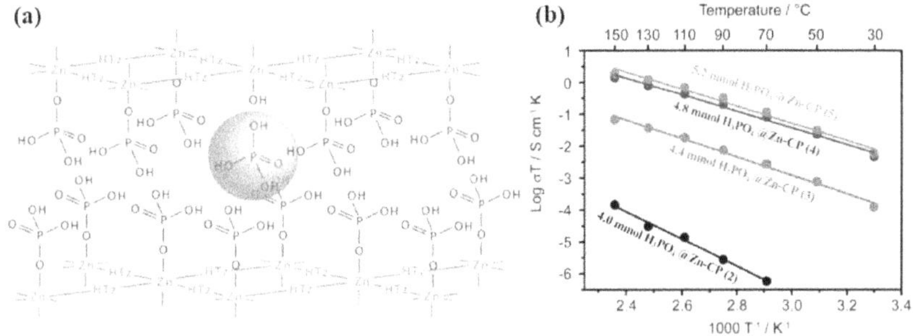

Figure 20. (a) Suggested structure of 2D Zn-MOF (5) with the OH$^-$ ion is coordinated to Zn rather than monodentate $H_2PO_4^-$, and H_3PO_4 occupying defect site in between the layers. (b) Arrhenius plot of Zn-MOF under N_2 with different acid content. "Reproduced with permission from Ref. [65]. Copyright 2016 American Chemical Society."

Despite the outstanding PC of MOFs over a wide temperature range, owing to the crystalline nature is difficult to process and offers an extra resistance component compromising the performance of solid-state electrolytes.[66] MOFs with reversible phase transition from hard crystals to liquids and glasses provide the solution to realize grain boundary-free monolith to improve solid-solid interface contact.[67] Contrary to the crystalline state of MOFs, the glassy states are highly disordered and defect-rich states with highly dynamic structures.[28] The glassy state possesses properties such as transparency, relatively softer mechanical properties, and network dynamics, typically absent in crystalline states.[68, 69] The most effective and often used method for creating glassy MOFs from their crystals is melt quenching.[28, 66] However, obtaining the glassy state of MOFs is highly challenging due to limited knowledge of melting and glass formation in MOFs, small temperature window (T_m-T_d gap) available to process MOF glass, stability of glassy state or avoidance of recrystallization (also known as glass-forming ability (GFA).

Horike and co-workers reported a glassy MOF [{$Zn_2(HPO_4)_2(H_2PO_4)$}$(ClbimH^+)_2 \cdot (H_2PO_4) \cdot (MeOH)]_n$ that was obtained by quenching the molten MOF rapidly. Unlike the crystalline counterpart, which has no PC capabilities under similar conditions, this glassy state of 2D MOF has a considerable PC value of 1.2×10^{-4} $S \cdot cm^{-1}$ (25 °C, 98% RH).[69] The disordered phosphates and Clbim in the glassy state in addition to adsorbed water molecules are responsible for efficient proton transfer. The same group also utilized a mechanical milling route to make MOF glasses formulated as [Cd-$(H_2PO_4)_2(HTz)_2]_n$ with increasing amounts of defects.[28] The PC values were found to increase with an increasing amount of defects (Ball milling time) introduced in the MOF structure (40 min milling = 10^{-6} $S \cdot cm^{-1}$, 240 min milling = 10^{-5} $S \cdot cm^{-1}$, 500 min milling = 10^{-4} $S \cdot cm^{-1}$).

5.7 MOF-Composite Materials

As highlighted earlier, almost all proton-conducting MOFs are synthesized as microcrystalline powder which is challenging to process in the form of defect-free films or membranes for electrochemical applications. To overcome this issue and to further enhance the ionic conductivities, researchers employed the MOFs-polymer composites approach to combine the best of both materials in one.

Hou *et al.* reported a Zn-MOF {[$ZnL(bpe)_{0.5}] \cdot 0.5(H_2bpe) \cdot 3(H_2O)$}$_n$ (L = 3,5-Dicarboxy-1-(3,5-dicarboxybenzyl)pyridin-1-ium chloride and bpe and used as a filler for Nafion to prepare a Zn-MOF/Nafion hybrid membrane (Fig. 21).[70] The PC measurements revealed 1.87 times higher PC value for Zn-MOF/Nafion as compared to the pristine Nafion membrane. Furthermore, the NH_3-water vapor loading yielded a 5.47 times higher value (2.13×10^{-2} $S \cdot cm^{-1}$) than that of a pure Nafion membrane.

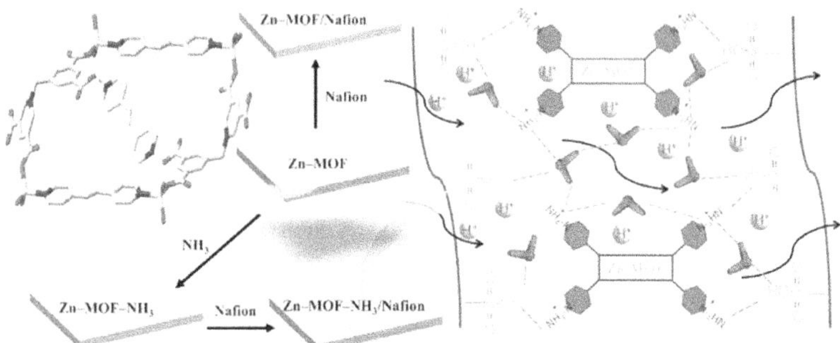

Figure 21. Synergistic H-bond interactions in the Zn-MOF and Nafion in presence of water molecules are shown schematically. "Reproduced with permission from Ref. [70]. Copyright 2021 American Chemical Society."

The conductivity of Nafion decreases drastically as the temperature exceeds 80 °C due to loss of conducting media (H_2O) thus limiting operating temperature. To overcome this limitation, very recently, Wang *et al.* reported loading of 3, 6 and 9 wt % of 3D Cd-MOF, $[Cd_2(pdc)(H_2O)(DMA)_2]_n$ to SPPO polymer substrate and prepared a series of MOF composite materials Cd-MOF@SPPO-3, Cd-MOF@SPPO-6, and Cd-MOF@SPPO-9.[71]. Interestingly, the hybrid membrane with 6 wt% MOF content (MOF@SPPO-6) showed a pronounced increase and reached a PC value of 2.64×10^{-1} S·cm⁻¹ (70 °C, 98% RH, E_a = 0.68 eV) as against parent MOF (1.15×10^{-3} S·cm⁻¹, 90 °C, 98% RH). The E_a value indicates the vehicular mechanism of the proton transport (Fig. 22).

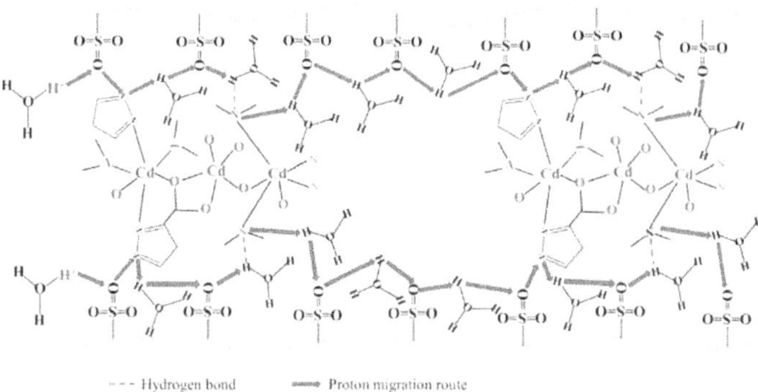

Figure 22. Suggested proton transfer pathway in Cd-MOF@SPPO hybrid film. "Reproduced with permission from Ref. [71]. Copyright 2021 American Chemical Society."

The synergistic effects between the functional groups of the fillers and MOFs exhibit superior functioning as compared to individual constituents, in addition to efficient conduction.[72] Graphene oxide (GO) with carboxyl, epoxy, and hydroxyl functional groups on the surface and at the edge, is a prominent filler material and has been utilized in the making of MOF composites for achieving highly efficient PC.

In addition to MOF alone, it has been found that metered insertion of MOFs/GO into another polymer matrix considerably improves the ionic conductivity of prepared composites. Accordingly, Sun and co-workers prepared a set of SPEEK/UiO-66-SO_3H@GO-X (X = 5, 10, and 15 are weight ratios of the UiO-66@GO to SPEEK) composite membranes using a UiO-66 MOF with $-SO_3$Na functional group and graphene oxide (GO) (Fig. 23).[73] The PC values for the SPEEK/UiO-66@GO-10 membrane were recorded to be 0.268 S·cm^{-1} (70 °C, 95% RH) and 16.57 mS·cm^{-1} (100 °C, 40% RH), which are 2.6 and 6.0 times higher than that of the SPEEK polymer alone. Proton transport is attributed to ionic nanochannels formed at the interfaces between UiO-66@GO and the SPEEK matrix *via* H_2O molecules and $-SO_3$H, as well as natural Bronsted acidic sites in the MOF framework.

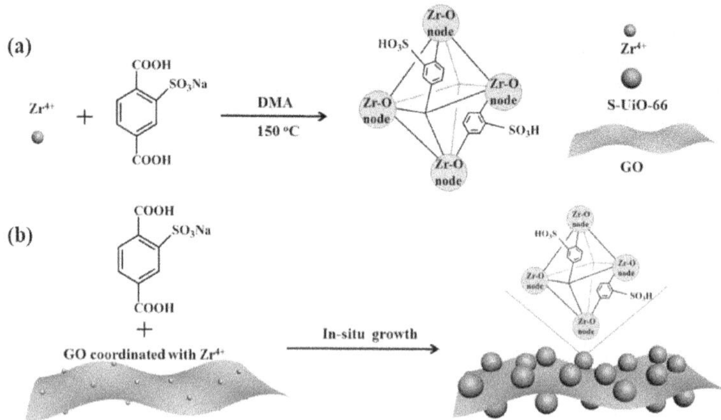

Figure 23. (a) Synthetic scheme of S-UiO-66 MOF. (b) The *in-situ* growth of S-UiO-66@GO hybrid nanosheets. "Reproduced with permission from Ref. [73]. Copyright 2017 American Chemical Society."

The optimal filler-polymer combination, uniform distribution, prudent selection of MOFs with appropriate functional groups and H-bond chains to aid proton transport in a polymer matrix and ease material processing.

6. Applications of Proton Conducting MOFs

6.1 Fuel Cell

The FCs are currently seen as a promising technology for generating clean and efficient power in the twenty-first century. Only a few proton-conducting MOFs have been employed for membrane-electrode assembly (MEA) so far to probe the practicality of MOF electrolytes. In this regard, Shimizu and co-workers sought to successfully integrate β-PCMOF2(Tz)$_{0.45}$ MOF into the FC assembly and measured their proton-conducting performance under real-world conditions.[35] A MEA based on β-PCMOF2(Tz)$_{0.45}$ MOF was built for this purpose, and the electromotive force of the H$_2$/air cell was measured. The MOF membrane was found to be gas-tight and to have a stable open circuit voltage (OCV) of 1.18 V at 100 °C for 72 hours, consequently demonstrating that β-PCMOFs have the potential to be used in FC devices. Further, as mentioned in the earlier section, Kitagawa and co-workers obtained PC $\geq$ ca. 10 mS cm^{-1} in defect rich Zn based MOF ([Zn-(H$_2$PO$_4$)$_2$(HTz)$_2$]$_n$), which is a reasonably high value for the utilization as a solid electrolyte in FC. The MEA comprising Zn-MOF was prepared by sandwiching the MOF in between two platinum-loaded carbon electrodes to obtain a CPFC pellet (CPFC = coordination polymer fuel cell) (Fig. 24a).[65] Then platinum meshes were connected to both sides of the platinum electrodes. A sealant gas obstacle was strongly connected to the CPFC pellet to separate air and H$_2$ gas in the FC experiment. The measurement was done under dry N$_2$ flow. A sealant gas barrier was attached tightly to the CPFC pellet to isolate air and H$_2$ gas during the experiment. The highest open-circuit voltage of 0.88 V was obtained and was retained for an hour (Fig. 24b). The maximum power density attained was 2.8 mW cm^{-2} (Fig. 24c). The OCV value is similar to that of standard polymer electrolyte fuel cells, although being somewhat less than the calculated value of 1.16 V. Fabrication of MOF-based membrane assemblies and evaluation of FC performance should be a frequent practice in the future, in addition to having good conductivity and durability.

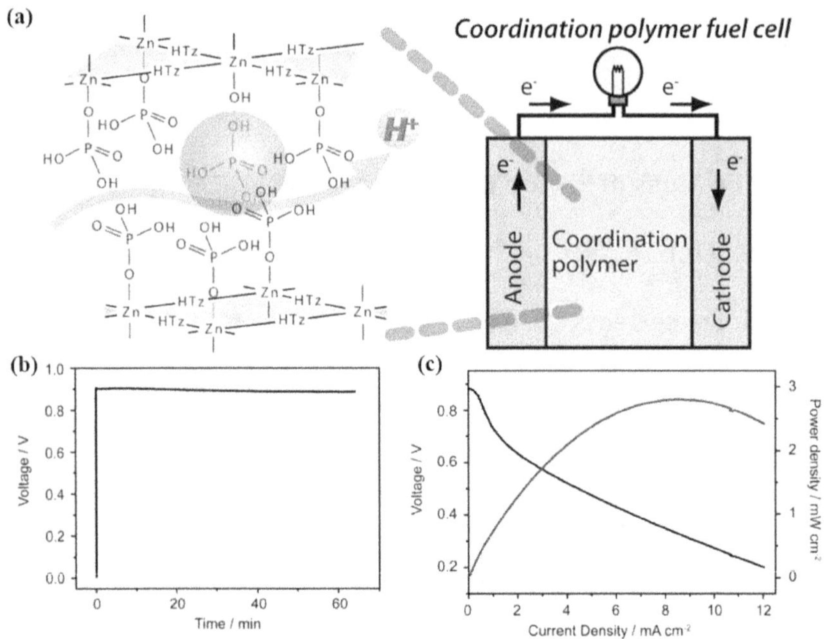

Figure 24. (a) Cartoon representation of FC assembly with MOF as a solid electrolyte. (b) Open circuit voltage plot of 2D Zn-MOF [Zn-$(H_2PO_4)_2(HTz)_2]_n$ membrane in an H_2/O_2 FC (120 °C, anhydrous condition). (c) Performance of H_2/O_2 FC with $[Zn-(H_2PO_4)_2(HTz)_2]_n$ as the electrolyte under a 26% RH and 120 °C. (black line = current-voltage curve and blue lines = current–power curve). "Reproduced with permission from Ref. [65]. Copyright 2016 American Chemical Society."

6.2 Bio-inspired Light Harvesting

The human eye converts the light energy to a proton gradient thus creating a chemical response. The gradient is then translated to the image we '*see*'. This proton gradient across the membrane can be harvested to convert it to electrical energy. Kitagawa *et al.* employed the solid-liquid phase transition in 1D MOF $[Zn(HPO_4)(H_2PO_4)_2](ImH_2)_2$ crystal to achieve bio-inspired photo electric conversion.[74] For this purpose, the 1D MOF was melted and in a molten state, the pyranine, a photoacid molecule was doped to accomplish the light-induced proton gradient in the MOF. On light irradiation, the pyranine molecule reversibly produces the mobile acidic protons as well as the local defects in the glass MOF leading to bulk switchable PC (Fig. 25a). On light irradiation, the quantity of mobile protons increases, resulting in improved PC and the formation of the defects by doped acid. When the doped MOF placed between the two ITO electrodes was irradiated with the light only on one side while the other was left dark, the electrodes revealed a creation potential difference (Fig. 25b). This is ascribed to the creation of asymmetric proton gradient in MOFs, similar to the human eye. It is worth noting here that, as against the membrane-based systems, the present system is all solid-state and does not require sophisticated instrumentations or techniques.

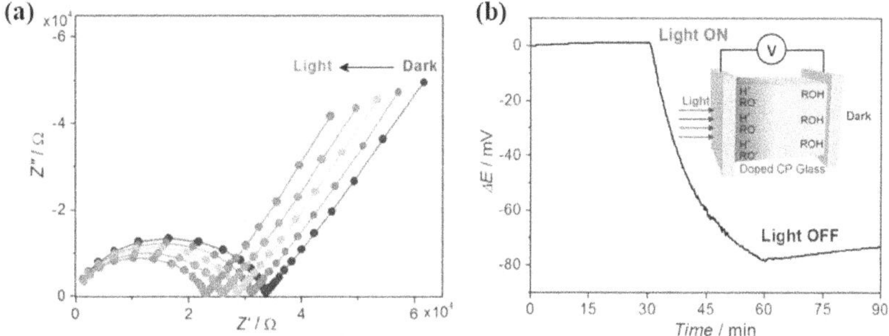

Figure 25. (a) Light irradiation reduces the resistance to conduction in pyranine doped $[Zn(HPO_4)(H_2PO_4)_2](ImH_2)_2$ MOF glass. (b) In the absence and presence of light irradiation, the formation of potential difference (ΔE_{cell}) changes among the electrodes concerning time for 1D MOF. The experimental setup used to investigate the potential difference between the irradiation and dark sides of the MOF is illustrated in the inset. "Reproduced with permission from Ref. [74]. Copyright 2017 Wiley-VCH."

6.3 Proton Batteries

While the application of typical liquid electrolytes is inconceivable at higher temperatures due to safety issues, the all-solid-state H^+ batteries with high PC and moldability are highly desirable. In view of this, for the first time, Horike *et al.* developed all-solid-state proton batteries using new proton conductive MOF glass $[Zn_3(H_2PO_4)_6(H_2O)_3](BTA)$.[75] The melt quenched glassy-state of MOF (denoted as 1g) is free of grain boundaries as compared to its parent polycrystalline form which contains grain boundaries and flaws leading to interface issues (Fig. 26a). The low melting temperature (T_m) of 114 °C, prevented the electrode materials from degrading through the fabrication procedure. The molten state of MOF (denoted as 1m) is loaded on a carbon fiber (CF) electrode by pressing and quenched to RT (CF–1g). SEM images of the electrode-electrolyte interface (CF–1g) (Fig. 26b), indicated the absence of grain boundary. The position of the CF electrode (C), as well as uniformly scattered Zn, P, and O elements was further resolved by element mapping (Fig. 26D-G). The proton dynamics of both phosphate and BTA were boosted in a glassy state, resulting in a PC value of 8.0×10^{-3} S·cm^{-1} (120 °C). It was discovered that the proton transference number (t_+) was 1.0, indicating single ion conduction. Furthermore, utilizing MoO_3 and $Cu^{II}[Fe^{III}(CN)_6]_{2/3} \cdot 4H_2O$ as model cathode and anode, the author investigated glass MOF (1g) as a solid-state electrolyte for H$^+$ batteries. Figure 26H displays the charge-discharge profiles from 0 to 1.2 V. The performance of the H$^+$ battery was also tested at a high temperature in an inert atmosphere (Fig. 26I). At 10 mAg^{-1}, the greatest discharge capacity of 55.4 mAh g^{-1} was measured. This is the first example of a rechargeable all-solid-state proton battery that uses the glassy-state MOF and can operate at temperatures ranging from 25 to 110 °C.

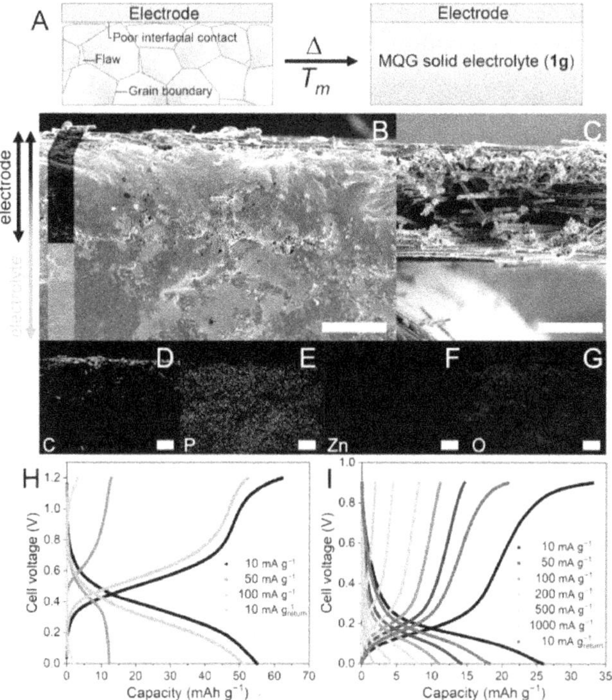

Figure 26. (A) Illustrative comparison of the connections or deformities in the microcrystalline and MQG solid electrolyte. (B) SEM images of the electrode-electrolyte connection (CF–1g) and (C) CF electrode. (D-G) Elemental mapping. (H-I) Proton battery charge-discharge curve by employing 1g as a solid-state. "Reproduced from Ref. [75] with permission from the Royal Society of Chemistry."

6.4 Sensing

MOFs have widely been used for sensing applications owing to their porous structure and change in the physical/chemical properties in response to incoming guests. Typically, changes in luminescence, electrical conductivity, colour, or magnetic properties in response to the analyte are used in MOF-based sensors. As discussed earlier, the incorporation of protic guests in the MOF framework leads to an increase in PC as probed by impedance analysis. Conversely, the change in the conductivity value can also be used as a measure to detect the guest molecules. Following this, Ruan *et al.* employed Titanium-based MOF NH_2-MIL-125(Ti) to detect the humidity.[76] The humidity sensor was devised by coating MOF nanoparticles on inter-digitated electrodes and showed linearly increasing PC response with a short response time when exposed to RH of 11-95 %. The increase in PC was assigned to an increased proton movement within the cavity establishing the potential of MOF as a humidity sensor.

Recently, Li and co-workers employed a similar mechanism to detect NH_3 under humid conditions and at very low concentrations. A 2D copper-based MOF $[Cu(p-IPhHIDC)]_n$ showed an optimum PC value of 1.5 10^{-3} S·cm^{-1} at 100 °C and 98% RH.[77] MOF exhibited a considerable rise in PC values when exposed to NH_3 gas under wide RH (68–98% RH) and temperature (25–40 °C) parameters. The greater E_a of 0.5 eV indicates that the increased PC on NH_3 exposure is related to the migration of exotic water and ammonia molecules inside the layers followed by NH_4^+ or $H^+(H_2O)_n$. The average shift in impedance value toward 130 ppm NH_3 was as high as 8620%. Without interference from seven distinct gases, the detection limit was determined to be 2 ppm. The MOF-based impedance gas sensor is planned to be employed in agricultural and animal husbandry to monitor air quality.

7. Future Perspective

In recent decades, Metal-Organic Frameworks (MOFs) have witnessed explosive development after immense efforts from the scientific community globally and are rising every year for realizing ultra-high PC of >10^{-1} S·cm^{-1}. So far, the development in the proton conducting MOFs was more focused on FC applications, and the emergence of novel proton-based electrochemical energy storage and conversion technologies such as proton batteries, supercapacitors, and wearables have further expanded the scope of the research area in the domain of clean energy. This chapter presents the key features of MOFs as designer solid-state electrolytes in context with underlying operating mechanisms, strategies of designing the MOFs, their chemical and physical modification, and novel applications with representative examples.

From a research perspective, investigation and improvement of the chemical and redox stability of the MOFs are required for real-world applications. While acidic guest molecules impart high ionic conductivity the leaching of these guest molecules should be avoided to inhibit corrosion of electrodes and current collectors during use. MOFs with smaller pore windows, stronger interaction with the framework, or chemical immobilization strategies should be explored. The pellet-based measurement significantly deviated from the intrinsic conductivity values, thus measurements using single-crystal or monoliths should be reported where ever possible. As practically observed, when E_a is ~0.4 eV, the conduction mechanism could not be assigned to either of the mechanisms. Also, the effect of local viscosity change on proton mobility cannot be ruled out. Thus, for a comprehensive understanding of the structure-property correlation use of diverse chemical, physical and computational tools would be valuable. As all these electrochemical processes occur at the electrode/electrolyte interface significant attention should be provided to interface engineering and mechanical properties. In this respect, glassy or liquid states of MOFs or MOF-composites could be worthwhile to tune and address the solid-solid interface issue. The specific application-oriented development criteria to be achieved are avoiding fuel cross-over during FC operation and low-temperature PC for space and related proton battery applications.[78]

References

1. Yue, M., Lambert, H., Pahon, E., Roche, R., Jemei, S., and Hissel, D. (2021). Hydrogen energy systems: A critical review of technologies, applications, trends and challenges. *Renewable and Sustainable Energy Reviews*, **146**, 111180.
2. Sorrell, S. (2015). Reducing energy demand: A review of issues, challenges and approaches. *Renewable and Sustainable Energy Reviews*, **47**, 74–82.
3. Avtar, R., Tripathi, S., Aggarwal, A. K., and Kumar, P. (2019). Population–Urbanization–Energy Nexus: A Review. *Resources*, **8**(3), 136.
4. Bernstein, J. A., Alexis, N., Barnes, C., Bernstein, I. L., Nel, A., Peden, D., Diaz-Sanchez, D., Tarlo, S. M., Williams, P. B. (2004). Health effects of air pollution. *Journal of Allergy and Clinical Immunology*, **114**(5), 1116–1123.
5. Winter, M., and Brodd, R. J. (2004). What Are Batteries, Fuel Cells, and Supercapacitors? *Chemical Reviews*, **104**(10), 4245–4270.
6. Atkins, P., and Paula, J. de. (2009). *Atkins' Physical Chemistry*, 9ᵗʰ edition, Oxford University Press, UK.
7. Acres, G. (2001). Recent advances in fuel cell technology and its applications. *Journal of Power Sources*, **100**(1–2), 60–66.
8. Kirubakaran, A., Jain, S., and Nema, R. K. (2009). A review on fuel cell technologies and power electronic interface. *Renewable and*

Sustainable Energy Reviews, **13**(9), 2430–2440.

9. Wang, Y., Chen, K. S., Mishler, J., Cho, S. C., and Adroher, X. C. (2011). A review of polymer electrolyte membrane fuel cells: Technology, applications, and needs on fundamental research. *Applied Energy*, **88**(4), 981-1007.

10. Gao, H., and Lian, K. (2014). Proton-conducting polymer electrolytes and their applications in solid supercapacitors: A review. *RSC Advances*, **4**, 33091-33113.

11. Xu, Y., Wu, X., and Ji, X. (2021). The Renaissance of Proton Batteries. *Small Structures*, **2**(5), 2000113.

12. Zhao, Q., Stalin, S., Zhao, C.-Z., and Archer, L. A. (2020). Designing solid-state electrolytes for safe, energy-dense batteries. *Nature Reviews Materials*, **5**(3), 229–252.

13. Yuan, S., Feng, L., Wang, K., Pang, J., Bosch, M., Lollar, C., Sun, Y., Qin, J., Yang, X., Zhang, P., Wang, Q., Zou, L., Zhang, Y., Zhang, L., Fang, Y., Li, J., and Zhou, H.-C. (2018). Stable Metal-Organic Frameworks: Design, Synthesis, and Applications. *Advanced Materials*, **30**(37), 1704303.

14. Dietzel, P. D. C., Besikiotis, V., and Blom, R. (2009). Application of metal–organic frameworks with coordinatively unsaturated metal sites in storage and separation of methane and carbon dioxide. *Journal of Materials Chemistry*, **19**(39), 7362.

15. Ranocchiari, M., and Bokhoven, J. A. van. (2011). Catalysis by metal–organic frameworks: fundamentals and opportunities. *Physical Chemistry Chemical Physics*, **13**(14), 6388.

16. Mínguez Espallargas, G., and Coronado, E. (2018). Magnetic functionalities in MOFs: from the framework to the pore. *Chemical Society Reviews*, **47**(2), 533–557.

17. Keskin, S., and Kızılel, S. (2011). Biomedical Applications of Metal Organic Frameworks. *Industrial & Engineering Chemistry Research*, **50**(4), 1799–1812.

18. Kreno, L. E., Leong, K., Farha, O. K., Allendorf, M., Van Duyne, R. P., and Hupp, J. T. (2012). Metal–Organic Framework Materials as Chemical Sensors. *Chemical Reviews*, **112**(2), 1105–1125.

19. Qiu, T., Liang, Z., Guo, W., Tabassum, H., Gao, S., and Zou, R. (2020). Metal–Organic Framework-Based Materials for Energy Conversion and Storage. *ACS Energy Letters*, **5**(2), 520–532.

20. Kanda, S., Yamashita, K., and Ohkawa, K. (1979). A Proton Conductive Coordination Polymer. I. [N , N ′-Bis(2-hydroxyethyl)dithiooxamido]copper(II). *Bulletin of the Chemical Society of Japan*, **52**(11), 3296–3301.

21. Kitagawa, H., Nagao, Y., Fujishima, M., Ikeda, R., and Kanda, S. (2003). Highly proton-conductive copper coordination polymer, H2dtoaCu (H2dtoa=dithiooxamide anion). *Inorganic Chemistry Communications*, **6**(4), 346–348.

22. Joarder, B., Lin, J. Bin, Romero, Z., and Shimizu, G. K. H. (2017). Single Crystal Proton Conduction Study of a Metal Organic Framework of Modest Water Stability. *Journal of the American Chemical Society*, **139**(21), 7176–7179.

23. Xu, G., Otsubo, K., Yamada, T., Sakaida, S., and Kitagawa, H. (2013). Superprotonic Conductivity in a Highly Oriented Crystalline Metal–Organic Framework Nanofilm. *Journal of the American Chemical Society*, **135**(20), 7438–7441.

24. Agmon, N. (1995). The Grotthuss mechanism. *Chemical Physics Letters*, **244**(5–6), 456–462.

25. Ramaswamy, P., Wong, N. E., and Shimizu, G. K. H. (2014). MOFs as proton conductors – challenges and opportunities. *Chem. Soc. Rev.*, **43**(16), 5913–5932.

26. Gupta, A., Goswami, S., Elahi, S. M., and Konar, S. (2021). Role of Framework–Carrier Interactions in Proton-Conducting Crystalline Porous Materials. *Crystal Growth & Design*, **21**(3), 1378–1388.

27. Kreuer, K.-D., Rabenau, A., and Weppner, W. (1982). Vehicle Mechanism, A New Model for the Interpretation of the Conductivity of Fast Proton Conductors. *Angewandte Chemie International Edition*, **21**(3), 208–209.

28. Horike, S., Nagarkar, S. S., Ogawa, T., and Kitagawa, S. (2020, April 20). A New Dimension for Coordination Polymers and Metal–Organic Frameworks: Towards Functional Glasses and Liquids. *Angewandte Chemie - International Edition*, **59**(17), 6652-6664.

29. Sadakiyo, M., Yamada, T., and Kitagawa, H. (2009). Rational designs for highly proton-conductive metal-organic frameworks. *Journal of the American Chemical Society*, **131**(29), 9906–9907.

30. Taylor, J. M., Dawson, K. W., and Shimizu, G. K. H. (2013). A Water-Stable Metal–Organic Framework with Highly Acidic Pores for Proton-Conducting Applications. *Journal of the American Chemical Society*, **135**(4), 1193–1196.

31. Mallick, A., Kundu, T., and Banerjee, R. (2012). Correlation between coordinated water content and proton conductivity in Ca–BTC-based metal–organic frameworks. *Chemical Communications*, **48**(70), 8829.

32. Elahi, S. M., Chand, S., Deng, W., Pal, A., and Das, M. C. (2018). Polycarboxylate-Templated Coordination Polymers: Role of Templates for Superprotonic Conductivities of up to 10 −1 S cm −1. *Angewandte Chemie International Edition*, **57**(22), 6662–6666.

33. Sikdar, N., Dutta, D., Haldar, R., Ray, T., Hazra, A., Bhattacharyya, A. J., and Maji, T. K. (2016). Coordination-Driven Fluorescent J-Aggregates in a Perylenetetracarboxylate-Based MOF: Permanent Porosity and Proton Conductivity. *The Journal of Physical Chemistry C*, **120**(25), 13622–13629.

34. Bureekaew, S., Horike, S., Higuchi, M., Mizuno, M., Kawamura, T.,

Tanaka, D., Yanai, N., and Kitagawa, S. (2009). One-dimensional imidazole aggregate in aluminium porous coordination polymers with high proton conductivity. *Nature Materials*, **8**(10), 831–836.

35. Hurd, J. A., Vaidhyanathan, R., Thangadurai, V., Ratcliffe, C. I., Moudrakovski, I. L., and Shimizu, G. K. H. (2009). Anhydrous proton conduction at 150 °C in a crystalline metal–organic framework. *Nature Chemistry*, **1**(9), 705–710.

36. Horike, S., Umeyama, D., Inukai, M., Itakura, T., and Kitagawa, S. (2012). Coordination-Network-Based Ionic Plastic Crystal for Anhydrous Proton Conductivity. *Journal of the American Chemical Society*, **134**(18), 7612–7615.

37. Nagarkar, S. S., Unni, S. M., Sharma, A., Kurungot, S., and Ghosh, S. K. (2014). Two-in-one: Inherent anhydrous and water-assisted high proton conduction in a 3D metal-organic framework. *Angewandte Chemie - International Edition*, **53**(10), 2638–2642.

38. Lim, D.-W., and Kitagawa, H. (2020). Proton Transport in Metal–Organic Frameworks. *Chemical Reviews*, **120**(16), 8416–8467.

39. Biradha, K., Goswami, A., Moi, R., and Saha, S. (2021). Metal–organic frameworks as proton conductors: strategies for improved proton conductivity. *Dalton Transactions*, **50**(31), 10655–10673.

40. Ye, Y., Gong, L., Xiang, S., Zhang, Z., and Chen, B. (2020). Metal–Organic Frameworks as a Versatile Platform for Proton Conductors. *Advanced Materials*, **32**(21), 1907090.

41. Ponomareva, V. G., Kovalenko, K. A., Chupakhin, A. P., Dybtsev, D. N., Shutova, E. S., and Fedin, V. P. (2012). Imparting High Proton Conductivity to a Metal–Organic Framework Material by Controlled Acid Impregnation. *Journal of the American Chemical Society*, **134**(38), 15640–15643.

42. Li, X. M., Dong, L. Z., Li, S. L., Xu, G., Liu, J., Zhang, F. M., Lu, L. S., and Lan, Y. Q. (2017). Synergistic Conductivity Effect in a Proton Sources-Coupled Metal-Organic Framework. *ACS Energy Letters*, **2**(10), 2313–2318.

43. Le Bideau, J., Viau, L., and Vioux, A. (2011). Ionogels, ionic liquid based hybrid materials. *Chem. Soc. Rev.*, **40**(2), 907–925.

44. Sun, X.-L., Deng, W.-H., Chen, H., Han, H.-L., Taylor, J. M., Wan, C.-Q., and Xu, G. (2017). A Metal-Organic Framework Impregnated with a Binary Ionic Liquid for Safe Proton Conduction above 100 °C. *Chemistry - A European Journal*, **23**(6), 1248–1252.

45. Wang, Z., and Cohen, S. M. (2009). Postsynthetic modification of metal–organic frameworks. *Chemical Society Reviews*, **38**(5), 1315.

46. Yin, Z., Wan, S., Yang, J., Kurmoo, M., and Zeng, M.-H. (2019). Recent advances in post-synthetic modification of metal–organic frameworks: New types and tandem reactions. *Coordination Chemistry Reviews*, **378**, 500–512.

47. Yamada, T., Sadakiyo, M., and Kitagawa, H. (2009). High Proton

Conductivity of One-Dimensional Ferrous Oxalate Dihydrate. *Journal of the American Chemical Society*, **131**(9), 3144–3145.

48. Jeong, N. C., Samanta, B., Lee, C. Y., Farha, O. K., and Hupp, J. T. (2012). Coordination-Chemistry Control of Proton Conductivity in the Iconic Metal–Organic Framework Material HKUST-1. *Journal of the American Chemical Society*, **134**(1), 51–54.

49. Liu, S.-S., Han, Z., Yang, J.-S., Huang, S.-Z., Dong, X.-Y., and Zang, S.-Q. (2020). Sulfonic Groups Lined along Channels of Metal–Organic Frameworks (MOFs) for Super-Proton Conductor. *Inorganic Chemistry*, **59**(1), 396–402.

50. Ye, Y., Guo, W., Wang, L., Li, Z., Song, Z., Chen, J., Zhang, Z., Xiang, S, and Chen, B. (2017). Straightforward Loading of Imidazole Molecules into Metal–Organic Framework for High Proton Conduction. *Journal of the American Chemical Society*, **139**(44), 15604–15607.

51. Zhang, F.-M., Dong, L.-Z., Qin, J.-S., Guan, W., Liu, J., Li, S.-L., Lu. M., Lan, Y.-Q., Su, Z.-M., and Zhou, H.-C. (2017). Effect of Imidazole Arrangements on Proton-Conductivity in Metal–Organic Frameworks. *Journal of the American Chemical Society*, **139**(17), 6183–6189.

52. Wong, N. E., Ramaswamy, P., Lee, A. S., Gelfand, B. S., Bladek, K. J., Taylor, J. M., Spasyuk, D. M., and Shimizu, G. K. H. (2017). Tuning Intrinsic and Extrinsic Proton Conduction in Metal–Organic Frameworks by the Lanthanide Contraction. *Journal of the American Chemical Society*, **139**(41), 14676–14683.

53. Kundu, T., Sahoo, S. C., and Banerjee, R. (2012). Alkali earth metal (Ca, Sr, Ba) based thermostable metal–organic frameworks (MOFs) for proton conduction. *Chemical Communications*, **48**(41), 4998.

54. Yang, F., Huang, H., Wang, X., Li, F., Gong, Y., Zhong, C., and Li, J.-R. (2015). Proton Conductivities in Functionalized UiO-66: Tuned Properties, Thermogravimetry Mass, and Molecular Simulation Analyses. *Crystal Growth & Design*, **15**(12), 5827–5833.

55. Tu, T. N., Phan, N. Q., Vu, T. T., Nguyen, H. L., Cordova, K. E., and Furukawa, H. (2016). High proton conductivity at low relative humidity in an anionic Fe-based metal–organic framework. *Journal of Materials Chemistry A*, **4**(10), 3638–3641.

56. Shigematsu, A., Yamada, T., and Kitagawa, H. (2011). Wide Control of Proton Conductivity in Porous Coordination Polymers. *Journal of the American Chemical Society*, **133**(7), 2034–2036.

57. Park, S. S., Rieth, A. J., Hendon, C. H., and Dincă, M. (2018). Selective Vapor Pressure Dependent Proton Transport in a Metal-Organic Framework with Two Distinct Hydrophilic Pores. *Journal of the American Chemical Society*, **140**(6), 2016–2019.

58. Otake, K., Otsubo, K., Komatsu, T., Dekura, S., Taylor, J. M., Ikeda, R., Sugimoto, K., Fujiwara, A., Chou, C.-P., Sakti, A. W., Nishimura, Y., Nakai, H., and Kitagawa, H. (2020). Confined water-mediated high

proton conduction in hydrophobic channel of a synthetic nanotube. *Nature Communications*, **11**(1), 843.

59. Phang, W. J., Jo, H., Lee, W. R., Song, J. H., Yoo, K., Kim, B., and Hong, C. S. (2015). Superprotonic Conductivity of a UiO-66 Framework Functionalized with Sulfonic Acid Groups by Facile Postsynthetic Oxidation. *Angewandte Chemie*, **54**(17), 5142–5146.

60. Mukhopadhyay, S., Debgupta, J., Singh, C., Sarkar, R., Basu, O., and Das, S. K. (2019). Designing UiO-66-Based Superprotonic Conductor with the Highest Metal–Organic Framework Based Proton Conductivity. *ACS Applied Materials & Interfaces*, **11**(14), 13423–13432.

61. Meng, X., Wang, H.-N., Wang, L.-S., Zou, Y.-H., and Zhou, Z.-Y. (2019). Enhanced proton conductivity of a MOF-808 framework through anchoring organic acids to the zirconium clusters by post-synthetic modification. *CrystEngComm*, **21**(20), 3146–3150.

62. Wei, Y.-S., Hu, X.-P., Han, Z., Dong, X.-Y., Zang, S.-Q., and Mak, T. C. W. (2017). Unique Proton Dynamics in an Efficient MOF-Based Proton Conductor. *Journal of the American Chemical Society*, **139**(9), 3505–3512.

63. Garai, A., Kumar, A. G., Banerjee, S., and Biradha, K. (2019). Proton-Conducting Hydrogen-Bonded 3D Frameworks of Imidazo-Pyridine-Based Coordination Complexes Containing Naphthalene Disulfonates in Rhomboid Channels. *Chemistry – An Asian Journal*, **14**(23), 4389–4394.

64. Taylor, J. M., Komatsu, T., Dekura, S., Otsubo, K., Takata, M., and Kitagawa, H. (2015). The Role of a Three Dimensionally Ordered Defect Sublattice on the Acidity of a Sulfonated Metal–Organic Framework. *Journal of the American Chemical Society*, **137**(35), 11498–11506.

65. Inukai, M., Horike, S., Itakura, T., Shinozaki, R., Ogiwara, N., Umeyama, D., Nagarkar, S., Nishiyama, Y., Malon, M., Hayashi, A., Ohhara, T., Kiyanagi, R., and Kitagawa, S. (2016). Encapsulating Mobile Proton Carriers into Structural Defects in Coordination Polymer Crystals: High Anhydrous Proton Conduction and Fuel Cell Application. *Journal of the American Chemical Society*, **138**(27), 8505–8511.

66. Bennett, T. D., and Horike, S. (2018). Liquid, glass and amorphous solid states of coordination polymers and metal–organic frameworks. *Nature Reviews Materials*, **3**(11), 431–440.

67. Qiao, A., Bennett, T. D., Tao, H., Krajnc, A., Mali, G., Doherty, C. M., Thorton, A. W., Mauro, J. C., Greaves, G. N., and Yue, Y. (2018). A metal-organic framework with ultrahigh glass-forming ability. *Science Advances*, **4**(3), eaao6827.

68. Ogawa, T., Takahashi, K., Nagarkar, S. S., Ohara, K., Hong, Y., Nishiyama, Y., and Horike, S. (2020). Coordination polymer glass

from a protic ionic liquid: proton conductivity and mechanical properties as an electrolyte. *Chemical Science*, **11**(20), 5175–5181.

69. Inukai, M., Nishiyama, Y., Honjo, K., Das, C., Kitagawa, S., and Horike, S. (2019). Glass-phase coordination polymer displaying proton conductivity and guest-accessible porosity. *Chemical Communications*, **55**(59), 8528–8531.

70. Wang, H., Zhao, Y., Shao, Z., Xu, W., Wu, Q., Ding, X., and Hou, H. (2021). Proton Conduction of Nafion Hybrid Membranes Promoted by NH 3 -Modified Zn-MOF with Host–Guest Collaborative Hydrogen Bonds for H 2 /O 2 Fuel Cell Applications. *ACS Applied Materials & Interfaces*, **13**(6), 7485–7497.

71. Wang, F. D., Su, W. H., Zhang, C. X., and Wang, Q. L. (2021). High Proton Conductivity of a Cadmium Metal-Organic Framework Constructed from Pyrazolecarboxylate and Its Hybrid Membrane. *Inorganic Chemistry*, **60**(21), 16337–16345.

72. Xu, W., Wu, Z., and Tao, S. (2016). Recent progress in electrocatalysts with mesoporous structures for application in polymer electrolyte membrane fuel cells. *Journal of Materials Chemistry A*, **4**(42), 16272–16287.

73. Sun, H., Tang, B., and Wu, P. (2017). Rational Design of S-UiO-66@GO Hybrid Nanosheets for Proton Exchange Membranes with Significantly Enhanced Transport Performance. *ACS Applied Materials & Interfaces*, **9**(31), 26077–26087.

74. Nagarkar, S. S., Horike, S., Itakura, T., Le Ouay, B., Demessence, A., Tsujimoto, M., and Kitagawa, S. (2017). Enhanced and Optically Switchable Proton Conductivity in a Melting Coordination Polymer Crystal. *Angewandte Chemie International Edition*, **56**(18), 4976–4981.

75. Ma, N., Kosasang, S., Yoshida, A., and Horike, S. (2021). Proton-conductive coordination polymer glass for solid-state anhydrous proton batteries. *Chemical Science*, **12**(16), 5818–5824.

76. Zhang, Y., Chen, Y., Zhang, Y., Cong, H., Fu, B., Wen, S., and Ruan, S. (2013). A novel humidity sensor based on NH2-MIL-125(Ti) metal organic framework with high responsiveness. *Journal of Nanoparticle Research*, **15**(10), 2014.

77. Sun, Z., Yu, S., Zhao, L., Wang, J., Li, Z., and Li, G. (2018). A Highly Stable Two-Dimensional Copper(II) Organic Framework for Proton Conduction and Ammonia Impedance Sensing. *Chemistry - A European Journal*, **24**(42), 10829–10839.

78. Ye, Y., Zhang, L., Peng, Q., Wang, G.-E., Shen, Y., Li, Z., Wang, L., Ma, X., Chen, Q.-H., Zhang, Z., and Xiang, S. (2015). High Anhydrous Proton Conductivity of Imidazole-Loaded Mesoporous Polyimides over a Wide Range from Subzero to Moderate Temperature. *Journal of the American Chemical Society*, **137**(2), 913–918.

Abbreviations

PC	Proton conductivity
H2dtoa	N, N′-disubstituted dithiooxamide
adp	Adipic acid
Tz	1,2,4-triazole
ox	oxalate
H5L	1,2,4,5-tetrakisphosphonomethylbenzene
DMA	Dimethylacetamide
BTC	benzene tri-carboxylate
bpy	4,4'-Bipyridine
fdc	furan dicarboxylate
btec	benzene tetra-carboxylate
1,4-ndc	1,4-naphthalene-dicarboxylate
1,4-bdc or BDC	1,4-benzene dicarboxylate
2,6-ndc or NDC	naphthalene-2,6-dicarboxylate
TTFTB	tetrathiafulvalene-tetrabenzoate
dach	(1R,2R)-(-)-1,2 diaminocyclohexane
Clbim	5-chloro-1H-benzimidazole
bpe	1,2-di(pyridin-4-yl) ethylene)
pdc	3,5-pyrazoledicarboxylic acid
DMA	Dimethylamine
SPEEK	Sulfonated poly(ether ether ketone)
ImH2	Monoprotonated imidazole
pIPhHIDC	2-(p-N-imidazol-1-yl)-phenyl-1 H-imidazole-4,5-dicarboxylic acid
EIMS	1-(1-ethyl-3-imidazolium) propane-3-sulfonate
HTFSA	N,N-bis(trifluoromethanesulfonyl) amide
Im	Imidazole
SBBA	4,4′-sulfobisbenzoic acid
K_a	Dissociation constant
PSA	3-pyridinesulfonic acid
PESA	2-(4- pyridyl)ethanesulfonic acid
PTC	Perylenetetracarboxylate
BTA	1,2,3-benzotriazole
PIP	Unsymmetrical olefinic ligand
NDS	1,5-napthalenedisulphonate
SPPO	Sulfonated poly(phenylene oxide))

8

Conversion of CO$_2$ to Valuable Chemicals by Metal-Organic Frameworks

Tapan K. Pal
Department of Chemistry, Pandit Deendayal Energy University,
Gandhinagar, Gujarat, India

1. Introduction

Quite well-known the CO$_2$ is a major contributor to the global warming and its concentration is gradually increases [1,2]. To alleviate this situation much endeavour has been given to captivate, storage and sequestration of CO$_2$ [3,4]. There into, the capture/storage of CO$_2$ followed by its conversion not only help to reduce the CO$_2$ concentration, also can produce the several value added chemicals [5,6]. The friendly characteristics nature of CO$_2$ like non-flammable, highly abundant, non-toxic and easy to handle are the additional advantage of its uses. However, the conversion of CO$_2$ is challenging as high energy is required to break the strong C=O bond (bond energy +805 kJ/mol) [7]. This can be circumvented by choosing proper reaction condition or by using efficient catalyst.

Currently, for energy, we are largely depends on the fossil fuels, and combustion of its releases large amount of CO$_2$ into the atmosphere. Thus, the transformation of CO$_2$ into value added products/fuels are vigorously pursuing in the research laboratory. Regarding value added products the CO$_2$ fixation with epoxide to cyclic carbonate is very attractive as the cyclic carbonates are very useful in various industrial purpose such as dye, polymers, pharmaceuticals etc.[8, 9]. Several homogenous catalysts have been employed in the cyclic carbonate synthesis. But the difficulty in the separation of product as well as the reactant, preclude us to follow the homogeneous catalytic reaction. Another important catalyst is the ionic liquid which contains number of active sites and are very effective in the cyclic carbonate synthesis [10]. But, the drawback is associated with this process is the leaching of the catalyst into the reaction system. Recently, Metal organic frameworks (MOFs) are attractive porous materials and behave as an excellent heterogeneous catalysis. The main constituents in MOFs are the metal nodes (or cluster) and the organic ligand. By judicial selection of the organic linker and metal ion, numerous MOF can be constructed. The inherent characteristics features such as highly stable [11], porous nature [12], large surface are [13], structure flexibility [14,15] and pore size modulations [16] endow them as a potential heterogeneous catalyst [17]. Of note MOFs are the hybrid of homogeneous and heterogeneous catalyst: homogeneous in sense that the orderly distribution of the catalytic site throughout the network with its easy accessibility and regenerability illustrates its heterogeneous nature. The catalytic activity of MOF was recognized earlier [18] and now MOFs have been emerged as a heterogeneous catalyst in numerous organic reaction [19]. It has been found that in the last decades, the MOFs comes out to be a potential candidate in the chemical fixation of CO_2 with epoxide to cyclic carbonate (**Figure 1**).

On the other hand, MOFs have the potency for photochemical reduction of CO_2 to various C1 commodity chemicals (CH_4, CH_3OH, CO and HCOOH) (**Figure 1**). In this regard, an alternative sustainable process, the uses of free and plentiful solar energy for the photocatalytic conversion of CO_2 is very alluring [20]. Enthused from the nature photosynthetic process, researchers have motivated to study the reduction of CO_2 with the assistance of solar energy. Literature survey reveals that the light driven photocatalytic conversion of CO_2 is taking place through certain fundamental ladders [21,22]. For convertion the solar energy into the chemical energy, few steps are involved. First is the light absorption ability: photocatalysts must absorb the maximum sunlight (especially visible light) for better efficiency towards the yield of chemical production. This is because the ultraviolet (UV) light accounts for about 4% of the solar energy whereas, the visible light occupies of about 53% solar energy [23]. Second is the generation of reactive centre: upon absorption of light by photocatalyst, an electron (photoinduced electron) will be produced that subsequently move onto the surface of photocatalyst. Further, the displacement of these generated electron from the surface creates a heterojunction or redox centres and concomitantly inhibit the reconciliation of the charges [24]. Third is the surface active site interact strongly with the adsorbed CO_2 molecule for its reduction and it has been found that the reduction efficacy increase manifold with the increase in the number of reactive sites [25].

Moreover, the large CO_2 adsorption ability and the enrichment of CO_2 molecule around the active sites is help to enhance the yield of the photocatalytic product [26]. Keeping all the above facets in mind, huge endeavour have been devoted in order to enhance the visible light absorption capacity and increase the efficiency for the photo reduction of CO_2 [27]. Several semiconductor based photocatalysts like TiO_2, Bi_2WO_6, $Fe_2V_4O_{13}$, $Sr_2Nb_2O_6$, HNb_3O_8, etc. have been employed for CO_2 reduction under irradiation of light [28]. Different methodologies have been developed to increase the light absorption range and efficiency of the reaction for converting solar energy to chemical energy [29]. In spite of the several progress have been made, for optimization of the photocatalytic reaction, the efforts should be given for searching of new materials with the following advantages like easily accessible catalytic site, controllable light absorption capability, ability to prevent the remerging of the photo excited ion pairs, and finally exceptional CO_2 uptake capacity. Quite interestingly, in this context MOFs are the excellent promising catalysts owing to their inimitable features like potency in broad range of light harvesting capability, generation of separate electron hole, distribution of catalytic site throughout the network along with their easy accessibility. Further, the well understanding of the structure property relationship which help to establish the mechanism of the catalytic reaction. Additionally, the excellent CO_2 uptake capacity of MOF confers their merit in the photocatalytic transformation of CO_2 in to C1 chemicals by enriching the CO_2 around the catalytic site. In this chapter, we have discussed the role of MOFs for the synthesis of cyclic carbonate through cycloaddition reaction between CO_2 with epoxide and the photochemical reduction of CO_2 to C1 fuels with their mechanism of formation.

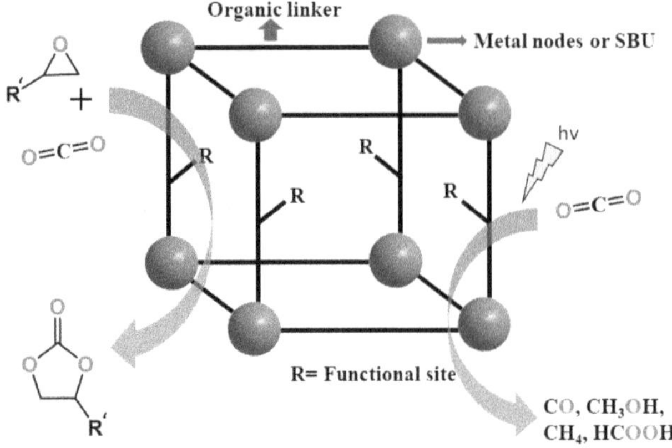

Figure 1. MOF as catalyst for the conversion of CO_2 to cyclic carbonate and its photochemical reduction to various C1 chemicals.

2. The Cycloaddition Reaction between CO_2 and Epoxide Promoted by MOFs

Due to the versatile important industrial application of cyclic carbonate, several catalysts are used to achieve the synthesis of cyclic carbonate. Sometimes the high pressure and temperature is require to accomplish the reaction, and thus the alternative catalytic system is essential that can perform the synthesis of cyclic carbonate under milder conditions. Recently, MOFs comes out to be a potential candidate in this regard. The careful structural investigation of MOFs and CO_2 fixation mechanism, indicated that the catalytically active sites are present within the MOFs are responsible for the conversion of CO_2. These catalytically active sites are (a) Lewis acidic behaviour from open metal sites or the existence of Brönsted acid counterpart, (b) the synergistic catalytic effect bybase and acid moeity (c) the presence of ionic environment in the MOFs and (d) sometimes the defects in the MOF perform the CO_2 cycloaddition reaction. Involvement of these catalytically active sites in the CO_2 fixation reactions are presented in the following sections.

2.1 Fixation of CO_2 Encouraged by Acidic Site in MOF

As discussed, the exposed metal site behave as a Lewis acid which can activate the epxoide ring to induce the cycloaddition reaction. Sometimes, to enhance the yield of the cyclic carbonate under milder condition the co-catalyst is used [30]. A general mechanism is presented in **scheme 1**. The oxygen atom of the expoxide ring is coordinated to the open metal site and gets activate. Next, the nucleophile from co-catalyst attack on the less hindered site of the activated epoxide ring to form alkoxy derivatives which is then reacted with CO_2 to form alkoxy carbonate. Finally this compound undergoes intramolecular nucleophilic attack on the carbon bearing halide leaving group to form cyclic carbonate which subsequently release from the MOF and rejuvenate the main catalyst, MOF and co-catalyst again. Apart from the presence of the open metal site, the existence of acidic part within MOF, the cyclo addition reaction with epoxide can be efficiently performed in presence of co-catalyst.

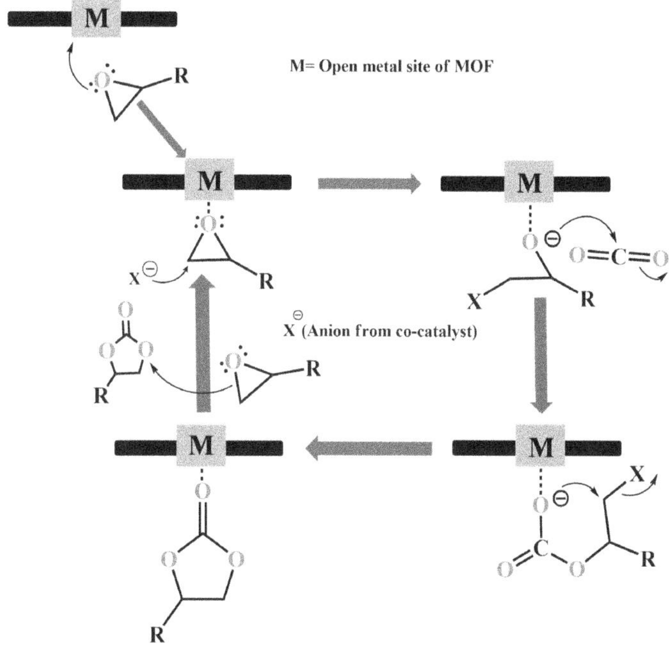

Scheme 1. The activation of epoxide followed by its reaction with CO_2 by the acidic site (open metal site is shown here) present in the MOF with the assistance of TBAB (co-catalyst).

Based on this concept Zhao and co-worker reported 3D MOF that contains open metal Ni^{2+} ion which help to accomplish cycloaddition reaction between CO_2 (0.1 MPa) with diverse epoxide at 70 °C for 12h with the help of TBAB [31]. Although the yield of the carbonate was good to excellent but the large amount of TBAB was employed. Another interesting approached was delivered by Ma and co-worker where the 1D tubular channels of 3D porous MOF contains open metal sites (In^{3+}) and the coordinating halide ions [32]. They activate CO_2 and epoxide very strongly for the fixation reaction (**Figure 2**). However, the the product yield was further enhanced upon addition of co-catalyst, TBAB.

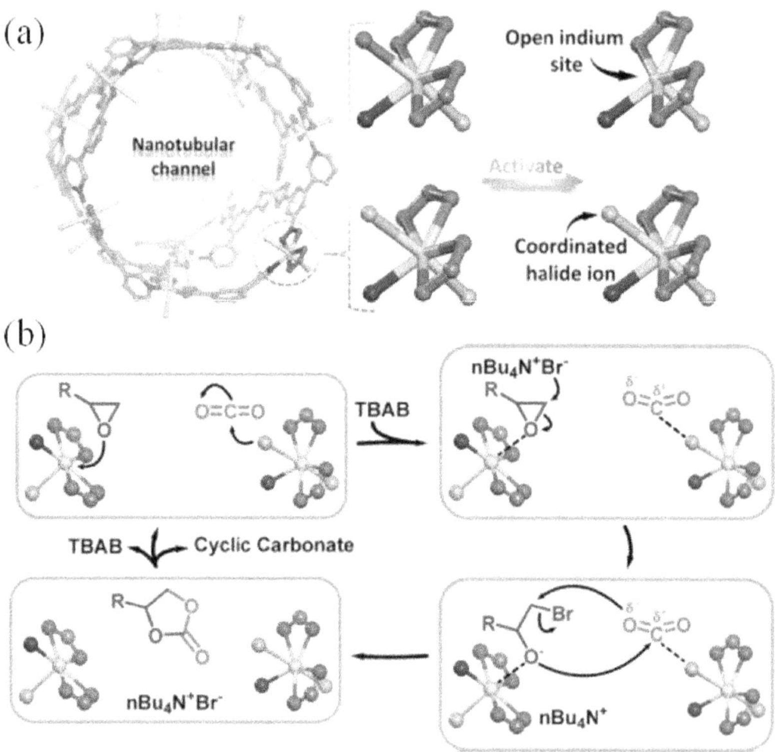

Figure 2. (a) The nanotubular channel containing two different types of indium metal ion. (b) The fixation of CO_2 with epoxide promoted by Indium and coordinated halide ion. Reproduced with permission from [32] American Chemical Society.

The mechanistic study reveal that the coordination of the epoxide oxygen to metal ion helps to activate the epoxide and hence it is anticipating that the more the oxophilic character of the metal ion, the faster will be the reaction rate. In this regard, the zirconium metal ion is preferred as the Zr ion is oxophilic in nature and in addition the MOF constructed from Zr metal ions are generally highly stable and robust which are ideal for CO_2 cycloaddition reaction [33]. The corollary on oxophilic concept was further support when the CO_2 cycloaddition reaction was performed with Hf-NU-100 [34]. The reaction kinetics between CO_2 with styrene epoxide by Hf-NU-100 was superior to some well-known MOF, gea-MOF-1 [35] and Cr-MIL-101 [36]. The Hf cluster is more acidic and exerts more oxophilic than Zr based cluster. In addition, the Hf cluster contains –OH (Brønsted acid) functionality which creates positive impact in the enhancement of CO_2 conversion rate. The idea on by using free Brønsted acid functionality for the CO_2 fixation reaction can achieve very efficiently [37]. One of the interesting examples was reported by Gándara and co-worker [38]. They have designed a ligand which contains a number of carboxylate groups to anticipate one or more –COOH group may remain free in the MOF. As expected the synthesized 3D porous MOFs contains free –COOH group in its cavity. These MOFs showed the CO_2 (1 atm) fixation reaction with epoxide at 80 ºC very efficiently under solvent free condition with the assistance of TBAB. Here the free –COOH group is activating the epoxide ring in a similar way by open metal sites (**Figure 3**). The other notable example based on this methodology also realised in recent reports [39,40,41].

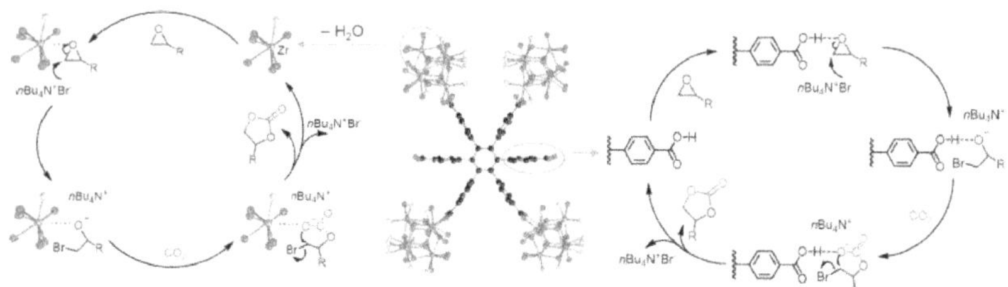

Figure 3. The CO_2 fixation mechanism with epoxide promoted by Brønsted acid (-COOH) present within MOF in presence of TBAB. Reproduced with permission from [38] American Chemical Society.

It is not always feasible to have the free open metal sites in the as synthesised MOFs. Generally, the labile solvents are binds to the metal centre. By thermal instigation the solvent molecules can be removed to generate open metal sites. But in this case we have to careful on the thermal stability of MOFs. Sometimes, the PSM process is allow us to incorporated active metal sites in the framework to perform the CO_2 cycloaddition reaction. An interesting example was reported by Lin and co-worker [42]. They have synthesized 3D porous zinc MOF (1-Zn) and it contains two dissimilar SBUs viz, tetrameric Zn_4O SBU and bimetallic paddle wheel SBU. The zinc metal ion in paddle wheel SBU underwent metal exchange reaction with Cu and Co to produce 1-Cu and 1-Co. These two new MOFs were able to perform the CO_2 cycloaddition reaction with propylene oxide with TBAB as a co-catalyst. However, the catalytic reactivity of 1-Zn was much better than the daughter frameworks and this may be due to the lowering of HOMO-LUMO energy of epoxy propane-CO_2 of 1-Zn as revealed by theoretical investigation. Another important PSM process was demonstrated by Zhang and co-worker. They have introduced alkali metal cations (M= Li^+ to Cs^+) within the cavity of well-known MOF, UTSA-16 to construct M-UTSA-16. The potassium ion is exist in the cavity of UTSA-16 is underwent the metal exchange reaction with these alkali metal cations (M). All the daughter frameworks showed the synthesis of cyclic carbonate from CO_2 and epoxide. But the CO_2 fixation reactivity of Li- UTSA-16 was superior than the other daughter frameworks and this may be due to the high charge density of lithium which activate the CO_2 molecule very proficiently (**Figure 4**) [43]. Recently, Huang and co-worker used well known porphyrin based MOF, PCN-224 and post-synthetically incorporated the bismuth (acidic site) within the core of the porphyrin to afforded Bi-PCN-224. Daughter framework beautifully performed CO_2 cycloaddition reaction with epoxide with the aid of TBAB under milder condition [44]. An interesting report where instead of simple metal ion, by introducing the metal nano-particle inside the MOF cavity the CO_2 cycloaddition reaction with epoxide was achieved very smoothly [45].

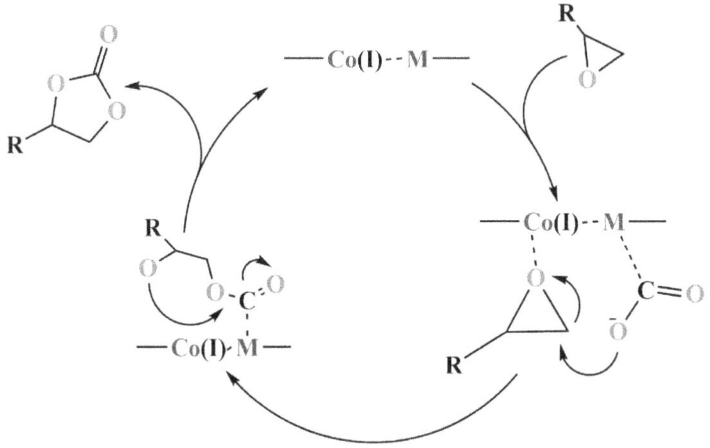

Figure 4. Mechanism of CO_2 cycloaddition with epoxide by M-UTSA-16.

However, the judicial designing of the organic ligand can embedded the metal ions in to the linker back bone of MOFs. This open metal site is nicely exposed to the substrate for smooth execution of the CO_2 fixation reaction. In this regard, metallo-salen ligand is very suitable to meet this requirement [46]. Various metallosalen linkers **(Figure 5)** containing metal ions in its core are used to construct four porous zirconium based frameworks (ZSF-1-4) [47]. Among them the CO_2 fixation activity of ZSF-1 is highest and this is due to the presence of unsaturated Ni^{2+} metal centre which can nicely interact with the adsorbed CO_2 molecule and also efficiently activate the epoxide ring. The catalysts were delivered a very good yield of the desired product except with the larger substrate. This is because the large substrate cannot easily enter into the pore of the framework and hence difficulty for them to interact with the catalytically active site.

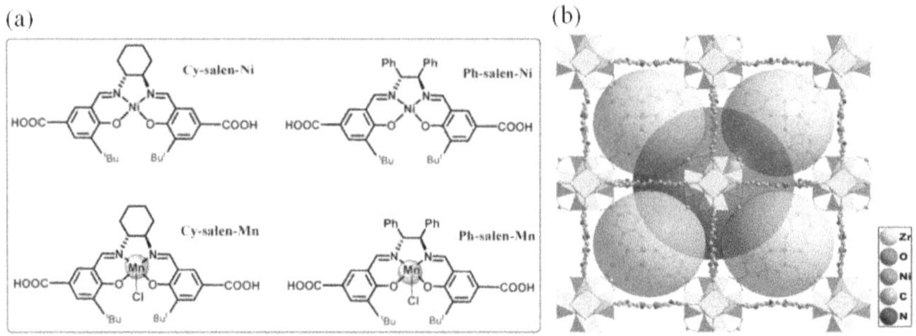

Figure 5. (a) The representation of the salen based linkers and (B) the network view of ZSF-1. Reproduced with permission from [47] American Chemical Society.

Ma and co-workers reported the four fold interpenetrating porous MOF, MMPF-18 which contains Zn_4O SBU and the metalloligand contains Zn^{2+} metal ion (**Figure 6**). This four-fold interpenetration offered high-density metalloporphyrin centers that help to enhance high CO_2 uptake and facile cyclic carbonate synthesis with variety of epoxide with CO_2 (1 atm) for 2 days at RT in presence of TBAB [48]. The yield of the reaction was excellent with smaller size epoxide and when the sterically cumbered epoxide was used then yield of the desired carbonate decreases. This is due to the difficulty in diffusion of the epoxide into the pore of the framework.

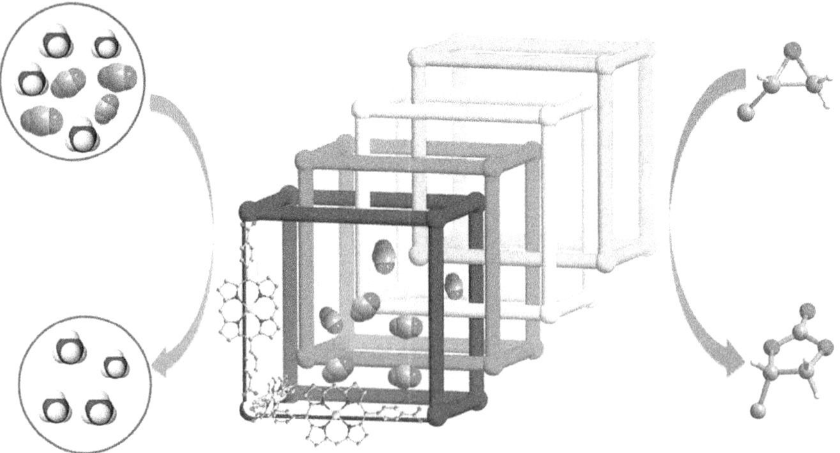

Figure 6. Porphyrine metallo ligand form a Zn based MOF along with exposed Zn^{2+} ion in the porphyrine core to perform CO_2 fixation reaction. Reproduced with permission from [48] American Chemical Society.

It has been noticed that the attendance of heterometallic sites in the MOF network leads to the very good CO_2 fixation activity. In this regard, under solvothermal condition two MOFs CPM-200-In and CPM-200-In/Mg were synthesized. The porous nature of the MOF endorse high CO_2 uptake capacity. The later MOF consist more number of active site (exposed Lewis acidic metal site) and hence it shows the high CO_2 fixation activity than the former MOF with the help of TBAB [49].

2.2 Bi-functional Catalytic Activity from Both Lewis Acid and Lewis Base

It becomes clear if any MOF can uptake large amount of CO_2 then the concentration of CO_2 will be more around the catalytically active site and may enhance the CO_2 fixation rate. Several functional Lewis basic sites such as pyrazole [50], amine [51], imidazole [52] etc. are installed within the MOF to enhance the CO_2 uptake capacity. The presence of Lewis basic group group in the MOF not only enhance the CO_2 uptake capacity also can accomplish the CO_2 fixation reaction [53,54]. Therefore, the role of ligand is very vital because by judicial designing of ligand it is easy to install these functionalities in MOF. In the previous section we have seen that the existence of acidic component in the framework can enhance the CO_2 fixation rate. Therefore, it may anticipated that the existence of both acidic and basic functionality within framework may reinforce the CO_2 cycloaddition reaction rate under milder conditions. A general mechanism for CO_2 fixation reaction is presented in **scheme 2** where the involvement of these functionalities are greatly contribute to activate the epoxide ring and the adsorbed CO_2 molecule... The presence of TBAB is also helping to enhance the yield of the desired carbonate product.

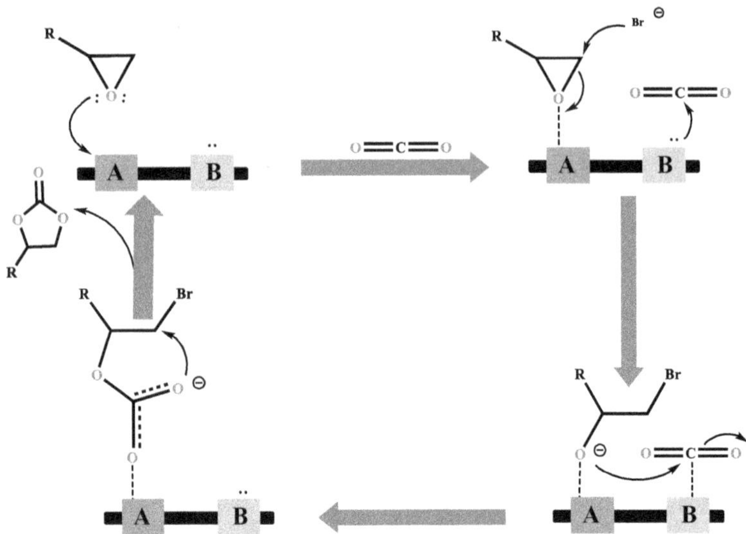

Scheme 2. The combinatory effect of acidic and basic functionalities for the CO_2 fixation reaction.

Based on this concept Bharadwaj and co-worker synthesized a linear linker containing tetracarboxylic acid as a metal binding site and the pendant -NH_2 group on the aromatic back bone for the guest interaction site (**Figure 7a**). The Cu based MOF constructed from the amino ligand that performed CO_2 (20 bar) fixation reaction with epoxide at 100 ºC with the help of TBAB.. However the conversion of styrene epoxide was ~ 50 % [55]. Later on, the same group further designed the bent tetracarboxylic ligand containing –NH_2 group (**Figure 7b**) and the Cu-MOF constructed from this ligand efficiently accomplished the CO_2 fixation reaction with diverse epoxide in presence same co-catalyst under mild reaction condition. Interestingly, the atmospheric CO_2 was converted to the cyclic carbonate by using this Cu-MOF [56]. This indicates that the relative position of the amino group in the ligand with respect to the open metal sites and carboxylic acid group is very crucial to impart in the CO_2 fixation activity.

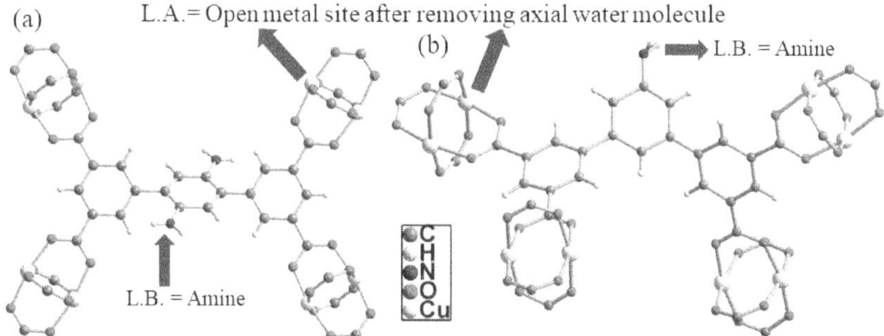

Figure 7. (a) A tetracobyxlic linear linker used to form Cu-MOF and (b) a bent tetracobyxlic linker is used to form Cu-MOF. In both the cases there is a presence of free amino group in the MOF cavity.

Li and co-worker synthesized four isostructural highly porous co-balt based MOFs (BUT-127-X, Where X= Bromine, amine, hydroxyl, and hydrogen) for the CO_2 cycloaddition reaction with epoxide [57]. All the synthesized MOF showed the very good CO_2 adsorption capacity. Interestingly, the CO_2 fixation rate with epoxide is shown by BUT-127-NH_2 is highest with respect to other three MOFs and the lowest performance was seen for BUT-127-Br. The presence of Lewis acid Co(II) and Lewis base -NH_2 in the framework not only help to adsorbed the large quantity of CO_2 but also activate both the epoxide and CO_2 efficiently for fixation between them to cyclic carbonate. In case BUT-127-Br, the existence of large bromine atom creates difficulty in diffusion for the substrate molecule into the pore of the MOF. As a result the interaction between the catalytically active site and the substrate is very poor, results low CO_2 fixation activity. An interesting work reported by Ye and co-worker where they have systematically tune the linker back bone from amine to urea and then constructed Cu(II) based porous MOFs (**Figure 8**) [58]. All of them contains the open metal Cu(II) site, but the CO_2 fixation reactivity with epoxide by urea based Cu(II)-MOF is superior than the –NH_2 based Cu(II)-MOF in presence of TBAB. The mechanistic study reveals that the urea form a strong hydrogen bonding interaction with the epoxides compare to amino group. Therefore, the urea based MOF showed the higher catalytic efficiency than the –NH_2 based MOF.

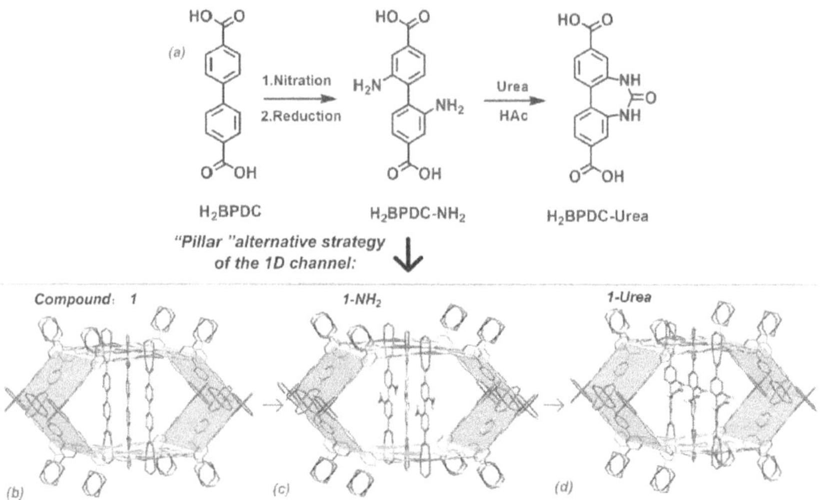

Figure 8. (a) The tuning of functional group: non-functionality to amine and then urea, (b), (c) and (d) are the Cu(II)-MOF having no functional group, amine functional group and urea functional gropu on their linker backbone respectively. Reproduced with permission from [58] American Chemical Society.

Yin and co-worker used a two dimensional metal organic layer (MOL) which was coating on the PCN-222(Co) to synthesized hybrid composite, PCN-222(Co)@MTTB for the conversion of CO_2 to cyclic carbonate (**Figure 9**) [59]. The MOL and the PCN-222(Co) are consisting exposed Zr^{4+} and Co^{2+} cations respectively. Additionally, the hybrid composite is oriented with the basic amino moiety. Therefore, the presence of heterometallic exposed metal site and the Lewis basic amino moieties, the hybrid composite displayed brilliant CO_2 conversion to cyclic carbonate. Importantly, in this case no TBAB was used.

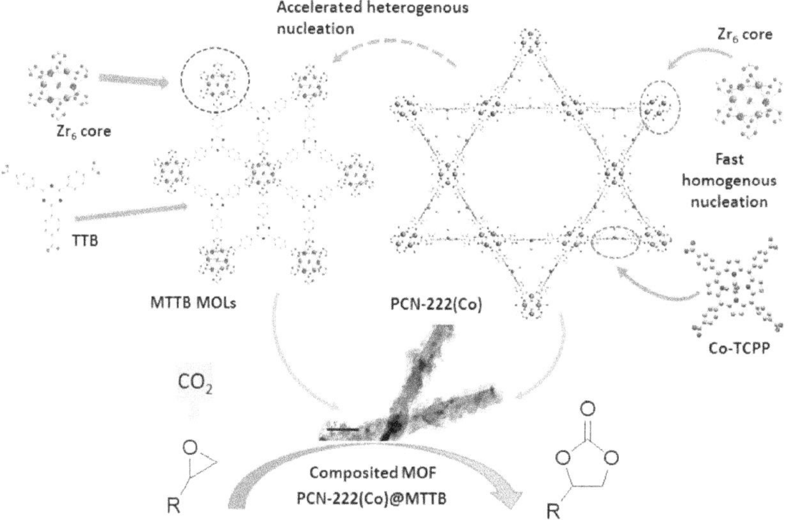

Figure 9. The synthesis of hybrid composite PCN-222(Co)@MTTB for the utilization of its in the conversion of CO_2 to cyclic carbonate. Reproduced with permission from [59] American Chemical Society.

Duan and co-workers synthesized a homochiral MOF, ZnW-PYI1 which contains free amino group and open zinc metal sites [60]. Initially the framework showed epoxidation of styrene to styrene epoxide (79 % ee, 92% yield) at 50 ºC for 5 days in presence of TBHP. Interestingly, the framework was able to showed the asymmetric tandem reaction to the generation of cyclic carbonate from alkene via the formation of alkene epoxide in presence of CO_2 (5 bar), TBHP, 96 h and 50 ºC. This excellent asymmetric CO_2 fixation activity was due to presence of polyoxometalate as oxidant, chiral organo catalyst, open Zn^{2+} site and $-NH_2$ which are helping to increase the CO_2 concentration around the catalytically active site as well as strongly activated them to expedite the tandem reaction (**Figure 10**).

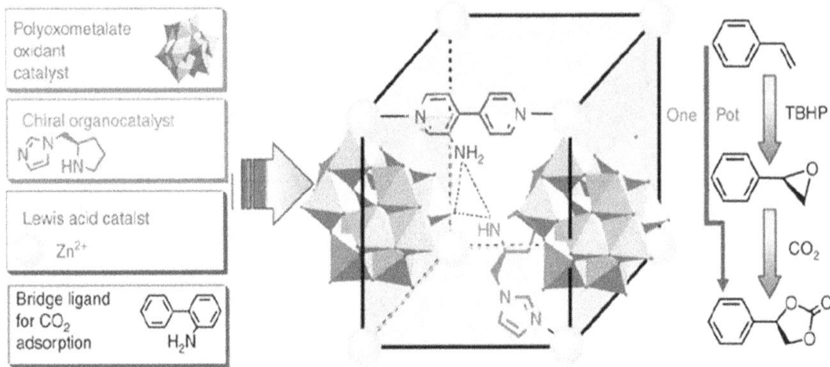

Figure 10. The CO_2 cycloaddition reaction with epoxide to cyclic carbonate executed by different catalytic components (left) those are present in the MOF (center).

Recently, Neogi and co-workers installed the Brønsted acidic moiety (-OH) on the linker backbone to performed CO_2 cyclo addition reaction [61]. To do this a tricarboxylic acid ligand H_3TCA and bpg linker were reacted with cadmium metal ion to delivered 3D porous Cd-MOF. The framework contains open cadmium metal site and Brønsted acidic functional moiety (-OH) which are activating both CO_2 and epoxide to form cyclic carbonate in presence of TBAB. Nagaraja and co-worker judiciously used NHC based MOF that underwent PSM process for the incorporation of catalytically active Cu(I) site on NHC component to construct Cu(I)@NHC–MOF [62]. The NHC is having affection for CO_2 and the Cu(I) site is capable to activate the triple bond of proparygyl alcohols. Thus, these two cooperative factors rendered the Cu(I)@NHC–MOF to performed the one pot three-component CO_2 fixation reaction between propargyl alcohols, amine and CO_2 to oxazolidinones in attendance of base DBU at RT (**Figure 11**).

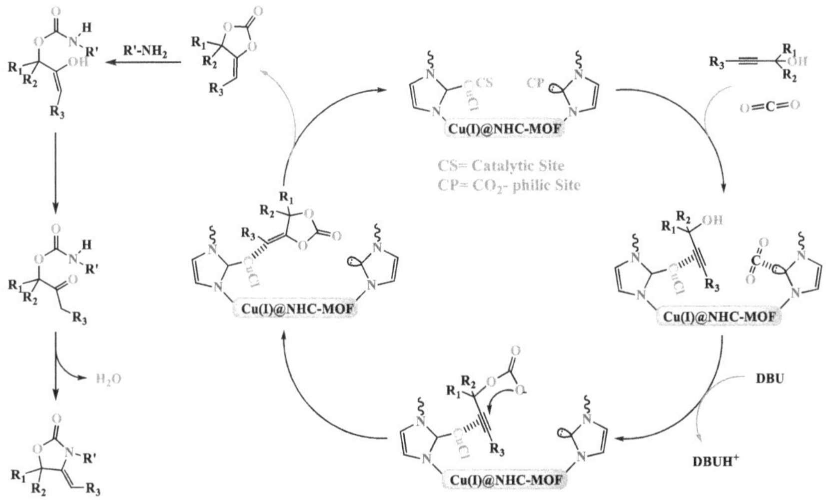

Figure 11. The formation of oxazolidinones from propargyl alcohols, CO_2 and amine by using Cu(I)@NHC–MOF.

2.3 CO_2 Fixation Activity by the Presence of Ionic Environment in the MOFs

So far, we have seen that the presence of co-catalyst in the catalytic system can enhance the rate of the reaction. But the addition of co-catalyst (TBAB) can be avoided if we can make a modification in the organic linker or metal node (or SBU) to bring the ionic liquid nature within the framework. This ionic liquid contains the counter anion as nucleophile which will be extra beneficiary for the catalytic system to get rid of from the uses of external co-catalyst TBAB. In this regard, the designing of the organic linker is very crucial as the organic linker is leverage us to easy perform of the PSM process to generate ionic environment or can preconceived the ionic liquid environment to establish ionic network in the MOF which will contains the nucleophilic anion. The role of anions in ionic liquid of MOF is exactly same as the role of anion from external co-catalyst (**scheme 1**). Based on this concept, Dong and co-worker reported a zirconium based framework, UiO-67-IL-Br which contains imidazolium ion for the source of nucleophilic Br⁻ ion (**Figure 12**) [63]. Quite interestingly, the framework was able to execute the conversion of CO_2 to variety of cyclic carbonates without using TBAB as an external co-catalyst.

Figure 12. An ionic liquid based MOF for CO_2 fixation reaction. Reproduced with permission from [63] American Chemical Society.

Recently, Gao and co-worker demonstrated sequential PSM process through click chemistry reaction of a stable MOF, MIL-101 to form ionic MOF followed by anion exchange to bring Br⁻ ions in its cavity (**Figure 13**) [64]. This daughter MOF effectively showed the conversion of CO_2 with variety of epoxide under suitable reaction condition with good to excellent yield.

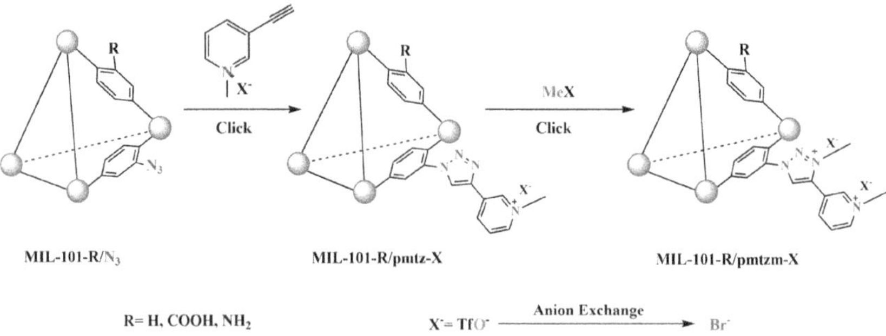

Figure 13. Stepwise PSM process brings highly anionic co-catalyst in the MOF to perform CO_2 cycloaddition reaction with epoxide. Reproduced with permission from [121] American Chemical Society, Copyright 2021.

Apart from the PSM process on the ligand to generate ionic liquid environment, the same can be achieved through the PSM process on the SBU. Bury and co-worker used well known MOF, NU-1000(M) (M=Zr, Hf) which underwent double step PSM process on SBU to engender nucleophilic anion Br^-/I^- within the MOF network (**Figure 14**) [65]. Quite interestingly, the daughter MOF deliberately completed CO_2 fixation reaction with numerous epoxide.

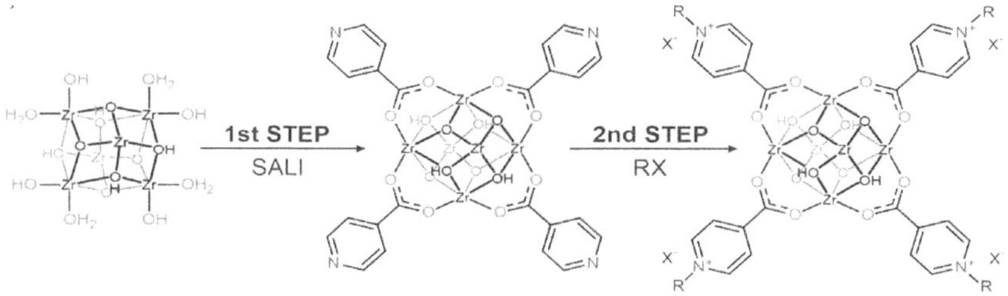

Figure 14. The PSM process on the SBU to bring the ionic environment within MOF. $X^-=Br^-/I^-$. Reproduced with permission from [65] American Chemical Society.

2.4 Fixation of CO_2 Promoted by Defects in MOF

The contribution of defects in MOF for CO_2 fixation with epoxide was first realized with MOF-5 [66]. The catalyst was successfully converted propylene oxide to cyclic carbonate (96 % yield) with CO_2 (6 Mpa) for 6 h at 323 K in existence of TBAB. The metal nodes are completely saturated and the ligand does not contain any catalytic site, thus the defect in the framework is responsible for such high catalytic activity. Afterwards, several defect MOFs as catalytic systems were exposed to demonstrated the conversion of CO_2 to cyclic carbonates with diverse epoxides. Recently, Fischer and co-workers demonstrated linker strategy to generate defects in MOF for CO_2 conversion fixation reaction [67]. A copper based MOF, NOTT-100 alone can constructed from ligand A. But, when three ligands A, B and C were employed at a time then defect Cu-MOF was fabricated (**Figure 15**). The presence of defects along with the easy accessible Brønsted acidic (-COOH) and Lewis basic (nitrogen) sites in the MOF delivered the facile CO_2 fixation reaction.

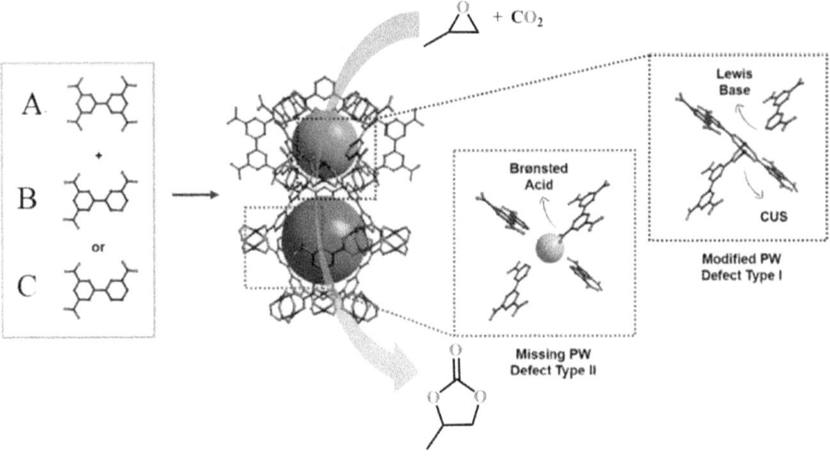

Figure 15. The combination of linkers A, B and C lead to the formation defects in NOTT-100. Reproduced with permission from [67] American Chemical Society.

3. Photoreduction of CO_2 by MOFs

Usually, a suitable catalyst with appropriate band gap would absorb certain amount of visible or UV light which is then reduce the CO_2 to valuable chemicals. In this particular reaction the electrode potential of CO_2 to the corresponding chemicals are very important. Here in **Table 1** the electrode potentials are presented in aqueous medium at pH=7 with respect to the normal hydrogen electrode [68].

Table 1. The numbers of electrons are involved for CO_2 reduction to the respective chemicals and their corresponding E° values are shown here

Reactions	E^0
$CO_2 + 2H^+ + 2e^- = HCOOH$	-0.61 V
$CO_2 + 2H^+ + 2e^- = CO + H_2O$	-0.53 V
$CO_2 + 4H^+ + 4e^- = HCHO + H_2O$	-0.48 V
$CO_2 + 6H^+ + 6e^- = CH_3OH + H_2O$	-0.24 V
$CO_2 + 8H^+ + 8e^- = CH_4 + 2H_2O$	-0.24 V
$2H^+ + 2e^- = H_2$	-0.41 V

Upon light illumination, the photoactive MOFs could generate photoinduced electron/ positive holes which facilitate the oxidation and reduction reactions. But some MOFs are not photoactive and even if some are active, their semiconducting property is not up to the mark with respect to the traditional semiconductor. It has been found that the poor conductivity of MOF is due the insufficient arrangements of energy levels prohibit the long range electron movements [69]. Garcia and co-worker, showed for the first time, the light absorption capacity, semiconducting property and charge separation process occurred within the MOF-5 [70]. Since then, tremendous efforts have been devoted towards the exploration of semiconductor MOFs [71]. The photoinduced electron transfer (PET) from ligand to metal nodes illustrated as a LMCT [72]. But in some MOFs containing Fe-O metal clusters the excitation of electron takes place from the bridging oxygen (negative entity) and it moves to the metal ions in the cluster (positive entity) (**Figure 16**) [73].

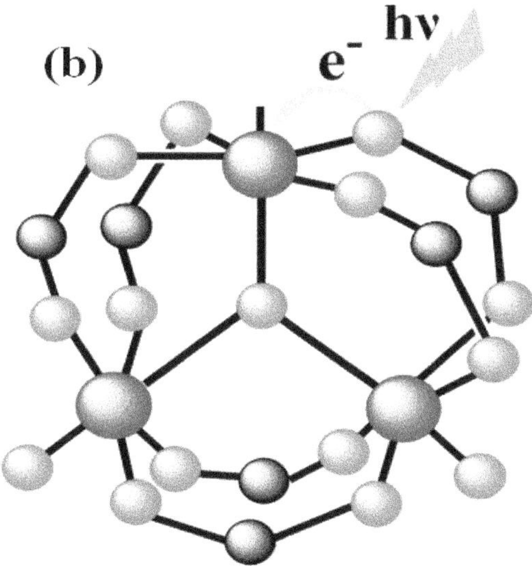

Figure 16. Charge transfer to metal ion from the oxygen (bridge position) through direct excitation in Fe-O clusture of MIL-101. The colour code at the metal cluster: Green (Fe), red (oxygen), gray (carbon). Reproduced with permission from [73] American Chemical Society respectively.

In general, the conducting materials contain two important parts: valence band and conduction band. Upon absorption of suitable energy, the excitation of electron takes place from the valence to the conduction band, leads to the formation of charge separated state [74]. But in case of MOF, the valence band and conduction bands are represented as HOMO and LUMO respectively, which alternatively indicates that organic linker can be recognized as HOMO and metal/metal clusters as a LUMO of the MOF [75]. Consequently, oxidation and reductions will take place on the organic ligand and metal nodes respectively. The MOF as a photocatalyst must fulfil the following criteria: (a) the HOMO-LUMO energy gap would be analogous to the absorbed UV-visible energy, (b) PET from HOMO (organic linker) to LUMO (metal cluster) and thus the HOMO of ligand should be higher in energy than the LUMO of metal cluster, (c) the standard reduction potentials of the CO_2 to renewable fuels must be greater than the HOMO-LUMO separation. In principal, the photoreduction of CO_2 is monitored by the position of HOMO-LUMO energy level in MOF. Several techniques have been adopted to calculate the HOMO-LUMO energy gap in MOFs. The HOMO-LUMO energy gap (3.4 eV) of MOF-5 was calculated by correlation between radiation and reflectance energy [76]. The UV-visible spectroscopy is also a helpful technique to calculate the HOMO-LUMO energy gap and the calculation is similar to the band gap calculation of semiconductors [77]. The Mott-Schottky plot and DFT study are also effective methodology for determining the energy gap of n-type semiconductor based MOFs [78]. The metal ions are capable of showing variable oxidation state and the functionalization in the organic linkers can reduce the HOMO–LUMO energy gap which will help to enlarge the light absorption ability [79]. Therefore, through proper design, MOFs can exhibit photocatalytic activity in the visible range which can induce the high light consumption efficiency.

On the other hand, MOFs having no photocatalytic activity can be converted to the effective photocatalysts by coupling with semiconductors or photosensitizers. In addition, the special characteristic features of MOFs (high porosity, surface area, huge CO_2 adsorption etc.) are inherited into the new composite to perform the better photocatalytic activity [80]. Furthermore, the existence of highly dense catalytically reactive sites which are homogeneously distributed in the composite, have a direct effect on catalytic efficiency. Due to the large CO_2 uptake capacity of MOFs, the catalytic centre will be highly concentrated with CO_2. This will allow the feasible interaction of CO_2 with the reactants and facilitate the formation of desired products. The unsaturated metal centres in MOF accelerate both the reaction rate and CO_2 uptake capacity. Therefore, by making a new composite with other materials, MOFs with no catalytic activity can be made as active photocatalyst for superior CO_2 photo-reduction.

The stability (photostability and water stability) of MOFs is an important issue regarding the photocatalytic effectiveness (reaction rate, turnover number and quantum efficiency). Generally, MOFs stability is controlled by Lewis acidity of metal clusters and the pKa values of organic ligand [81]. The MOFs with poor water stability is difficult to use for photoreduction of CO_2. For long duration of catalytic reaction, the water molecules can attack on the metal coordination site and replace the attached ligand, resulting in decrease of porosity as well as crystallinity and finally break-down of the framework [82]. Thus, non-aqueous solvent has also been used to carry out the photocatalytic reduction of CO_2. However, it is essential to develop water-stable MOFs for CO_2 photoreduction. On the other hand, for prolong photocatalytic reaction the photocatalyst may undergo photodecomposition reaction (or photo degradation of the organic ligand) leading to the breakdown of the framework [83]. The stability of the photocatalytic MOFs can also be accomplished by maintaining the surface charge balance. In this regard foreign compounds e.g., TEA and TEOA, have been used to exhaust the photogenerated holes. Therefore, MOFs stability need to be monitored by PXRD, surface area analysis and other spectroscopic techniques. Herein, the role of different components in pure MOFs (Sections 3.1 and 3.2) or MOF based hybrid composite (Section 3.3) will be discussed for the photoreduction of CO_2.

3.1 Photoreduction of CO_2 by Ligand

The functionalization in highly conjugate organic ligands is not only increases the light absorption capabilities but also facilitate the distributing the charge holes which prevents the charge recombination. The aromatic ligands with carboxylate binding sites are considered to be a good linkers as it can easily produce photoinduced electrons when exposed to light due to its superfluous electron density. In addition, polycarboxylate binding sites lead to the formation of metal clusters that helps to increase the stability of MOFs which is an important requirement for photocatalytic reaction and sometimes cluster electronically plays a vital role.

3.1.1 Presence of Amino Group and its Modification

As discussed earlier the MOFs having -NH$_2$ functional group would increase the CO$_2$ uptake ability. Also, the amino group in the MOF would increase the electron density, thereby reducing the HOMO-LUMO energy gap. The shifting of absorption maximum to the higher wavelengths and in some cases to the visible range is very effective for CO$_2$ photoreduction. Influenced from the photocatalytic application of the MIL-125(Ti), Li's group synthesized Ti based framework, NH$_2$-MIL-125(Ti) from NH$_2$-BDC instead of BDC [84].CO$_2$ uptake capacity of NH$_2$-MIL-125(Ti) is enhanced to 132.2 cc g^{-1} due the presence of amino group in the ligand whereas for non-amino framework MIL-125(Ti) it was 98.6 cc/g. Besides, the absorption maximum also changed: 350 nm for MIL-125(Ti) and 550 nm for NH$_2$-MIL-125(Ti). Thus, NH$_2$-MIL-125(Ti) nicely performed the CO$_2$ photoreduction to formate under visible light in acetonitrile solvent with the help of sacrificial reagent, TEOA. The production rate of HCOO$^-$ is 16.28 μmol h^{-1} g^{-1} by NH$_2$-MIL-125(Ti) is significantly higher than MIL-125(Ti). The framework exhibited photochromic effect and existence of ionic state was confirmed by ESR study. The probable mechanism is occurred with the following steps: Upon irradiation, the photoinduced electron transfer is occurred from ligand to Ti^{4+} ion in Ti-O cluster that induced thereduction of Ti^{4+} to Ti^{3+}. A charge separated species were generated in the MOF that enhanced the performance of photocatalytic reduction of CO$_2$ by amino based MIL-125(Ti) (**Figure 17**).

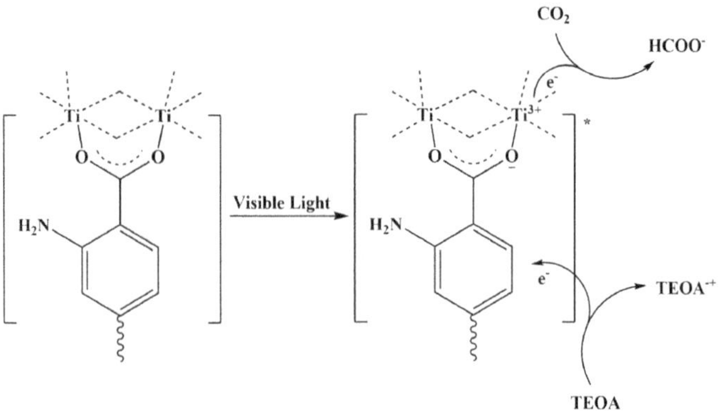

Figure 17. CO$_2$ reduction catalysed by NH$_2$-MIL-125(Ti) under visible light.

The efficiency of CO_2 photoreduction by introducing amino base linker in the MOF had been demonstrated by Li and co-worker with UiO-66(Zr) which contained the BDC linker. This BDC linker was replaced by NH_2-BDC to afford NH_2-UiO- 66(Zr) [85]. The following change noticed: (i) the maximum wavelength for visible light absorption was observed at 460 nm for NH_2-UiO- 66(Zr) whereas for UiO-66(Zr) it was 430 nm, (ii) gas uptake capacity increased to 71 cc g^{-1} (for pristine MOF it was 68 cc/g), (iii) most importantly, after 10 h of light irradiation, the amount of HCOO$^-$ produced was 20.7 µmol and 13.2 µmol for UiO-66(Zr)). On continuation to work on amino functionality Uribe-Romo and co-worker functionalized the amino group of NH_2-BDC ligand and subsequently formed a series of Ti based MOF (**Figure 18a**) [86]. The functionalization of $-NH_2$ was carried out by adding primary (methyl to n-heptyl), secondary and cyclic substituent's. Different amino based functional groups are interacting differently, results various amount of CO_2 Uptake. It had been observed that with increase in the chain length, the adsorption capacity of the MOF decreases. Thus, MIL-125-NH_2 adsorbed 221 cc g^{-1} whereas MIL-125-NHhep adsorbed 19.1 cc g^{-1}. Interestingly, the HOMO-LUMO energy gap of the MOFs followed the same trend. Upon illumination of all the MOFs under blue LED light in the existence of mesitylene and TEOA, the conversion of CO_2 to formate took place and the yield of the product increased with increase in the chain length of the substituents on amino group. The quantum yield of CO_2 photoreduction by MOFs having secondary substituent on the amino group was higher compared to that of the MOFs having primary group substituent. This was attributed to the lower HOMO–LUMO energy gap and extended excited-lifetime of the former group of MOFs than the latter group of MOFs.

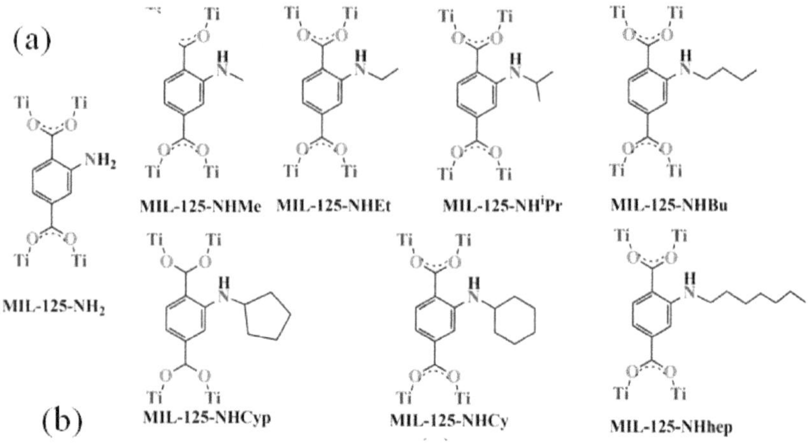

Figure 18. Series of ligand derivatives (NH$_2$-BDC) was used for synthesis of series of MIL-125(Ti).

3.1.2 Installation of Photosensitizer

Photocatalytic CO_2 reduction by transition metal based inorganic complexes containing carbonyl groups are fairly reported in the literature [87]. These complexes are acting as photosensitizers and simultaneously reduces the CO_2. In this context, a series of "Re" based inorganic complexes (light sensitizers) of the formula Re^I(CO)$_3$(bpy)X (where, X=halide ion) are reported as excellent photocatlysts for CO_2 photoreduction. But they were unstable and degraded when irradiated with light. The non-photosensitizing MOFs can be made to photosensitizers by incorporating the inorganic photosensitizers. The instability of inorganic photosensitizers can be improve by coupling with stable MOFs. The resulting MOF could display high light harvesting efficiency. The synthetic procedure for the construction of the photosensitized MOFs are illustrated in the **scheme 3**.

Method 1 (*inclusion of photosensitizer by direct method*)

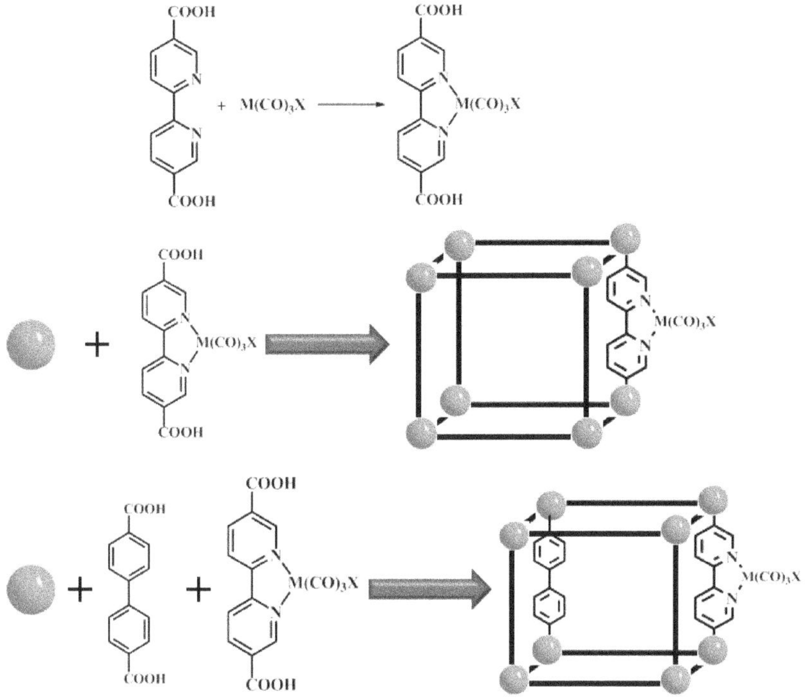

Scheme 3. Schematic representation for the direct synthesis of the photosensitizer to produced photocatalyst MOF.

Method 1: The chelating *N,N'*-site present at the bpy units of dcbpy linker coordinated with the metal carbonyl complexes ($Re^I(CO)_5Cl$ and $Ru^{II}(CO)_2Cl_2$ etc.) leading to the formation of precursors like $Ru^{II}(4,4'\text{-}dcbpy)_3$, $Re^I(5,5'\text{-}dcbpy)(CO)_3Cl$, etc. and finally these were utilized for the MOFs formation. In the metal-ligand complexes, the MLCT was possible (metal to π^* of bpy) that led to enlargement of the light absorption range (from UV to visible region) and increased the photocatalytic efficiency for the CO_2 reduction. Lin and co-worker synthesized a zirconium based MOF, $Zr_6O_4(OH)_4(BPDC)_{6-x}[Re(5,5'\text{-}dcbpy)(CO)_3Cl]x$ by reacting BPDC and $Re^I(5,5'\text{-}dcbpy)(CO)_3Cl)$ with Zr^{4+} ion [88]. This MOF can be reduce CO_2 to CO under visible light for 20 h in acetonitrile solvent and in presence TEA as a sacrificial reagent. The TON for this reaction was 7.0 and TON of $Re^I(5,5'\text{-}dcbpy)(CO)_3Cl$ was 3.5. The catalyst degradation was responsible for such low TON. Luo and co-worker synthesized a photosensitized MOF, ($[Ir(ppy)_2(dcbpy)]_2[OH]$) (Ir-CP) which showed the photoreduction of CO_2 to formate [89]. The framework exhibited very broad light absorption range in the visible region and the framework had two important features: a light harvesting unit, $[Ir(ppy)_2(dcbpy)]^-$ and a catalytic CO_2 reduction site. When the compound was subjected to the visible light it produced 25 μmol $HCOO^-$ after 6 h and was reusable for five cycles without significant changes in the yields, indicating its photostability. The probable mechanism put forth was as follow: upon visible light irradiation $[Ir(ppy)_2(dcbpy)]^-$ unit reached the upper state and then the reduction of CO_2 took place in presence of sacrificial reagent TEOA which donated electron to the excited photosensitizer (**Figure 19**).

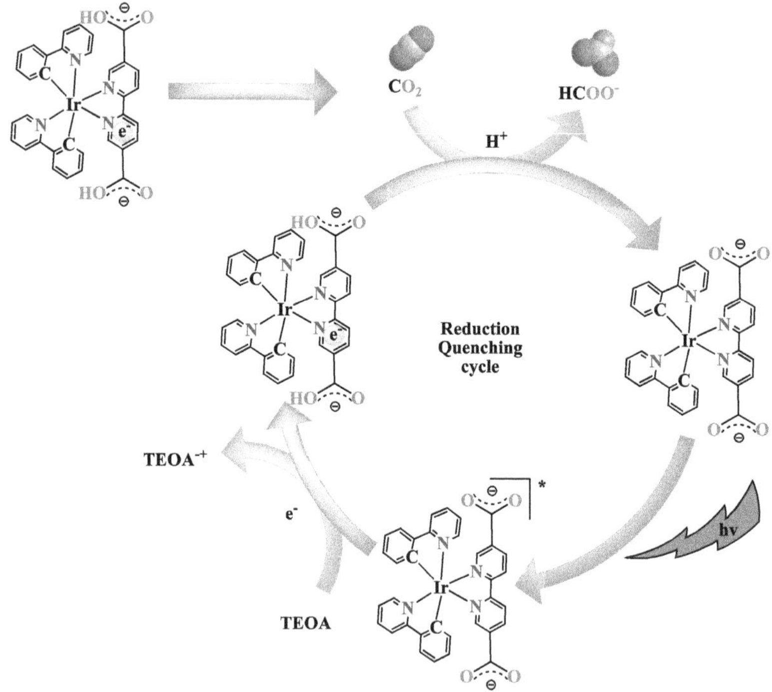

Figure 19. Probable photocatalytic mechanism by Ir-CP.

Method 2: Unlike previous method, the present one is based on the insertion of photosensitizer (metal carbonyl complex) within MOF through post-synthetic procedure (**scheme 4**). Cohen and co-workers reported for the synthesis of Zr based MOF (UiO-67-dcbpy) by the solvothermal reaction between 5,5′-dcbpy and $ZrCl_4$ in DMF solvent for 24 h at 120 °C. Then the compound $Mn^I(CO)_5Br$ got inserted at the *N,N'*-functional site of bpy unit and afforded a new MOF, UiO-67-Mn(bpy)(CO)$_3$Br (**Figure 20**) [90]. This MOF was able to perform photocatalytic CO_2 reduction to $HCOO^-$ when exposed to visible light for 20 h and the TON was 110 ± 13 in presence of $[Ru(dmb)_3]^{2+}$ and BNAH. The mechanism for the reduction of CO_2 to formate was suggested as given in **Figure 20**. The photosensitizer accepted an electron from sacrificial reductant (BNAH to BNAH$^{·+}$) and initiated the reduction and then the monovalent Mn^+ ion present in the MOF, accepted the electron from $[Ru(dmb)_3]^{2+}$ and got reduced to Mn^0 state. In this state it could bind a CO_2 molecule and with the help of another reagent, TEOA the metal bound CO_2 was converted to $HCOO^-$.

Method 2 (inclusion of photosensitizer by post synthetic method)

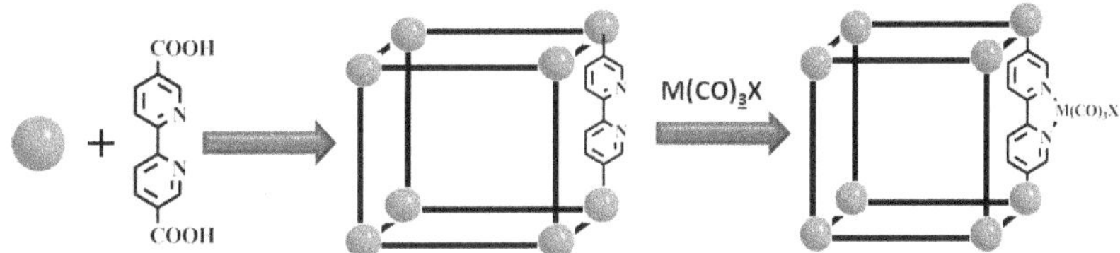

Method 3 (inclusion of photosensitizer by PSLE)

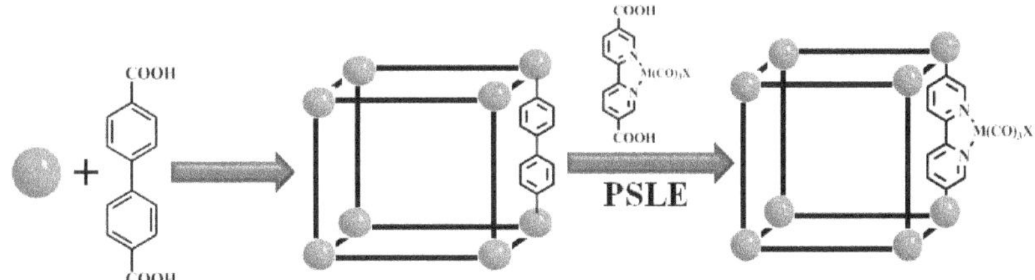

Scheme 4. Schematic representation of the insertion of photosensitizers via post synthetic method.

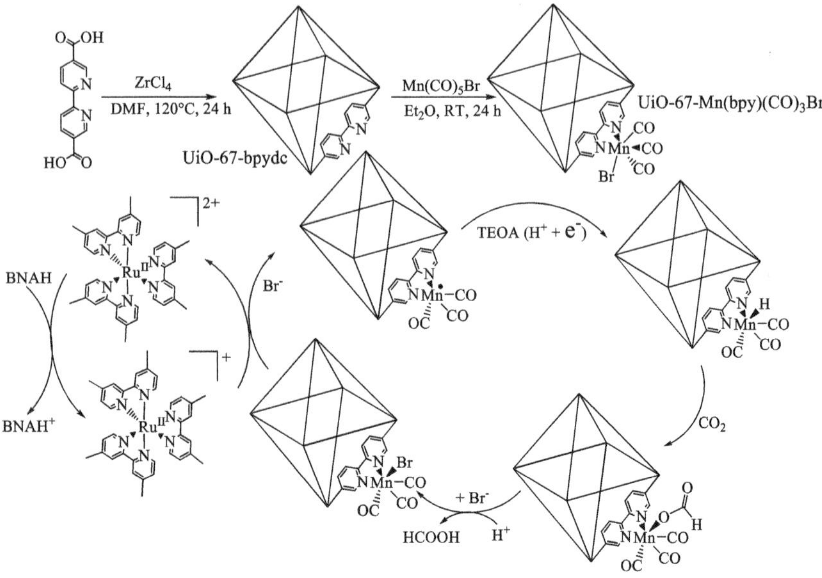

Figure 20. (a) Synthetic scheme of UiO-67-Mn(bpy)(CO)₃Br and possible Co2 photoreduction mechanism.

Method 3: Method 3 is quite similar to method 2. In this method, the organic linker in the MOF is partially replaced through PSLE process with a new ligand that having the light harvesting ability (**scheme 4**). Following this technique, Marc Fontecave and co-workers reported [91] CO_2 reduction to formate under visible light. They integrated Cp*Rh(5,5′-dcbpy)Cl within UiO-67 through PSLE process in aqueous medium at RT for 24 h that afforded Cp*Rh@UiO-67 (**Figure 21**). This modified MOF exhibited photocatalytic CO_2 reduction to formate with a TON of 47 and it was greater than the TON of Cp*Rh(5,5′- dcbpy)Cl which was 42. Althhough the enhancement in TON value was marginal, the MOF showed greater selectivity and higher stability compared to the Cp*Rh(5,5′- dcbpy)Cl. Kitagawa and co-workers introduced the photosensitizer, H_2RuCO in the Zr-bpdc MOF (UiO-67) via PSLE process to afford Zr-bpdc/RuCO MOF [92]. Under visible light the resulting composite reduced CO_2 to formic acid in presence of TEOA as a sacrificial reagent.

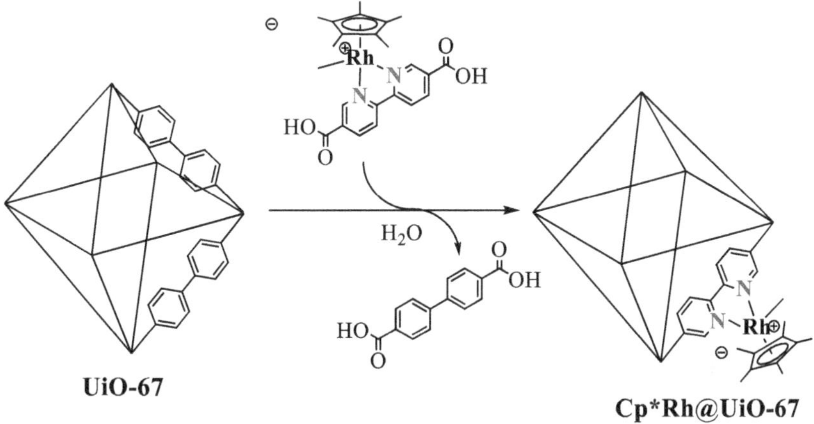

Figure 21. The PSM process for the synthesis of Cp*Rh@UiO-67.

3.1.3 Miscellaneous

It has been observed that the highly conjugated electronically rich organic linkers can modulate the photocatalytic behaviour of MOF. The advantages to have such linkers are as follows: (a) the highly polarized surface of ligands can interact proficiently with CO_2 (b) easy to generate photoinduced electron and concomitantly it is moving to the metal cluster, and (c) the redox potential energy of such systems will be comparable with the standard reduction potential of CO_2. In this regard, the porphyrin ligand would be very suitable. In highly conjugated porphyrins system the pyrrole moieties are interconnected by carbon bridges and hence the π electron can freely move over the ligand backbone. Additionally, the ligand enriched with basic nitrogen elements and the presence of aromatic surface, the interaction with the CO_2 molecule will be very effective. Therefore, Jiang and co-worker synthesized Zr based porphyrinic n-type semiconductor MOF, PCN-22 by the reaction of H_2TCPP ligand with $ZrCl_4$ salt [93]. The synthesized MOF nicely participated in the photocatalytic reduction of CO_2 to formate. The standard reaction condition such as MeCN as solvent, under visible light and with the attendance of TEOA, the production rate of formate found to be 60 µmol h^{-1} g^{-1}. Under the same reaction the activity of ligand was very insignificant. The production rate of formate by ligand was found to be 4.8 µmol h^{-1} g^{-1}. Upon illumination with light a photo induced electron is produced from the ligand which is then reduced the Zr^{4+}

to Zr^{3+} in Zr-O cluster. The adsorbed CO_2 is reduced to formate by the Zr^{3+} in Zr-O cluster. Importantly, the generation of Zr^{3+} in Zr-O cluster was confirmed by ESR analysis (**Figure 22**).

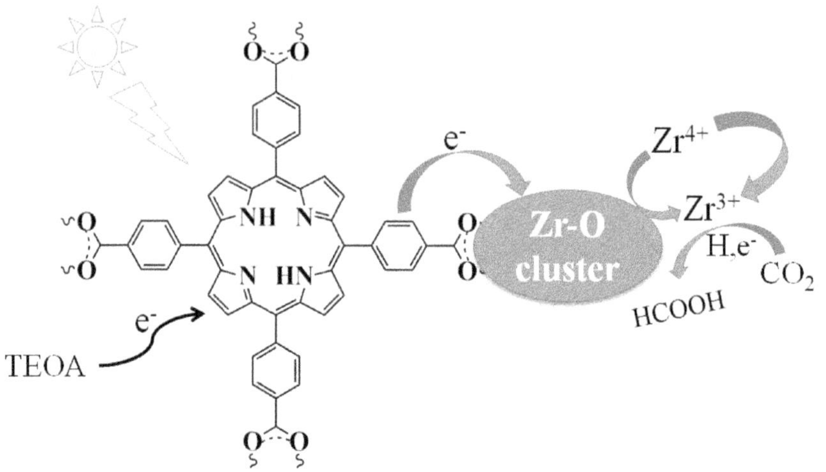

Figure 22. Illustration of CO_2 photoreduction mechanism by PCN-22.

After successful photoreduction of CO_2 by porphyrin ligand based MOF, next the investigation was focused on the searching of highly conjugated linker. In this respect the anthracene would be remarkable candidate. Su and co-worker synthesized an anthracene ligand based zirconium MOF, NNU-28 for the photocatalytic reduction of CO_2 to formate under visible light for 10 h [94]. The production rate or formate was 52.8 μmolh^{-1}g^{-1}. The ESR study was employed to investigate the reaction mechanism. It was noticed that in visible light as well as in dark the ligand displayed ESR peak at g=2.003 which indicating the generation of radical entity. The as synthesized MOF did not show any ESR peak under dark condition. However, when the MOF was exposed to the light for a long time two new peaks at 2.009 and 2.030 were produced. Such peaks were not shown by the ligand. As the HOMO-LUMO energy gap in NNU-28 is 5 eV which is very high and that ruled out the possibility of direct electronic transition of Zr-O cluster. Therefore, the generations of these two new ESR peaks are due to the LMCT process between ligand and Zr-O cluster.

3.2 Photoreduction of CO_2 by Metal Cluster

A detailed literature survey revealed that the MOFs containing oxo-metal cluster of Fe(III), Zr(IV) and Ti(IV) are acting as a good photocatalyst. The combination of these metals clustered with the amino functionality in the organic ligand would be an efficient photocatalyst.

3.2.1 Pure Metal Cclusters (M-O, M= Fe and Ti)

Most of the MOFs are constructed from 3d transition metal ions (divalent and trivalent), lanthanides and 3p metals with organic ligand containing nitrogen and carboxylate as binding sites. The metal with variable oxidation states like Fe^{3+}/Fe^{2+} and Ti^{4+}/Ti^{3+} etc. are generally preferred to form a MOF with the carboxylate base ligand. Such combinations allow to form a metal-oxo cluster (Fe–O, Ti–O and Zr–O cluster) that are able to catch the photoexcited electrons and get reduced to their corresponding lower oxidation states. This results in the formation of charge separated species which is an ideal situation for photocatalytic reaction [95]. Iron is a highly earth abundant element, non-toxic and importantly it can exhibit variable oxidation states namely +III and +II (Fe^{3+}/Fe^{2+}). In Fe-based MOF, the iron in Fe-O cluster is in +III state (Fe^{3+}-O) and upon light irradiation the photo excited electron transfer from the ligand or oxygen (O^{2-}) to the Fe^{3+}ion. Therefore, the Fe-O cluster behave as photocatalytic active centre for the photocatalytic reaction [96]. Li and co-workers explored the photocatalytic activity of Fe based MIL series (MIL-88B(Fe), MIL-53(Fe) and MIL- 101(Fe)) for the photo reduction of CO_2 to $HCOO^-$. The reactions were carried out in acetonitrile solvent in attendance of TEOA under visible light and the yield of the formate are 9.0 μmol, 29.7 μmol, and 59.0 μmol respectively after 8 h [73]. The framework MIL- 101(Fe) contains open metal sites which not only enhance the CO_2 adsorption (26.4 g cm^{-3}) capacity also reduce the CO_2 efficiently. Such open metal sites are absent in other two MIL series (MIL-88B(Fe) and MIL-53(Fe)). The photocatalytic activity of MIL-53(Fe) was greater than MIL-88B(Fe) due to higher CO_2 uptake capacity of the former (13.5 g cm^{-3}) than the latter (10.4 g cm^{-3}). This assumption is further proved by the amino functionalized MOFs. The superior photocatalytic activities and the yields are 30 μmol, 46.5 μmol, and 178 μmol respectively under the same reaction condition. The enhancement in yield is due to the high CO_2 uptake by amino functionalize ligand along with the excitation of electron from the amino functionalize ligand to the iron centre as well as the direct excitation of electron from the Fe-O cluster (**Figure 23**).

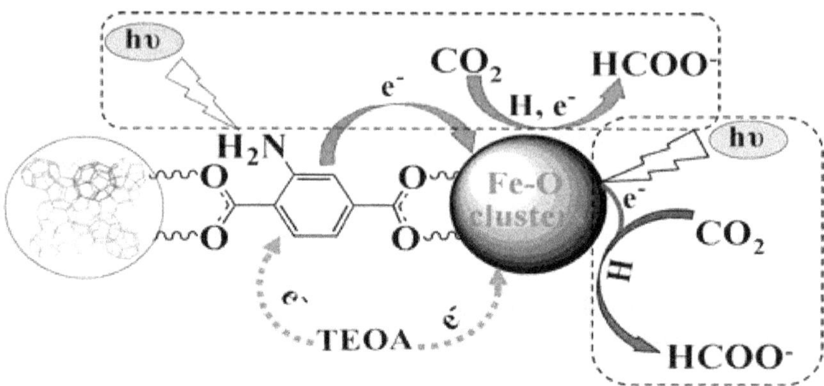

Figure 23. Represents the dual excitations of electron i.e., transfer of photoinduced electron from amino functionalized ligand to metal cluster and also from the bridging oxygen to metal rendered the CO_2 reduction. Reproduced with permission from [96] American Chemical Society.

Like iron, the Titanium can also form a variable charge cations (Ti^{4+}/Ti^{3+}) [97]. The HOMO-LUMO energy gap of MIL-125(Ti) in UV range was found to be 3.6eV [98]. The MIL-125(Ti) is exhibited excellent photochromic effect by changing its colour from white to purple under UV light, indicating the existence of long lived charge separated state which was originated by the transfer of photoinduced electron from the organic ligand to the Ti^{4+}ion in Ti-O cluster concurrently +IV state is reduce to +III state (Ti^{3+}) [99]. The photocatalytic CO_2 reduction of MIL-125(Ti) was completed in acetonitrile solvent and TEOA as a sacrificial reagent under at 365 nm and the yield of HCOO⁻ was 2.41 µmol after 10 h [84]. When the BDC linker was replaced by the NH_2-BDC the yield of the HCOO⁻ under same experimental condition was much higher at 8.14 µmol. Compare to MIL-125(Ti), the enhance photoreduction efficiency for NH_2-MIL-125 (Ti) was due to presence of amino group in the NH_2-BDC linker increase CO_2 adsorption capability and expedite the LMCT from linker/oxygen to Ti^{4+} centre in Ti-O cluster afforded long-lived charge separated state.

3.2.2 Hybrid Metal Clusters

Though some MOFs contain metal ions with variable oxidation states but they may not able to perform as photocatalyst for CO_2 reduction. This is due to their inconsistency in potential energy levels. However, they can be converted to photocatalyst by introducing a different metal cation into the metal cluster to bring the redox potential at reasonable energy level. This foreign metal ion can be introduced through post-synthetic metal exchange (PSME) process. Due to this partial substitution, the presence of two different metal ions in the metal cluster leads to the different redox system with the possibility of MMCT that helps in the photocatalytic activity. In a recent work by Li and co-workers, the Ti^{4+} ion was partially embedded into the Zr-O clusters of NH_2-UiO-66(Zr) via PSME process by keeping NH_2-UiO-66(Zr) in DMF solution of $TiCl_4(THF)_2$ that afforded two different MOFs, NH_2-UiO-66(Zr/Ti)-100-4 and NH_2-UiO-66(Zr/Ti)-120-16 respectively [100]. Introduction of Ti^{4+} ion in the SBU enhanced the gas uptake capacity and increase the number of catalytic active sites. Thus, both of these factors reinforced the catalytic performance. The proposed mechanism for the photocatalytic CO_2 reduction was the transfer of photoinduced electron from NH_2-BDC to Ti^{4+} ion rather than to Zr^{4+} ion in the SBU, $(Ti^{4+}/Zr^{4+})_6O_4(OH)_4$ affording a new excited SBU, $Ti^{3+}/Zr^{4+})_6O_4(OH)_4$. Then, Ti^{3+} ion donated the electron to the Zr^{4+} leading to the formation of $Ti^{4+}-O-Zr^{3+}$ and this induce the charge transfer from the organic linker to the Zr-O cluster endowed with superior photo catalytic activity of NH_2-UiO-66(Zr/Ti) (**Figure 24**).

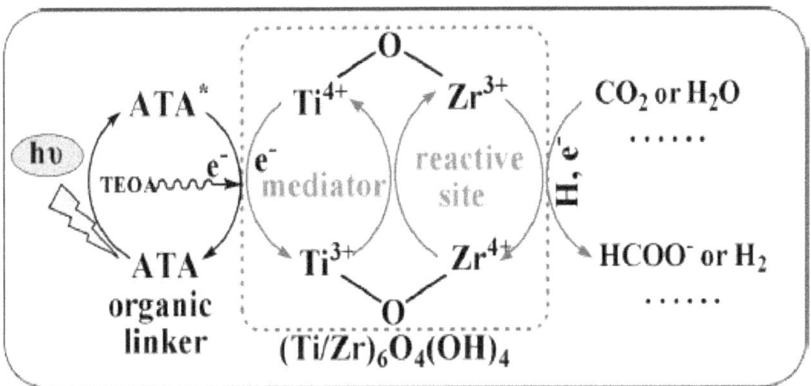

Figure 24. Probable mechanism for the CO_2 photoreduction.

3.3 Photoreduction of CO_2 by MOF based Hybrid Composite

The modification in the metal cluster and in the organic linker, a non-photocatalytic MOF can transform into photocatalytic active MOF. The combinations of MOFs with other materials so called the new composites can have the ability to form heterojunctions that empowers charge transfer from one to the other material helping facile charge separation and enhance the photocatalytic activity.

3.3.1 Composite with Semiconductor

Most of the MOFs are not conducting and hence cannot behave as photocatalysts. When a MOF is combined with a semiconductor, the new composite material can function as a photocatalyst. In this respect, the MOFs with huge CO_2 uptake capacity can be integrated with semiconductor to anticipate better CO_2 photoreduction ability. Based on this concept, Wang and co-worker synthesized a MOF composite between Co-ZIF-9 and CdS, and the new composite reduced CO_2 to CO [101]. The duo played an effective role for the photo reduction of CO_2 to CO and the TON was 50.4 µmol and its catalytic activity was much more superior to the semiconductor itself. The formation of CO from CO2 is occurred through the following steps: upon visible light illumination, the generation of electron hole pairs occurred on the semiconductor CdS. The photoinduced electron jumped to the CB of CdS and finally with the help of bpy linker it moved to Co-ZIF-9 where reduction of adsorbed CO_2 to CO took place (**Figure 25**). The positive hole on CdS was quenched by the TEOA by the donation of the electron.

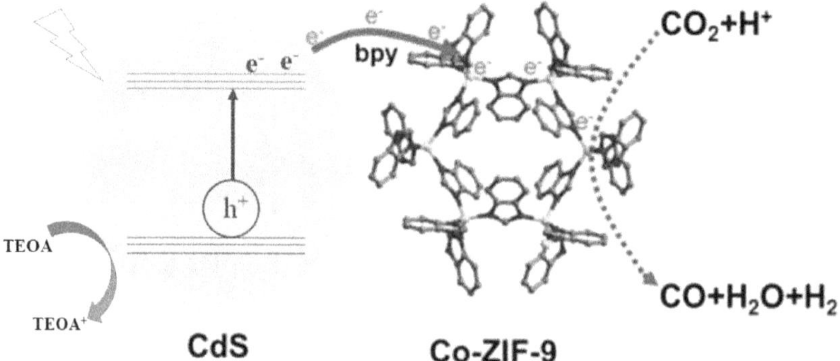

Figure 25. (a) The CO formation with respect to time of CdS+bpy+ Co-ZIF-9 and (c) the possible mechanism for CO2 reduction to CO.

Xiong and co-worker performed a photocatalytic CO_2 reduction into CH_4 by making a new MOF composite between TiO_2 and the HKUST. In a systematic way, they grew the TiO_2 nanoparticle on the surface of octahedral microcrystals of HKUST [102]. The newly formed nano composite exhibited high CO_2 uptake 49.17 cc/ g at 298 K and 1 atm and showed facile photoreduction of CO_2 to CH_4 under ultraviolet irradiation in water for 4 h. The ultrafast transient absorption and photoexcited kinetics of both TiO_2 and $Cu_3(BTC)_2@TiO_2$ illustrated that in the hybrid nano structure, TiO_2 was excited and produced the electron pair which moved to the $Cu_3(BTC)_2$. Thus, creation of two interface charge states between TiO_2 and $Cu_3(BTC)_2@TiO_2$ that opposed the charge recombination resulting in high yield of CH_4.

3.3.2 Composite with Photosensitizer

It is also possible to make a new composite between photosensitizer and MOF for developing new photocatalytic materials. In this case the light harvesting capability and CO_2 adsorption capacity of photosensitizer and MOF respectively play their active role in the photoreduction of CO_2. In this context, Wang and co-workers synthesized a new MOF composite between Co-ZIF-9 and $[Ru(bpy)_3]Cl_2 \cdot 6H_2O$. This new composite material was able to reduce CO_2 to CO under mild reaction conditions (visible light, 1 atm, 20 °C) in acetonitrile solvent and in presence of TEOA [103]. The composite converted CO_2 to CO at a rate of 41.8 μmol/30 min along with the production of hydrogen (29.9 μmol /30 min). Interestingly, no CO and H_2 were formed in absence of $[Ru(bpy)_3]Cl_2 \cdot 6H_2O$.

Again, three monometallic MOF, *viz.,* MOF-Ni, MOF-Co, and MOF-Cu had been reported which were able to perform as photocatalyst for the reduction of CO_2 to CO [104]. The LUMO of MOF-Ni is -0.94 V which is more negative compare to others two MOFs i.e. MOF-Cu (-1.08V) and MOF-Ni (-1.14 V). Mott–Schottky study revealed MOF-Ni was an n-type semiconductor and the HOMO-LUMO energy gap is 1.50 V which is small compared to others (MOF-Cu and MOF-Co are 1.77 V, 2.02V respectively). Thus, MOF-Ni reduced CO_2 to CO in visible light in a mixture of solvents (MeCN/H_2O (13:1)) in presence of photo sensitizer $[Ru(bpy)_3]Cl_2 \cdot 6H_2O$ and TIPA as an electron donor. The LUMO of PS was higher than the LUMO of MOF-Ni, so facile electron transfer took place from PS to MOF-Ni for CO_2 reduction **(Figure 26)**.

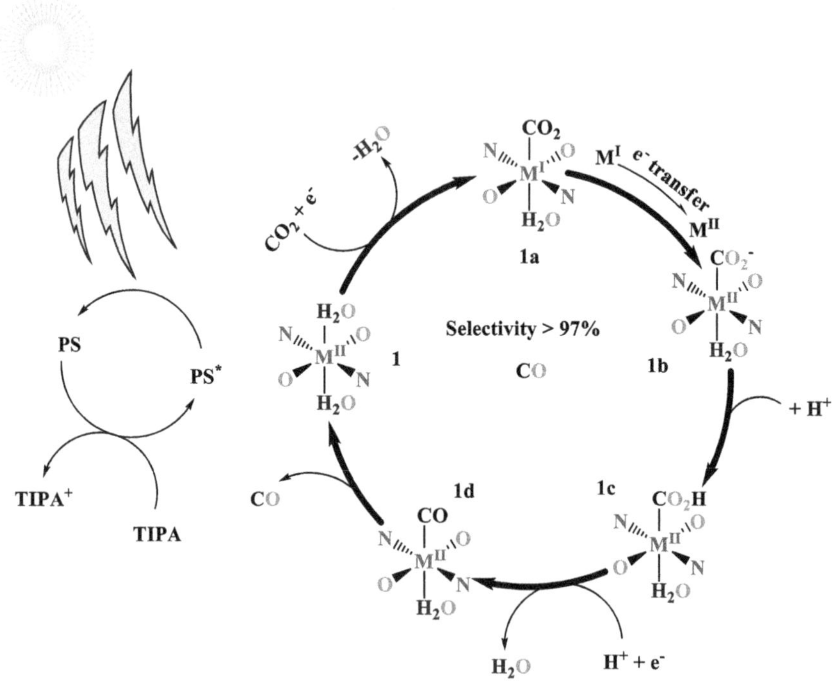

Figure 26. The mechanism of CO_2 photoreduction to CO by MOF-M (Where, M= Ni, Co and Cu).

3.3.3 Miscellaneous

The incorporation of the single metal and the metal carbonyl complexes (as photosensitizer), a MOF is able to perform as a suitable photocatalyst for the CO_2 photoreduction. The photosensitizer group present in MOF or composite is responsible for light harvesting and boost to enhance the product formation rate. Silver nanoparticles exhibits localized surface plasmon resonance (LSPR) making it a light sensitizer and useful in the photocatalytic system. A composite between semiconductor TiO_2 and Ag nanoparticles enhance the CO_2 photoreduction effectively. Due to the excitation of plasmonic Ag nanoparticles by light harvesting, a local strong electromagnetic field is generated which leads to the excitation of TiO_2

much greater than its excitation under normal light illumination. Consequently, the photocatalytic performance of the composite is considerably improved [105]. This sort of technique can be utilized to induce the photoactive nature in MOFs. Yaghi and co-workers synthesized MOF nanocomposite, Ag⊂Re$_n$-MOF where, n= Re^I(5,5′-dcbpy)(CO)$_3$Cl. This composite is having two types of functional sites (Re-complex as photosensitizers and Ag nanocubes) **(Figure 27a)** [106]. It was found that when n=3 and thickness was 16 nm, then highest photocatalysis was observed for Re$_3$-MOF. Thus, this Re$_3$-MOF coated on the Ag nanocubes to form Ag⊂Re$_3$-MOF-16. Importantly, UV–visible spectrum of the new composite, Ag⊂Re$_3$-MOF showed quadrupolar LSPR scattering peak which was same as that of Ag nanocubes at 480 nm, indicating the localized surface plasmon resonance of Ag-nanocubes retained in the composite **(Figure 27b)**. The photocatalytic reduction of CO_2 to CO by Ag⊂Re3-MOF-16 nm showed seven fold increments compare to Re$_3$-MOF. The photocatalytic activity of Re^I(5,5′-dcbpy)(CO)$_3$Cl complex was initially greater than the Ag⊂Re$_3$-MOF-16 nm. But after sometime, no further enhancement was observed. This is due to the decomposition of Re complex **(Figure 27c)**. The facile photocatalytic reaction of Ag⊂Re$_3$-MOF-16 nm was due to the transfer of photogenerated electrons from MOF to the metal nanoparticles. This charge transfer created long-lived charge separation and hindered their recombination which was suitable for the boosting of photocatalytic activity.

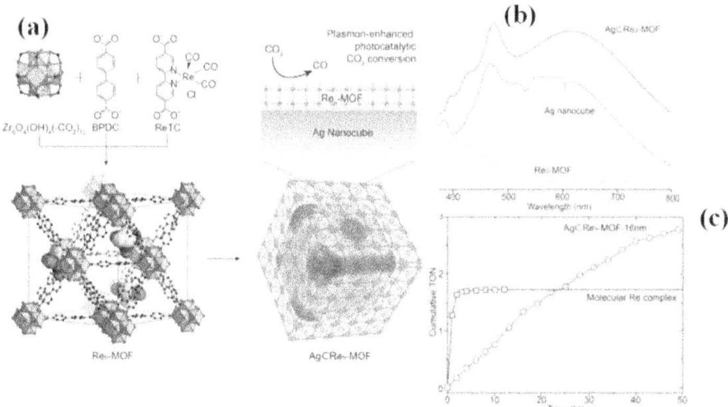

Figure 27. (a) Synthesis and structure of Re$_n$-MOF and Ag⊂Re$_n$-MOFrespectively, (b) the UV–visible spectra of the materials, (c) The photocatalytic stability of different materials. Reproduced with permission from [106] American Chemical Society.

4. Conclusion and Perspective

MOFs are the useful research materials and are immensely attracted by chemist or materials scientists for CO_2 capture and its sequestration to useful chemicals. It becomes easy to design or synthesis of specific MOF platform to achieve variety of CO_2 transformation such the cycloaddition of CO_2 with epoxide to cyclic carbonate and photochemical reduction of CO_2 to several C1 chemicals. The accurate structural determination of MOF is helping us to establish the mechanism of the reaction which opens up the fine correlation between local catalytic environment and catalytic performance. This output is really essential to design or construct next-generation MOF for better catalytic application. In this chapter, we have seen several outstanding MOFs have purposefully derived for CO_2 sequestration. Still there are huge opportunities for the creation of various MOF based heterogeneous catalyst that can accomplish the CO_2 sequestration reaction at milder reaction condition such as atmospheric pressure, low temperature and low CO_2 concentration. Also, more the CO_2 uptake capacity of MOF, then the CO_2 concentration around the catalytically active sites will be more and hence the yield of the product will be high. However, a deep, through investigation and development of new methodology is required to utilize these materials as heterogeneous catalyst in industrial scale.

Acknowledgements

TKP gratefully acknowledges the financial support from the SERB, India (TAR/2021/000090).

Abbreviations

TBAB= Tetrabutylammonium bromide
PSM= Post synthetic modification
DBU= Diazabicyclo[5.4.0]undec-7-ene
RT= Room temperature
TBHP= tert-Butyl hydroperoxide
PYI1= Pyrrolidine-2-yl-imidazole
H_3TCA= 4,4′,4″-tricarboxytriphenylamine
Bpg= –OH functionalized bipyridyl linker

SBU= Secondary building unit
CPM=crystalline porous materials
LMCT= Ligand to metal charge transfer
MLCT= Metal to ligand charge transfer
MMCT= Metal to metal charge transfer transition
HOMO = Highest occupied molecular orbital
LUMO=Lowest unoccupied molecular orbital
DFT= Density functional theory
MIL-125(Ti)= $Ti_8O_8(OH)_4$-$(O_2C$-C_6H_4-$CO_2)_6$
BDC=Benzene-1,4-dicarboxylic acid,
NH_2-MIL-125(Ti)= $(Ti_8O_8(OH)_4(NH_2$- BDC$)_6)$
NH_2- BDC= 2-Amino terephthalic acid
TEOA= Triethanolamine
TEA= Triethyl amine
ESR= Electron-spin-resonance
LED= Light-emitting diode
dcbpy= 2,2$'$-bipyridine dicarboxylic acid
BPDC= Biphenyl-4,4$'$-dicarboxylic acid
ppy= 2-phenylpyridine
dmb= 4,4$'$-dimethyl-2,2$'$-bipyridine
BNAH= 1-benzyl-1,4-dihydronicotinamide
Cp* = pentamethylcyclopentadiene
PSLE= Post-synthetic ligand exchange
PCN-22= $Zr_6(\mu_3$-OH$)_8(OH)_8(TCPP)_2$
H_2TCPP= 4,4$'$,4$''$,4$'''$-(porphyrin-5,10,15,20-tetrayl tetra benzoate
NNU-28= $[Zr_6O_4(OH)_4(L)_6]\cdot$6DMF).
BTC= 1,3,5-benzenetricarboxylate
HKUST= $Cu_3(BTC)_2$
TIPA= Triisopropanolamine
TON= Turn over number

References

1. Panwar, N., Kaushik, S., and Kothari, S. (2011) Role of renewable energy sources in environmental protection: A review. *Renewable Sustainable Energy Reviews*, **15**(3), 1513–1524.
2. Chu, S., Cui,Y., and Liu, N. (2017) The path towards sustainable energy. *Nat. Mater.*, **16**, 16–22.
3. Wang, W.-H., Himeda, Y., Muckerman, J. T., Manbeck, G. F., and Fujita, E. (2015) CO_2 Hydrogenation to Formate and Methanol as

an Alternative to Photo- and Electrochemical CO_2 Reduction. *Chem. Rev.*, **115**(23), 12936–12973.

4. Zou, Y. –H., Huang, Y. –B., Si, D. –H., Yin, Q., Wu, Q.-J., Weng, Z., and Cao, R. (2021) Porous Metal-Organic Framework Liquids for Enhanced CO_2 Adsorption and Catalytic Conversion. *Angew. Chem. Int. Ed.*, **60**(38), 20915-20920.

5. Guo, F., Su, C., Fan, Y., Shi, W., and Zhang, X. (2020) Design and Construction of a Porous Heterometallic Organic Framework Based on Cu_6I_6 Clusters and One-Dimensional TbIII Chains: Syntheses, Crystal Structure, and Various Properties. *Cryst. Growth Des.* **20**(6), 4135–4143.

6. Liu, J., Wu, D., Yang, G. –P., Wu, Y., Zhang, S., Jin, J., and Wang, Y. -Y. (2020) Rational Stepwise Construction of Different Heterometallic–Organic Frameworks (HMOFs) for Highly Efficient CO_2 Conversion. *Chem. Eur. J.* **26**(24), 5400-5406.

7. Ola, O., and Maroto-Valer, M. M. (2015) Review of material design and reactor engineering on TiO_2 photocatalysis for CO_2 reduction. *J. Photochem. Photobiol. C*, **24**, 16-42.

8. Clements, J. H. (2003) Reactive Applications of Cyclic Alkylene Carbonates. *Ind. Eng Chem Res.*, **42**(4), 663–674.

9. Lu, X. –B., and Darensbourg, D. J. (2012) Cobalt catalysts for the coupling of CO_2 and epoxides to provide polycarbonates and cyclic carbonates. *Chem Soc Rev.*, **41**(4), 1462–1484.

10. Chen, S. W., Kawthekar, R. B., and Kim, G. J. (2007) Efficient catalytic synthesis of optically active cyclic carbonates via coupling reaction of epoxides and carbon dioxide. *Tetrahedron Lett.*, **48**, 297-300.

11. Pal, T. K., De, D., Neogi, S., Pachfule, P., Senthilkumar, S., Xu, Q., and Bharadwaj, P.K. (2015) Significant Gas Adsorption and Catalytic Performance by a Robust Cu (II) -MOF Derived through Single-Crystal to Single-Crystal Transmetalation of a Thermally Less-Stable Zn (II) -MOF. *Chem. Eur. J.*, **21**(52), 19064-19070.

12. Li, H., Eddaoudi, M., O'Keeffe, M., and Yaghi, O. M. (1999) Design and synthesis of an exceptionally stable and highly porous metal-organic framework. *Nature*, **402**, 276–279.

13. Li, P. -Z., Su, J., Liang, J., Liu, J., Zhang, Y., Chen, H., and Zhao, Y. (2017) A highly porous metal–organic framework for large organic molecule capture and chromatographic separation. *Chem. Commun.*, **53**, 3434-3437.

14. Pal, T. K., De, D., Neogi, S., and Bharadwaj, P. K. (2015) Guest dependent reversible single-crystal to single-crystal structural transformation in a flexible Gd(III)-coordination polymer. *Inorg. Chem. Front.*, **2**, 395-402.

15. Pal, T. K. (2019) Syntheses, Structures and Topology Variations of Metal Organic Frameworks Built from a Semi-Rigid Tetracarboxylate Ligand. *Chemistry Select.*, **4**, 536-542.
16. Pal, T. K., Neogi, S., and Bharadwaj, P. K. (2015) Versatile Tailoring of Paddle-Wheel ZnII Metal–Organic Frameworks through Single-Crystal-to-Single-Crystal Transformations. *Chem. Eur. J.*, **21**(45), 16083-16090.
17. De, D., Pal, T. K., and Bharadwaj, P. K. (2016) A porous Cu (II)-MOF with proline embellished cavity: cooperative catalysis for the Baylis-Hillman reaction. *Inorg. Chem.*, **55**(14), 6842-6844.
18. Fujita, M., Kwon, Y. J., Washizu, S., and Ogura, K. (1994) Preparation, Clathration Ability, and Catalysis of a Two-Dimensional Square Network Material Composed of Cadmium (II) and 4, 4'-Bipyridine. *J. Am. Chem. Soc.*, **116**(3), 1151-1152.
19. Liu, J., Chen, L., Cui, H., Zhang, J., Zhang, and L., Su, C. -Y. (2014) Applications of metal–organic frameworks in heterogeneous supramolecular catalysis. *Chem. Soc. Rev.*, **43**, 6011-6061.
20. Li, X., Yu, J., Jaroniec, M., and Chen, X. (2019) Cocatalysts for Selective Photoreduction of CO_2 into Solar Fuels. *Chem. Rev.*, **119**(6), 3962–4179.
21. Chang, X., Wang, T., and Gong, J. (2016) CO_2 photo-reduction: insights into CO_2 activation and reaction on surfaces of photocatalysts. *Energy Environ. Sci.*, **9**, 2177-2196.
22. Gao, C., Wang, J., Xu, H., and Xiong, Y. (2017) Coordination chemistry in the design of heterogeneous photocatalysts. *Chem. Soc. Rev.*, **46**, 2799-2823.
23. Hisatomi, T., Kubota, J., and Domen, K. (2014) Recent advances in semiconductors for photocatalytic and photoelectrochemical water splitting. *Chem. Soc. Rev.*, **43**, 7520-7535.
24. White, J. L., Baruch, M. F., Pander III, J. E., Hu, Y., Fortmeyer, I. C., Park, J.E., Zhang, T. Liao, K., Gu, J., Yan, Y., Shaw, T. W., Abelev, E., and Bocarsly, A. B. (2015) Light-Driven Heterogeneous Reduction of Carbon Dioxide: Photocatalysts and Photoelectrodes. *Chem. Rev.*, **115**(23), 12888–12935.
25. Chen, X., Li, C., Grätzel, M., Kostecki, R., and Mao, S.S. (2012) Nanomaterials for renewable energy production and storage. *Chem. Soc. Rev.*, **41**, 7909–7937.
26. Linares, N., Silvestre-Albero, A. M., Serrano, E., Silvestre-Albero, J., and García-Martínez, J. (2014) Mesoporous materials for clean energy technologies. *Chem. Soc. Rev.*, **43**, 7681–7717.
27. Ma, Y., Wang, X., Jia, Y., Chen, X., Han, H., and Li, C. (2014) Titanium Dioxide-Based Nanomaterials for Photocatalytic Fuel Generations. *Chem. Rev.*, **114**, 9987– 10043.
28. Ran, J., Jaroniec, M., and Qiao, S. -Z. (2018) Cocatalysts in Semiconductor-based Photocatalytic CO_2 Reduction:

Achievements, Challenges, and Opportunities. *Advance materials*, **30**(7), 1704649.

29. Liu, L., Jiang, Y., Zhao, H., Chen, J., Cheng, J., Yang, K., and Li, Y. (2016) Engineering Coexposed {001} and {101} Facets in Oxygen-Deficient TiO_2 Nanocrystals for Enhanced CO_2 Photoreduction under Visible Light. *ACS Catal.*, **6**, 1097-1103.

30. Benedito, A., Acarreta, E., and Giménez, E. (2021) Highly Efficient MOF Catalyst Systems for CO2 Conversion to Bis-Cyclic Carbonates as Building Blocks for NIPHUs (Non-Isocyanate Polyhydroxyurethanes) Synthesis. *Catalysts*, 2021;**11**(5), 628.

31. Xu, H., Cao, C.–S., Hu, H.–S., Wang, S.–B., Liu, J.–C., Cheng, P., Kaltsoyannis, N., Li, J., and Zhao, B. (2019) High Uptake of ReO_4^- and CO_2 Conversion by a Radiation-Resistant Thorium–Nickle [$Th_{48}Ni_6$] Nanocage-Based Metal–Organic Framework. *Angew. Chem. Int. Ed.*, **131**(18), 6083-6088.

32. Yuan, Y., Li, J., Sun, X., Li, G., Liu, Y., Verma, G., and Ma, S. (2019) Indium–Organic Frameworks Based on Dual Secondary Building Units Featuring Halogen-Decorated Channels for Highly Effective CO_2 Fixation. *Chem Mater.*, **31**, 1084–1091.

33. He, T., Ni, B., Xu, X., Li, H., Lin, H., Yuan, W., Luo, J., Hu, W., and Wang, X. (2017) Competitive Coordination Strategy to Finely Tune Pore Environment of Zirconium-Based Metal–Organic Frameworks. *ACS Appl. Mater. Interfaces*, **9**(27), 22732–22738.

34. Beyzavi, H., Klet, R.C., Tussupbayev, S., Borycz, J., Vermeulen, N.A., Cramer, C.J., Stoddart, J. F., Hupp, J. T., and Farha, O. K. (2014) A Hafnium-Based Metal–Organic Framework as an Efficient and Multifunctional Catalyst for Facile CO_2 Fixation and Regioselective and Enantioretentive Epoxide Activation. *J. Am. Chem. Soc.*, **136**(45), 15861– 15864.

35. Guillerm, V., Weselin´ski, Ł.J., Belmabkhout, Y., Cairns, A.J., D'Elia, V., Wojtas, Ł., Adil, K., and Eddaoudi, M. (2014) Discovery and introduction of a (3, 18)-connected net as an ideal blueprint for the design of metal–organic frameworks. *Nat Chem.*, **6**, 673.

36. Zalomaeva, O.V., Maksimchuk, N. V., Chibiryaev, A.M., Kovalenko, K.A., Fedin, V. P., and Balzhinimaev, B. S. (2013) Synthesis of cyclic carbonates from epoxides or olefins and CO2 catalyzed by metal-organic frameworks and quaternary ammonium salts. *J. Energy Chem.*, **22**(1), 130–135.

37. Li, Y.-Z., Wang, H.-H., Yang, H.-Y., Hou, L., Wang, Y.-Y., and Zhu, Z. (2018) An Uncommon Carboxyl-Decorated MOF with Selective Gas Adsorption and Catalytic Conversion of CO_2. *Chem. Eur. J.*, **24**(4), 865–871.

38. Nguyen, P.T.K., Nguyen, H.T.D., Nguyen, H.N., Trickett, C.A., Ton, Q.T., Gutiérrez-Puebla, E., Monge, M. A., Cordova, K. E., and

Gándara, F. (2018) New Metal–Organic Frameworks for Chemical Fixation of CO_2. *ACS Appl Mater Interfaces*, **10**(1), 733–744.

39. Wang, G. –D., Li, Y. –Z., Shi, W. –J., Hou, L., Zhu, Z., and Wang, Y. -Y. (2020) A new honeycomb metal–carboxylate-tetrazolate framework with multiple functions for CO_2 conversion and selective capture of C_2H_2, CO_2 and benzene. *Inorg. Chem. Front.*, **7**, 1957-1964.

40. Yuan, R., and He, H. (2020) Constructing a 3D porous Co(II)-organic framework: Synthesis, characterization and chemical transformation of epoxide and CO2 into cyclic carbonate, *Inorg. Chem. Comm.*, **121**, 108235.

41. Taghavi, F., Khojastehnezhad, A., Khalifeh, R., Rajabzadeh, M., Rezaei, F., Abnous, K., Taghdisi, S. M. (2021) Design and synthesis of a new magnetic metal organic framework as a versatile platform for immobilization of acidic catalysts and CO_2 fixation reaction. *New J Chem.*, **45**, 15405-15414.

42. Sen, R., Saha, D., Koner, S., Brandão, P., and Lin, Z. (2015) Single Crystal to Single Crystal (SC-to-SC) Transformation from a Nonporous to Porous Metal–Organic Framework and Its Application Potential in Gas Adsorption and Suzuki Coupling Reaction through Postmodification. *Chem. Eur. J.*, **21**(15), 5962-5971.

43. Zhang, X., Chen, Z., Yang, X., Li, M., Chen, C., and Zhang, N. (2018) The fixation of carbon dioxide with epoxides catalyzed by cation-exchanged metal-organic framework. *Mesoporous Microporous*, **258**, 55–61.

44. Zhai, G., Liu, Y., Lei, L., Wang, J., Wang, Z., Zheng, Z., Wang, P., Cheng, H., Dai, Y., and Huang, B. (2021) Light-Promoted CO2 Conversion from Epoxides to Cyclic Carbonates at Ambient Conditions over a Bi-Based Metal–Organic Framework. *ACS Catal.*, **11**, 1988–1994.

45. Singh, M., Solanki, P., Patel, P., Mondal, A., and Neogi, S. (2019) Highly Active Ultrasmall Ni Nanoparticle Embedded Inside a Robust Metal–Organic Framework: Remarkably Improved Adsorption, Selectivity, and Solvent-Free Efficient Fixation of CO_2. *Inorg. Chem.*, **58**:8100–8110.

46. Xia, Q., Liu, Y., Li, Z., Gong, W., and Cui, Y. (2016) A Cr(salen)-based metal–organic framework as a versatile catalyst for efficient asymmetric transformations. *Chem Commun.*, **52**, 13167-13170.

47. Li, J., Ren, Y., Yue, C., Fan, Y., Qi, C., Jiang, H. (2018) Highly Stable Chiral Zirconium-Metallosalen Frameworks for CO 2 Conversion and Asymmetric C-H Azidation. *ACS Appl .Mater. Interfaces*, **10**, 36047–36057.

48. Gao, W.-Y., Tsai, C.-Y., Wojtas, L., Thiounn, T., Lin, C. -C., and Ma, S. (2016) Interpenetrating Metal–Metalloporphyrin Framework for

Selective CO2 Uptake and Chemical Transformation of CO_2. *Inorg Chem.*, **55**(15), 7291–7294.

49. Gu, Y., Choe, Y., Park, and D. -W. (2021) Catalytic Performance of CPM-200-In/Mg in the Cycloaddition of CO_2 and Epoxides. *Catalysts*, **11**, 430.

50. Jia, Y.Y., Liu, X. T., Feng, R., Zhang, S.Y., Zhang, P., He, Y.B., Zhang, Y.H., and Bu, X.H. (2017) Improving the Stability and Gas Adsorption Performance of Acylamide Group Functionalized Zinc Metal–Organic Frameworks through Coordination Group Optimization. *Cryst. Growth Des.*, **17**(5), 2584–2588.

51. Cabello, C.P., Berlier, G., Magnacca, G., Rumoria, P., and Palomino, G. T. (2015) Enhanced CO_2 adsorption capacity of amine-functionalized MIL-100(Cr) metal–organic frameworks. *CrystEngComm.*, **17**, 430-437.

52. Gupta, M., De, D., Tomar, K., and Bharadwaj, P. K. (2017) From Zn(II)-Carboxylate to Double-Walled Zn(II)-Carboxylato Phosphate MOF: Change in the Framework Topology, Capture and Conversion of CO_2, and Catalysis of Strecker Reaction. *Inorg. Chem.*, **56**(23), 14605–14611.

53. Li, X.-Y., Ma, L.-N., Liu, Y., Hou, L., Wang, Y.-Y., and Zhu, Z. (2018) Honeycomb Metal–Organic Framework with Lewis Acidic and Basic Bifunctional Sites: Selective Adsorption and CO_2 Catalytic Fixation. *ACS Appl. Mater. Interfaces.*, **10**, 10965–10973.

54. Guo, X., Zhou, Z., Chen, C., Bai, J., He, C., and Duan, C. (2016) New rht-Type Metal–Organic Frameworks Decorated with Acylamide Groups for Efficient Carbon Dioxide Capture and Chemical Fixation from Raw Power Plant Flue Gas. *ACS Appl. Mater. Interfaces.*, **8**(46), 31746–31756.

55. De, D., Pal, T.K., Neogi, S., Senthilkumar, S., Das, D., Gupta, S. S., and Bharadwaj, P. K. (2016) A Versatile CuII Metal–Organic Framework Exhibiting High Gas Storage Capacity with Selectivity for CO_2: Conversion of CO_2 to Cyclic Carbonate and Other Catalytic Abilities. *Chem. Eur. J.*, **22**(10), 3387–3396.

56. Sharma, V., De, D., Saha, R., Das, R., Chattaraj, P.K., and Bharadwaj, P. K. (2017) A Cu(ii)-MOF capable of fixing CO_2 from air and showing high capacity H_2 and CO_2 adsorption. *Chem. Commun.*, **53**, 13371-13374.

57. Bian, Z. -X., Zhang, Y. -Z., Tian, D., Zhang, X., Xie, L. -H., Zhao, M., Xie, Y., and Li, J.-R. (2020) Co7-Cluster-Based Metal–Organic Frameworks with Mixed Carboxylate and Pyrazolate Ligands: Construction and CO_2 Adsorption and Fixation. *Cryst. Growth Des.*, **20**(12), 7972–7978.

58. Wei, L. -Q, and Ye, B. -H. (2019) Efficient Conversion of CO_2 via Grafting Urea Group into a $[Cu_2(COO)_4]$-Based Metal–Organic

Framework with Hierarchical Porosity. *Inorg. Chem.,,***8**, 4385–4393.

59. Liu, E., Zhu, J., Yang, W., Liu, F., Huang, C., and Yin, S. (2020) PCN-222(Co) Metal–Organic Framework Nanorods Coated with 2D Metal–Organic Layers for the Catalytic Fixation of CO2 to Cyclic Carbonates. *ACS Appl. Nano Mater.*, **3**(4), 3578–3584.

60. Han, Q., Qi, B., Ren, W., He, C., Niu, J., and Duan, C. (2015) Polyoxometalate-based homochiral metal-organic frameworks for tandem asymmetric transformation of cyclic carbonates from olefins. *Nature Communications*, **6**, 10007.

61. Seal, N., Singh, M., Das, S., Goswami., R, Pathak, B., and Neogi, S. (2021) Dual-functionalization actuated trimodal attribute in an ultra-robust MOF: exceptionally selective capture and effectual fixation of CO2 with fast-responsive, nanomolar detection of assorted organo-contaminants in water. *Mater. Chem. Front.*, **5**, 979–994.

62. Das, R., and Nagaraja, C. M. (2021) Noble metal-free Cu(i)-anchored NHC-based MOF for highly recyclable fixation of CO_2 under RT and atmospheric pressure conditions. ***Green Chem.***, **23,** 5195-5204.

63. Ding, L.-G., Yao, B.- J., Jiang, W. -L., Li, J. -T., Fu, Q. -J., Li, Y. -A., Liu, Z. -H., Ma, J. -P., and Dong, Y. -B. (2017) Bifunctional Imidazolium-Based Ionic Liquid Decorated UiO-67 Type MOF for Selective CO_2 Adsorption and Catalytic Property for CO_2 Cycloaddition with Epoxides. *Inorg. Chem.*, **56**(4), 2337–2344.

64. Liu, W.–S., Zhou, L.–J., Li, G., Yang, S.–L., and Gao, E.–Q. (2021) Double Cationization Approach toward Ionic Metal–Organic Frameworks with a High Bromide Content for CO_2 Cycloaddition to Epoxides. *ACS Sus. Chem. Eng.*, **9**(4), 1880–1890.

65. Pander, M., Janeta, M., and Bury, W. (2021) Quest for an Efficient 2-in-1 MOF-Based Catalytic System for Cycloaddition of CO_2 to Epoxides under Mild Conditions. *ACS Appl. Mater. Interfaces.*, **13**(7), 8344–8352.

66. Song, J. L., Zhang, Z.F., Hu, S.Q., Wu, T.B., Jiang, T., and Han, B.X. (2009) MOF-5/n-Bu4NBr: an efficient catalyst system for the synthesis of cyclic carbonates from epoxides and CO_2 under mild conditions. *Green Chem.*, **11**, 1031–1036.

67. Fan, Z., Wang, J., Wang, W., Burger, S., Wang, Z., Wang, Y., Wöll, C., Cokoja, M., and Fischer, R. A. (2020) Defect Engineering of Copper Paddlewheel-Based Metal–Organic Frameworks of Type NOTT-100: Implementing Truncated Linkers and Its Effect on Catalytic Properties. *ACS Appl Mater. Interfaces.*, **12**(34), 37993–38002.

68. Kumar, B., Llorente, M., Froehlich, J., Dang, T., Sathrum, A., and Kubiak, C. P. (2012) Photochemical and Photoelectrochemical Reduction of CO_2. *Annu. Rev. Phys. Chem.*, **63**, 541-569.

69. Santaclara, J. G., Kapteijn, F., Gascon, J., and vanderVeen, M. A. (2017) Understanding metal–organic frameworks for photocatalytic solar fuel production. *CrystEngComm.*, **19**, 4118-4125.

70. Alvaro, M., Carbonell, E., Ferrer, B., iXamena, F. X. L., and Garcia, H. (2007) Semiconductor Behavior of a Metal-Organic Framework (MOF). *Chem. Eur. J.*, **13**(18), 5106-5112.

71. Li, R., Zhang, W., and Zhou, K. (2018) Metal–Organic-Framework-Based Catalysts for Photoreduction of CO_2. *Adv. Mater.*, **30**(35), 1705512.

72. Santaclara, J. G., Nasalevich, M. A., Castellanos, S., Evers, W. H., Spoor, F. C. M., Rock, K., Siebbeles, L. D. A., Kapteijn, F., Grozema, F., Houtepen, A., Gascon, J., Hunger, J., and van der Veen, M. A. (2016) Organic Linker Defines the Excited-State Decay of photocatalytic MIL-125(Ti)-Type Materials. *ChemSusChem.*, **9**, 388-395.

73. Wang, D., Huang, R., Liu, W., Sun, D., and Li, Z. (2014) Fe-Based MOFs for Photocatalytic CO_2 Reduction: Role of Coordination Unsaturated Sites and Dual Excitation Pathways. *ACS Catal.*, **4**(12), 4254-4260.

74. Ma, Y., Wang, X., Jia, Y., Chen, X., Han, H., and Li, C. (2014) Titanium Dioxide-Based Nanomaterials for Photocatalytic Fuel Generations. *Chem. Rev.*, **114**, 9987-10043.

75. Maina, W., Pozo-Gonzalo, C., Kong, L., Schütz, J., Hill, M., and Dumée, L. F. (2017) Metal organic framework-based catalysts for CO_2 conversion. *Mater. Horiz.*, **4**, 345-361.

76. Dhakshinamoorthy, A., Asiri, A. M., and Garcia, H. (2016) Metal–Organic Framework (MOF) Compounds: Photocatalysts for Redox Reactions and Solar Fuel Production. *Angew. Chem. Int. Ed.*, **55**, 5414-5445.

77. Xu, H.-Q., Hu, J., Wang, D., Li, Z., Zhang, Q., Luo, Y., Yu, S.-H., and Jiang, H.-L. (2015) Visible-Light Photoreduction of CO_2 in a Metal–Organic Framework: Boosting Electron–Hole Separation via Electron Trap States. *J. Am. Chem. Soc.*, **137**(42), 13440-13443.

78. Pham, H. Q., Mai, T., and Pham-Tran, N.-N. (2014) Engineering of Band Gap in Metal–Organic Frameworks by Functionalizing Organic Linker: A Systematic Density Functional Theory Investigation. *J. Phys. Chem. C*, **118**, 4567-4577.

79. Nguyen, H. L., Vu, T. T., Le, D., Doan, R. L. H., Nguyen, V. Q., and Phan, N. T. S. (2017) A Titanium–Organic Framework: Engineering of the Band-Gap Energy for Photocatalytic Property Enhancement. *ACS Catal.*, **7**(1), 338-342.

80. Schoedel, A., Ji, Z., and Yaghi, O. M. (2016) The role of metal–organic frameworks in a carbon-neutral energy cycle. *Nat. Energy*, **1**, 16034.

81. Lu, P., Wu, Y., Kang, H., Wei, H., Liu, H., and Fang, M. (2014) What can pKa and NBO charges of the ligands tell us about the water and thermal stability of metal organic frameworks? *J. Mater. Chem. A.*, **2**, 16250-16267.

82. Zhang, W., Hu, Y., Ge, J., Jiang, H. -L., and Yu, S.-H. (2014) A Facile and General Coating Approach to Moisture/Water-Resistant Metal–Organic Frameworks with Intact Porosity. *J. Am. Chem. Soc.*, **136**(49), 16978-16981

83. Silva, C. G., Corma, A., and García, H. (2010) Metal–organic frameworks as semiconductors. *J. Mater. Chem.*, **20**, 3141-3156.

84. Fu, Y., Sun, D., Chen, Y., Huang, R., Ding, Z., Fu, X., and Li, Z. (2012) An Amine-Functionalized Titanium Metal–Organic Framework Photocatalyst with Visible-Light-Induced Activity for CO_2 Reduction. *Angew. Chem. Int. Ed.*, **51**, 3364-3367.

85. Sun, D., Fu, Y., Liu, W., Ye, L,. Wang, D., Yang, L., Fu, X., and Li, Z. (2013) Studies on Photocatalytic CO2 Reduction over NH2-Uio-66(Zr) and Its Derivatives: Towards a Better Understanding of Photocatalysis on Metal–Organic Frameworks. *Chem. Eur. J.*, **19**, 14279-14285.

86. Logan, M. W., Ayad, S., Adamson, J. D., Dilbeck, T., Hanson, K., and Uribe-Romo, F. J. (2017) Systematic variation of the optical bandgap in titanium based isoreticular metal–organic frameworks for photocatalytic reduction of CO_2 under blue light. *J. Mater. Chem. A*, **5**, 11854-11863.

87. Morris, A. J., Meyer, G. J., and Fujita, E. (2009) Molecular Approaches to the Photocatalytic Reduction of Carbon Dioxide for Solar Fuels. *Acc. Chem. Res.*, **42**, 1983-1994

88. Wang, C., Xie, Z., deKrafft, K. E., and Lin, W. (2011) Doping Metal–Organic Frameworks for Water Oxidation, Carbon Dioxide Reduction, and Organic Photocatalysis. *J. Am. Chem. Soc.*, **133**(34), 13445-13454.

89. Li, L., Zhang, S., Xu, L., Wang, J., Shi, L.–X., Chen, Z.–N., Hongab, M., and Luo, J. (2014) Effective visible-light driven CO_2 photoreduction via a promising bifunctional iridium coordination polymer. *Chem. Sci.*, **5**, 3808-3813.

90. Fei, H., Sampson, M. D., Lee, Y., Kubiak, C. P., and Cohen, S. M. Photocatalytic CO_2 Reduction to Formate Using a Mn(I) Molecular Catalyst in a Robust Metal–Organic Framework. *Inorg. Chem.*, **54**(14), 6821-6828.

91. Chambers, M. B., Wang, X., Elgrishi, N., Hendon, C. H., Waksh, A., Bonnefoy, J., Canivet, J., Quadrelli, E. A., Farrusseng, D., Mellot-Draznieks, C., Fontecave, M. (2015) Photocatalytic Carbon Dioxide Reduction with Rhodium-based Catalysts in Solution and Heterogenized within Metal–Organic Frameworks. *ChemSusChem*, **8**, 603-608.

92. Kajiwara, T., Fujii, M., Tsujimoto, M., Kobayashi, K., Higuchi, M., Tanaka, K., and Kitagawa, S. (2016) Photochemical Reduction of Low Concentrations of CO_2 in a Porous Coordination Polymer with a Ruthenium(II)–CO Complex. *Angew. Chem. Int. Ed.*, **128**(8) 2747-2750.

93. Xu, H. -Q., Hu, J., Wang, D., Li, Z., Zhang, Q., Luo, Y., Yu, S.-H., and Jiang, H. -L. (2015) Visible-Light Photoreduction of CO_2 in a Metal–Organic Framework: Boosting Electron–Hole Separation via Electron Trap States. *J. Am. Chem. Soc.*, **137**(42), 13440-13443.

94. Chen, D., Xing, H., Wang, C., and Su, Z. (2016) Highly efficient visible-light-driven CO2 reduction to formate by a new anthracene-based zirconium MOF via dual catalytic routes. *J. Mater. Chem. A*, **4**, 2657-2662.

95. Windle, C. D., George, M. W., Perutz, R. N., Summers, P. A., Sun, X. Z., Whitwood, A. C. (2015) Comparison of rhenium–porphyrin dyads for CO_2 photoreduction: photocatalytic studies and charge separation dynamics studied by time-resolved IR spectroscopy. *Chem. Sci.*, **6**, 6847-6864.

96. Laurier, K. G. M., Vermoortele, F., Ameloot, Vos, R., D. E. D., Hofkens, J., and Roeffaers, M. B. J. (2013) Iron(III)-Based Metal–Organic Frameworks As Visible Light Photocatalysts. *J. Am. Chem. Soc.*, **135**(39), 14488-14491.

97. Assi, H., Mouchaham, G., Steunou, N., Devic, T., and Serre, C. (2017) Titanium coordination compounds: from discrete metal complexes to metal–organic frameworks. *Chem. Soc. Rev.*, **46**, 3431-3452.

98. Hendon, C. H., Tiana, D., Fontecave, M., Sanchez, C., D'arras, L., Sassoye, C., Rozes, L., and Mellot-Draznieks, C. (2013) A. Walsh, Engineering the Optical Response of the Titanium-MIL-125 Metal–Organic Framework through Ligand Functionalization. *J. Am. Chem. Soc.*, **135**(30) 10942-10945.

99. Dan-Hardi, M., Serre, C., Frot, T., Rozes, L., Maurin, G., Sanchez, C., and Férey, G.(2009) A New Photoactive Crystalline Highly Porous Titanium(IV) Dicarboxylate. *J. Am. Chem. Soc.*, **131**(31), 10857-10859.

100. Sun, D., Liu, W., Qiu, M., Zhang, Y., and Li, Z. (2015) Introduction of a mediator for enhancing photocatalytic performance via post-synthetic metal exchange in metal–organic frameworks (MOFs). *Chem. Commun.*, **51**, 2056-2059.

101. Wang, S., and Wang, X. (2015) Photocatalytic CO_2 reduction by CdS promoted with a zeolitic imidazolate framework. *Appl. Catal. B*, **162**, 494-500.

102. Li, R., Hu, J., Deng, M., Wang, H., Wang, X., Hu, Y., Jiang, H. -L., Jiang, J., Zhang, Q., Xie, Y., and Xiong, Y. (2014) Integration of an Inorganic Semiconductor with a Metal–Organic Framework: A Platform

for Enhanced Gaseous Photocatalytic Reactions. *Adv. Mater.*, **26**(28), 4783-4788.

103. Wang, S., Yao, W., Liu, J., Ding, Z., and Wang, X. (2014) Cobalt Imidazolate Metal–Organic Frameworks Photosplit CO_2 under Mild Reaction Conditions. *Angew. Chem. Int. Ed.*, **53**(4) 1034-1038.

104. Wang, X. –K., Liu, J., Zhang, L., Dong, L. –Z., Li, S. –L., Kan, Y. –H., Li, D. –S., and Lan, Y. –Q. (2019) Monometallic Catalytic Models Hosted in Stable Metal–Organic Frameworks for Tunable CO_2 Photoreduction. *ACS Catal.*, **9**(3), 1726-1732.

105. Hou, W., and Cronin, S. B. (2013) A Review of Surface Plasmon Resonance-Enhanced Photocatalysis. *Adv. Funct. Mater.*, **23**(13), 1612-1619.

106. Choi, K. M., Kim, D., Rungtaweevoranit, B., Trickett, C. A., Barmanbek, J. T. D., Alshammari, A. S., Yang, P., and Yaghi, O. M. (2017) Plasmon-Enhanced Photocatalytic CO2 Conversion within Metal–Organic Frameworks under Visible Light. *J. Am. Chem. Soc.*, **139**(1), 356-362.

www.ingramcontent.com/pod-product-compliance
Lightning Source LLC
Chambersburg PA
CBHW070741030726
47601CB00001B/98